Microbiology Monographs
Volume 8

Series Editor: Alexander Steinbüchel
Münster, Germany

Microbiology Monographs

Volumes published in the series

Inclusions in Prokaryotes
Jessup M. Shively (Editor)
Volume 1 (2006)

Complex Intracellular Structures in Prokaryotes
Jessup M. Shively (Editor)
Volume 2 (2006)

Magnetoreception and Magnetosomes in Bacteria
Dirk Schüler (Editor)
Volume 3 (2007)

Predatory Prokaryotes – Biology, Ecology and Evolution
Edouard Jurkevitch (Editor)
Volume 4 (2007)

Amino Acid Biosynthesis – Pathways, Regulation and Metabolic Engineering
Volker F. Wendisch (Editor)
Volume 5 (2007)

Molecular Microbiology of Heavy Metals
Dietrich H. Nies and Simon Silver (Editors)
Volume 6 (2007)

Microbial Linear Plasmids
Friedhelm Meinhardt and Roland Klassen (Editors)
Volume 7 (2007)

Prokaryotic Symbionts in Plants
Katharina Pawlowski (Editor)
Volume 8 (2009)

**Hydrogenosomes and Mitosomes:
Mitochondria of Anaerobic Eukaryotes**
Jan Tachezy (Editor)
Volume 9 (2008)

Uncultivated Microorganisms
Slava S. Epstein (Editor)
Volume 10 (2009)

Microbial Megaplasmids
Edward Schwartz (Editor)
Volume 11 (2009)

Katharina Pawlowski
Editor

Prokaryotic Symbionts in Plants

Springer

Editor
Dr. Katharina Pawlowski
Department of Botany
Stockholm University
10691 Stockholm
Sweden
e-mail: pawlowski@botan.su.se

Series Editor
Professor Dr. Alexander Steinbüchel
Institut für Molekulare Mikrobiologie und Biotechnologie
Westfälische Wilhelms-Universität
Corrensstraße 3
48149 Münster
Germany
e-mail: steinbu@uni-muenster.de

ISSN 1862-5576 e-ISSN 1862-5584
ISBN 978-3-540-75459-6 e-ISBN 978-3-540-75460-2
DOI 10.1007/978-3-540-75460-2
Springer Dordrecht Heidelberg London New York

Library of Congress Control Number: 2008940027

© Springer-Verlag Berlin Heidelberg 2009
This work is subject to copyright. All rights are reserved, whether the whole or part of the material is concerned, specifically the rights of translation, reprinting, reuse of illustrations, recitation, broadcasting, reproduction on microfilm or in any other way, and storage in data banks. Duplication of this publication or parts thereof is permitted only under the provisions of the German Copyright Law of September 9, 1965, in its current version, and permission for use must always be obtained from Springer. Violations are liable to prosecution under the German Copyright Law.
The use of general descriptive names, registered names, trademarks, etc. in this publication does not imply, even in the absence of a specific statement, that such names are exempt from the relevant protective laws and regulations and therefore free for general use.

Cover Design: WMXDesign GmbH, Heidelberg, Germany

Printed on acid-free paper

Springer is part of Springer Science+Business Media (www.springer.com)

Preface

This *Microbiology Monographs* volume provides a comprehensive and detailed source of information on endosymbioses between prokaryotes (eubacteria and cyanobacteria) and plants. Our first task is to thank all authors of the chapters in this volume. We greatly appreciate the investment in time and effort they have made to produce the comprehensive coverage of their topics.

The book comprises 12 chapters, authored by well-known scientists in the field. The first three chapters deal with nitrogen-fixing symbioses between rhizobia and legumes, which are the best-known nitrogen-fixing symbioses due to their agricultural importance. The area of genomics has led to great progress in the analysis of these symbioses in recent years. The discovery of beta-rhizobia has shed light on the extent of lateral transfer of symbiosis genes. Plant genetic studies have led to the elucidation of the perception of rhizobial signal factors by legumes, and the signal transduction steps that lead to nodule induction which involve functions recruited from plant symbioses with soil fungi.

The next three chapters concern nitrogen-fixing symbioses between actinomycetous soil bacteria of the genus *Frankia*, and mostly woody plants from eight different families, collectively called actinorhizal plants. These symbioses are important in agroforestry and soil reclamation. Being less accessible to molecular analysis than legume-rhizobia symbioses due to the woody nature of the host plants, and the fact that the microsymbionts are, so far, non-transformable, they have not been examined as closely as legume symbioses. However, transfer of knowledge obtained in legume research to actinorhizal research has recently shown that in both symbioses, the same signal transduction pathway is used in response to microsymbiont signal factors. The fact that *Frankia* strains cannot be transformed is in agreement with the lack, or rarity of, lateral transfer of symbiosis genes in this phylogenetic group, constituting a strong contrast with the situation in rhizobia.

The next five chapters deal with symbioses between cyanobacteria and mosses, ferns, gymnosperms and one angiosperm genus, *Gunnera*. As precursors of chloroplasts, filamentous heterocystous cyanobacteria are arguably the most successful endosymbionts in evolution (apart from the precursors of mitochondria), so it is not surprising that many symbioses exist between cyanobacteria and lower plants as well as gymnosperms. It is striking that only one cyanobacterial angiosperm symbiosis has evolved. The symbiosis between cyanobacteria and the water fern *Azolla* is of special importance, not only because of the role of *Azolla* as green manure in rice culture, but also because here, the cyanobacterium is an obligate symbiont, unable to survive outside the plant. A cameo chapter on tripartite lichens – symbioses between cyanobacteria, algae, and fungi – has been added.

The last chapter summarizes the current knowledge on endophytic diazotrophs. Endophytic bacteria, which do not cause any harm to the plant, but actually may support plant growth, are found in most plant species, but it has to be closely analyzed how this growth promotion is achieved, and whether under certain circumstances a usually advantageous endophyte may turn into a parasite.

Stockholm and Münster, April 2009   Katharina Pawlowski
Alexander Steinbüchel

Contents

Part I:
Rhizobia-Legume Symbioses

The Diversity and Evolution of Rhizobia
A. Dresler-Nurmi · D. P. Fewer · L. A. Räsänen · K. Lindström   3

Erratum to The Diversity and Evolution of Rhizobia
A. Dresler-Nurmi · D. P. Fewer · L. A. Räsänen · K. Lindström 43

Making Rhizobium-Infected Root Nodules
A. Untergasser · T. Bisseling · R. Geurts 45

Functional Genomics of Rhizobia
A. Becker 71

Part II:
Actinorhizal Symbioses

Evolution and Diversity of *Frankia*
P. Normand · M. P. Fernandez 103

Induction of Actinorhizal Nodules by *Frankia*
K. Pawlowski 127

Physiology of Actinorhizal Nodules
T. Persson · K. Huss-Danell 155

Part III:
Cyanobacterial Symbioses

Physiological Adaptations
in Nitrogen-fixing *Nostoc*-Plant Symbiotic Associations
J. C. Meeks 181

Why Does *Gunnera* Do It and Other Angiosperms Don't?
An Evolutionary Perspective on the *Gunnera–Nostoc* Symbiosis
B. Osborne · B. Bergman 207

Cyanobacteria in Symbiosis with Cycads
P. Lindblad 225

Structural Characteristics of the Cyanobacterium–*Azolla* Symbioses
W. Zheng · L. Rang · B. Bergman 235

Relations Between Cyanobacterial Symbionts in Lichens and Plants
J. Rikkinen 265

Part IV:
Diazotrophic Endophytes

Diazotrophic Bacterial Endophytes in *Gramineae* and Other Plants
M. Rothballer · M. Schmid · A. Hartmann 273

Subject Index 303

Part I
Rhizobia-Legume Symbioses

The Diversity and Evolution of Rhizobia

Aneta Dresler-Nurmi (✉) · David P. Fewer · Leena A. Räsänen · Kristina Lindström

Department of Applied Chemistry and Microbiology, Viikki Biocenter, University of Helsinki, PO Box 56 (Viikinkaari 9), 00014 Helsinki, Finland
aneta.dresler@helsinki.fi

1	Introduction	4
2	The Structure of the Rhizobial Genome	11
2.1	Chromosome and Plasmids	11
3	Species Concept Applicable to Rhizobia	11
4	Rhizobia-legume Symbiosis	13
4.1	Molecular Interactions Between Partners	13
4.2	Determinants of Host Specifity in Rhizobia	15
4.2.1	The NodD/Flavonoid Complex	16
4.2.2	Nod Factors	16
5	Diversity and Taxonomy of Rhizobia	18
5.1	Phenotypic and Genotypic Diversity	18
5.1.1	Diversity Based on "Core" Chromosomal Genes	20
5.1.2	Diversity Based on "Accessory" Symbiotic Genes	22
6	Phylogenetic and Biogeographic Inferences Based on the Analysis of Rhizobial Core and Accessory Genes	25
7	Taxonomy and Phylogeny of Legumes	27
8	Evolution of Interaction Between Rhizobia and Their Legume Hosts	28
8.1	Divergence Events in Rhizobia–Legume Symbiosis	29
8.2	Co-evolution of Rhizobia–Legume Symbiosis	31
9	Conclusions	32
References		32

Abstract Rhizobia are soil bacteria that are able to fix nitrogen in symbiosis with plants from the family Leguminosae. Rhizobia populate both soil and nodule niches. The rhizobial genome can be divided into a core of housekeeping genes and an accessory pool of non-essential genes. The accessory genome, together with factors acting within plants, confers symbiotic interaction between rhizobia and plant hosts. Rhizobia are distributed within the α- and β subdivisions of the *Proteobacteria* and intermingled with non-symbiotic photosynthetic and pathogenic relatives in the following genera: *Allorhizobium*, *Azorhizobium*, *Bradyrhizobium*, *Mesorhizobium*, *Rhizobium*, *Sinorhizobium* and *Methylobacterium* (α-rhizobia), as well as *Burkholderia* and *Cupriavidus* (β-rhizobia). Recently α-proteobacterial nitrogen-fixing isolates have been identified in the genera *Ochrobactrum, Devosia* and *Blastobacter*. Lateral gene transfer of the mobile accessory genome is the most likely explanation for the occurrence of rhizobial symbiotic loci in distantly related genera of proteobacteria. Consequently, similar symbiotic types can be found in different chromosomal backgrounds, and the same chromosomal background can harbour different symbiotic genotypes. Thus, comparisons of genes from core loci and the accessory gene pool reflect their separate evolutionary histories. Similarly, phylogenies from ribosomal sequences of rhizobia do not show parallel divergence with plant taxonomy. Phylogenies of closely evolving organisms might develop in parallel. In the case of rhizobia, only phylogenetic trees based on symbiotic genes show some correlation with host plant range, indicating that evolution of nodulation genes could develop under the functional constraint of the plant.

1
Introduction

Lightning accounts for the fixation of 10% of the globally available nitrogen. The remaining 90% of the 255 million tons of N_2 annually fixed on Earth is fixed through the action of microbes. Prokaryotes are the only organisms capable of utilizing molecular nitrogen. Symbiosis with leguminous plants alone accounts for 20% of the global biological nitrogen fixed annually. Bacteria and other micro-organisms that can convert atmospheric nitrogen to ammonia using the enzyme nitrogenase are called **diazotrophs** and the process is known as *biological nitrogen fixation (BNF)*.

The formula for BNF is:

$$N_2 + 8H^+ + 8\ e^- + 16ATP \rightarrow 2NH_3 + H_2 + 16ADP + 16P_i\ .$$

Diazotrophs are scattered across microbial taxonomic groups, mostly in the Bacteria but also in the Archaea (Leigh 2000). Within a diazotrophic species there are both nitrogen fixing and non-fixing strains (Postgate 1998).

Diazotrophs can be divided into three groups of i) non-symbiotic or free-living (e.g. *Azotobacter*); ii) associative symbiotic (e.g. *Azospirillum*); and iii) symbiotic diazotrophs (e.g. *Frankia* and rhizobia). The group of of non-symbiotic N_2 fixers comprises bacteria living independently of plant root systems. Within the second group, associative symbiotic bacteria live partly within the root and partly outside the plant. These bacteria-plant associations do not result in nodule formation. Only symbiotic diazotrophs such as the gram-positive filamentous actinomycete (*Frankia*) (Benson and Silvester 1993) and the gram-negative proteobacteria (rhizobia) are capable of infecting the plant root and inducing nodule-like structures.

Frankia species establish symbioses with non-leguminous actinorhizal plants belonging to eight families: Betulaceae, Casuarinaceae, Coriariaceae, Datiscaceae, Elaeagnaceae, Myricaceae, Rhamnaceae and Rosaceae.

The rhizobia establish symbioses mainly with a wide variety of plant species belonging to the family Leguminosae (Fabaceae), commonly known as legumes, including several important food and fodder plants such as beans, peas, soybean, clover and peanut. An exception is the non-legume *Parasponia* belonging to the family Ulmaceae, which also has the ability to form nodules with rhizobia (Appleby et al. 1983). In contrast to associative symbiotic nitrogen fixing microbes, most of the nitrogen fixed by rhizobia or *Frankia* species is transferred and assimilated by the plant and used for growth (Hirsch et al. 2001).

Legumes have the ability to enter into mutually beneficial symbioses with either rhizobia or mycorrhiza that provide the plant with water and minerals. Rhizobial and mycorrhizal invasion share a common plant-specified genetic programme controlling the early host interaction (Radutoiu et al. 2003).

Rhizobia are distinguished from other living and symbiotic nitrogen-fixing microorganisms by their ability to make Nod factor molecules. Legumes are also differentiated in their ability to recognize rhizobial specific signal molecules (Hirsch et al. 2001; Perret et al. 2000; Radutoiu et al. 2003).

Rhizobia populate both soil and nodule niches. Culture collections of most rhizobial species are dominated by nodulating strains because isolates are nearly always obtained from root nodules. However, they only represent a small part of the bacteria that carry symbiotic genes. The use of *nodD*-specific primers revealed a wider diversity of sequences in soil DNA than was found in nodule isolates (Zézé et al. 2001).

Rhizobia need to maintain sets of genes that enable them to function in these two very dissimilar environments. Rhizobia growing *ex planta* are able to survive in the rhizosphere for years, by leading a saprophytic life as non-symbiotic rhizobia. Such rhizobia may either exchange genetic information with other soil organisms or possibly, upon introduction of legumes, may acquire symbiotic genes from inoculant strains (Sullivan et al. 1996). Rhizobia cannot fix nitrogen independently in the soil but are able to do so in symbiosis with the plant host. However, adaptations that allow a life in the soil environment are poorly understood.

Thus, more information about rhizosphere bacteria and their genomes would be very welcome. This would allow a reliable hypothesis to be formed concerning the rhizosphere gene pool and its dynamics (Martinez-Romero et al. 2006).

Early classification of rhizobia was based on the cross-inoculation group concept, according to which rhizobia were grouped based on their ability to specifically infect and fix molecular nitrogen with a certain group of legumes (Fred et al. 1932). Host-based classification would be practical if rhizobial host-ranges mirrored legume taxonomy, but quite many rhizobia form symbioses across taxonomic divisions. Thus, it soon became obvious that nodulation ability spanned inoculation groups and could not be used alone in the taxonomic classification of rhizobia. However, the symbiotic phenotype has been always regarded as a very important trait in understanding the biology of rhizobia

Later, bacterial species started to be delineated comprehensively at phenotypic, genotypic and phylogenetic levels by polyphasic taxonomy (Vandamme et al. 1996; Martinez-Romero et al. 2006). The polyphasic approach has led to the continuous creation and distinction of new rhizobial genera and species (Vandamme et al. 1996; van Bercum and Eardly 1998).

According to the host plants, all rhizobia were first classified into one genus *Rhizobium*. Later, they were divided into fast-growing *Rhizobium* and slow-growing *Bradyrhizobium*. The validity of distinguishing fast and slow growers was confirmed by sequence divergences of small 16S rRNA (*rrs*) and large 23S rRNA (*rrl*) subunits of ribosomal RNA (Willems and Collins 1993; Yanagi and Yamasto 1993; Young et al. 1991). Young et al. (1991) additionally showed a clear division of rhizobia into three clusters: the fast or moderately growing genera *Rhizobium*, *Allorhizobium*, *Sinorhizobium* and *Mesorhizobium*, the slow-growing *Bradyrhizobium* and stem-nodulating *Azorhizobium*. Moreover, fast-growing rhizobia are closely related to non-symbiotic pathogenic or non-pathogenic bacteria from the *Agrobacterium* genus (Young and Haukka 1996).

Nowadays rhizobia do not form a single taxonomic cluster. They are instead distributed within distantly related lineages of α- and β- subdivisions of *Proteobacteria* (Table 1). Consequently, rhizobia are intermingled with non-rhizobial plant and animal pathogens and photosyntetic bacteria and many non-symbiotic bacteria are more closely related to rhizobia than rhizobia are to one another (Fig. 1; Table 1; Young and Haukka 1996).

Table 1 List of known rhizobial and non-rhizobial nodule forming bacteria species according to "List of bacterial names with standing in nomenclature" (http://www.bacterio.net/) by J. P. Euzéby

Species	Representative (type) strain[a]	Examples of host plant genera[b]
α-Rhizobia		
Bradyrhizobiaceae		
Blastobacter		
B. denitrificans (*Bradyrhizobium denitrificans*)[c]	IFAM 1005[T] (LMG 8443)	*Aeschynomene* (P)
Bradyrhizobium		
B. canariense	BTA-1[T]	*Chamaecytisus* (P), *Lupinus* (P)
B. elkanii	USDA 76[T]	*Glycine* (P), *Macroptilium* (P), *Vigna* (P)

Table 1 (continued)

Species	Representative (type) strain[a]	Examples of host plant genera[b]
B. japonicum	ATCC 10324[T]	Glycine (P), Macroptilium (P), Vigna (P)
B. liaoningense	2281[T]	Glycine (P)
B. yuanmingense	CCBAU 10071[T]	Lespedeza (P)
Brucellaceae		
Ochrobactum		
O. lupini	LUP21[T]	Lupinus (P)
Hyphomicrobiaceae		
Azorhizobium		
A. caulidonas	ORS 571[T]	Sesbania (P) stem nodules
A. doebereinerae	UFLA1-100[T]	Sesbania (P)
Devosia		
D. neptuniae	LMG 21357[T]	Neptunia (M)
Methylobacteriaceae		
Methylobacterium		
M. nodulans	ORS 2060[T]	Crotalaria (P)
Rhizobiaceae		
Allorhizobium		
A. undicola	ORS 992[T]	Neptunia (M), Medicago (P), Acacia (M), Lotus (P), Faidherbia (M)
Ensifer		
E. adhaerens	R-14065[T]	Sesbania (P), Leucaena (M), Pithecellobium (M), Medicago (P)
Rhizobium		
R. etli	CFN42[T]	
bv. mimosae		Phaseolus (P), Leucaena (M), Mimosa (M)
bv. phaseoli		Phaseolus (P)
R. gallicum	R602[T]	
bv. gallicum		Phaseolus (P), Leucaena (M), Macroptilium (P), Onobrychis (P), Phaseolus (P)
bv. phaseoli		
bv. orientale		Medicago (P), Gueldenstaedtia (P), Coronilla (P)
R. giardinii	HAMBI 152[T]	
bv. giardinii		Phaseolus (P), Leucaena (M), Macroptilium (P)
bv. phaseoli		Phaseolus (P)

Table 1 (continued)

Species	Representative (type) strain[a]	Examples of host plant genera[b]
R. galegae bv. officinalis bv. orientalis	HAMBI 540[T]	Galega (P)
R. hainanense	I66[T]	Acacia (M), Arachis (P), Centrosema (P), Desmodium (P), Macroptilium (P), Stylosanthes (P), Tephrosia (P), Uraria (P), Zornia (P)
R. huautlense	S02[T]	Sesbania (P), Leucaena (M)
R. indigoferae	CCBAU 71042[T]	Indigofera (P)
R. leguminosarum bv. trifolii	ATCC 10004[T]	Trifolium (P)
bv. viciae		Pisum (P), Vicia (P), Lathyrus (P), Lens (P)
bv. phaseoli		Phaseolus (P), Vigna (P), Glycine (P), Cajanus (P)
R. loessense	CCBAU 7190B[T]	Astragalus (P), Lespedeza (P)
R. lusitanum	P1-7	Phaseolus (P)
R. mongolense (R. gallicum bv. orientale)[d]	USDA 1844[T]	Medicago (P), Gueldenstaedtia (P), Coronilla (P), Phaseolus (P)
R. sullae	IS123[T]	Hedysarum (P)
R. tropici type A	CFN 299	Phaseolus (P), Leucaena (M), Amorpha (P)
type B	CIAT 899[T]	Phaseolus (P), Leucaena (M)
R. yanglingense (R. gallicum bv. orientale)[d]	SH 22623[T]	Gueldenstaedtia (P), Coronilla (P), Amphicarpaea (P)
Sinorhizobium		
S. americanum	CFNEI 156[T]	Acacia (M)
S. arboris	HAMBI 1552[T]	Acacia (M), Prosopis (M)
S. fredii	PRC 205	Glycine (P)
S. kostiense	HAMBI 1489[T]	Acacia (M), Prosopis (M)
S. kummerowiae	CCBAU 71042[T]	Kummerowia (P)
S. medicae	A321[T]	Medicago (P)
S. meliloti	LMG 6133[T]	Medicago (P), Melilotus (P), Trigonella (P)
S. morelense	Lc04[T]	Leucaena (M)
S. saheli	ORS 609[T]	Sesbania (P)
S. terangae bv. sesbaniae	ORS 1009[T]	Sesbania (P)
bv. acaciae		Acacia (M), Leucaena (M)

Table 1 (continued)

Species	Representative (type) strain[a]	Examples of host plant genera[b]
S. xinjiangense	ATCC 49357[T]	Glycine (P)
Phyllobacteriaceae		
Mesorhizobium		
M. amorhae	ATCC 19665[T]	Amorpha (P)
M. chacoense	CECT 5336[T]	Prosopis (M)[T]
M. ciceri	UPM-Ca7[T]	Cicer (P)
M. huakuii	CCBAU 2609[T]	Astragalus (P)
M. loti	NZP 2213[T]	Lotus (P), Anthyllis (P), Lupinus (P)
M. mediterraneum	LMG 14990[T]	Cicer (P)
M. plurifarium	ORS 1032[T]	Acacia (M), Prosopis (M)
M. septentrionale	SDW014[T]	Astragalus (P)
M. temperatum	SDW018[T]	Astragalus (P)
M. tianshanense	A-1BS[T]	Glycyrrhiza (P)
Phylobacterium		
P. trifolii	PETP02[T]	Trifolium (P), Lupinus (P)
β-Rhizobia		
Burkholderiaceae		
Burkholderia		
B. caribensis	LMG 18531[T]	Mimosa (M)
B. mimosarum	PAS 44[T]	Mimosa (M)
B. phymatum	LMG 21445[T]	Aspalathus (P), Machaerium (P)
B. tuberum	LMG 21444[T]	Aspalathus (P), Machaerium (P)
Cupriavidus		
C. taiwanensis	LMG 19424[T]	Mimosa (M)
Oxalobacteriaceae		
Herbaspirillum		
H. lusitanum	P6-12[T]	Phaseolus (P)

[a] USDA, US Department of Agriculture, Beltsville, MD; ATCC, American Type Culture, Rockville, MD; CCBAU, Culture Collection of Beijing Agricultural University, Beijing, China; ORS ORSTOM Collection, Dacar, Senegal; LMG, Collection of Bacteria of the Laboratorium voor Microbiologie, Ghent, Belgium; CFN, Centro de Investigacion sobre Fijacion de Nitrogeno, Universidad NacionalAutonoma de Mexico, Cuernavaca, Mexico; H, HAMBI Culture Collection of Division of Microbiology, University of Helsinki, Finland; CIAT, Rhizobium Collection, Centro International de Agricultura Tropical, Cali, Columbia; NZP, Colture Collection of the Department for Scientific and Industrial Research, Palmerston North, New Zeland.
[b] (M), (P) are abbreviations for two subfamilies of the Leguminosae family: Mimosoideae and Papilionoideae, respectively
[c] Strain reclassification according to van Bercum et al. (2006)
[d] Strain reclassification according to Silva et al. (2005)

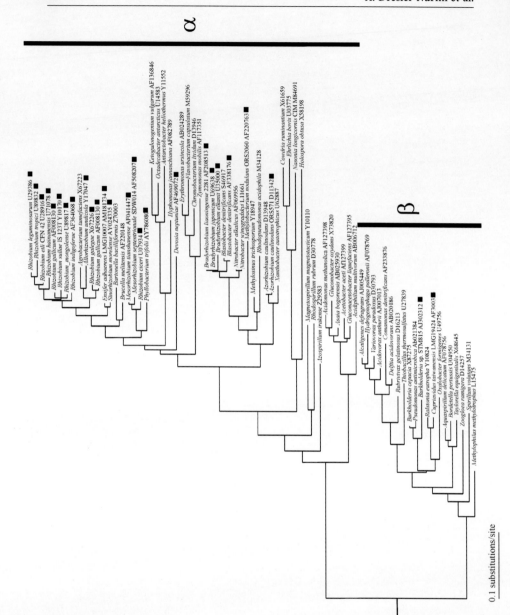

Fig. 1 A maximum likelihood tree based on the *rrs* gene from 75 taxa from α- and β- subdivisions of the proteobacteria. Representatives of species capable of forming nodules are marked with a *black box*

2
The Structure of the Rhizobial Genome

2.1
Chromosome and Plasmids

The rhizobial genome (5–8 Mb) is complex and physically partioned into circular chromosomes, smaller megaplasmids (e.g. *Sym* plasmids) and other plasmids (Martinez 1990). The number and size of plasmids and their gene composition vary between species and can also differ in other isolates of the same species (Young et al. 2006).

Daughter cells however are not always identical to the mother. Different strains of the same bacterial species are created through *horizontal gene transfer* (HGT) events in which an organism transfers genetic material to another cell that is not its offspring. In prokaryotes, horizontal gene transfer of accessory genomes is facilitated by three common mechanisms: i) transformation in which genetic material in the form of DNA or RNA is introduced, taken up, and expressed in a foreign cell; ii) transduction in which bacterial DNA is moved from one bacterium to another by a bacteriophage and iii) conjugation in which living bacterial cells transfer genetic material through cell-to-cell contact (Levin and Bergstrom 2000). Once the donor chromosome segment gets into the recipient cell, recombination events may occur. Changes in the genome are then heritable.

IS (insertion sequence) elements are mobile genetic elements often present in bacterial genomes. In rhizobia, they are parts of either the core or accessory gene pools and are commonly found flanking gene clusters (also the symbiotic compartments). Due to their mobility, IS elements promote genetic diversification through processes such as genomic rearrangement, recombination and transposition (Freiberg et al. 1997; Martinez 1990).

3
Species Concept Applicable to Rhizobia

An enormous potential for speciation, greater than in the sexual world of animals and plants, can be detected in the bacterial world (Cohan 2002). Defining the concept of a bacterial species is not easy and bacterial systematics has not yet reached a consensus for species definition that would satisfy all needs. The bacterial genus is even more difficult to define than the species (Cohan 2002).

Bacterial speciation is an everyday process. During evolution, bacteria on one hand diversify and on the other, form cohesive clusters; this process is called speciation. Speciation is an evolutionary process, which requires a sufficient amount of genetic difference and reproductive isolation between populations.

The species would thus be defined as a group whose recognition is based on significant biological discontinuities (barriers) among diverse populations. The speciation process may move along diverse routes. The simplest way to reproductively isolate a derived population from the ancestral population is to locate it in a separate geographic place (allopatric speciation) (Mayr 1957). If a population is physically and genetically isolated, even later gene flow from the ancestral population might not overcome the genetic difference. Among rhizobia non-geographic isolation can also be the cause of reproductive isolation. Then this isolation is then due to recombination events (sympatric speciation) (Vinuesa et al. 2005c).

The species can contain multiple distinct geographical entities (ecotypes) or symbiotic phenotype variants (biovars). Thus, it is very important to recognize these ecotypes or biovars as a component of rhizobial diversity. Contrary to Cohan's (2002) viewpoint on ecotypes or biovars exhibiting very important features of species, Silva et al. (2005) argued that rhizobial symbiotic phenotype populations should be recognized taxonomically at a subspecific rank (subspecies, biovar). Obviously, due to events of lateral transfer of plasmid to other bacteria of the same or different species or simply loss of plasmids or genomic islands, rhizobial symbiotic phenotypes cannot form the basis for defining a species (Lindström and Gyllenberg 2006; Martinez-Romero et al. 2006).

The use of estimates of evolutionary relationships inferred from sequence variation of the *rrs* gene has been widely used in the delineation of rhizobial genera. However, rhizobial genomes can contain divergent ribosomal copies due to gene conversion and recombination events. Therefore, analyses of other chromosomal 'core' housekeeping protein-coding genes have been found to be good markers for establishment of relationships between species at the genus and species level. The multilocus sequence analysis (MLSA) approach is regarded to be more robust than analyses done using only *rrn* operon markers (Gaunt et al. 2001; Martinez-Romero et al. 2006; Silva et al. 2005; Vinuesa et al. 2005a,b,c; Wernegreen and Riley 1999). The resolving power of such approaches is maximized when large rhizobial strain collections are used (Palys et al. 1997; Silva et al. 2005; Vinuesa and Silva 2004; Vinuesa et al. 2005b).

The core and accessory components of the rhizobial genome are useful for the definition of a rhizobial species (Lan and Reeves 2000, 2001; Martinez-Romero et al. 2006). A major factor in the maintenance of species specificity in the core genome is the existence of recombination barriers between species. There is no strong selection pressure for capture of new genes from the outside in the core genome. In contrast, selective pressure will lead to a capture of new genes from the outside by lateral transfer in the accessory genome.

The rhizobial accessory genome is related more to lifestyle and can determine different ecological niches on similar core backgrounds (Young et al.

2006). Conversely, the same symbiotic genotype can be found in divergent core genomes (Louvrier et al. 1996; Silva et al. 2005).

However, a bacterial species is not just a biological concept. The naming of bacteria is a social necessity, and a modern bacterial species concept takes the needs of the end-users into consideration. Lindström and Gyllenberg (2006) argued that the most useful species concept is a cross-disciplinary one. This concept is created from the interaction of philosophical, biological and social learning components.

4
Rhizobia-legume Symbiosis

An effective symbiosis between rhizobia and leguminous plant requires among others several bacterial nodulation genes (*nod, nol* and *noe*) and nitrogen fixation genes (*nif* and *fix*). Nodulation genes are essential for the first steps of infection and nodule formation in the rhizobia-legume symbiosis. The *nod* genes responsible for production of Nod factors, lipo-chitin oligosaccharide (LCOs), can be divided into three groups: the regulatory, common and host-specific genes (Kondorosi et al. 1984; Martinez 1990). The *nif* and *fix* genes are responsible for encoding and controlling the nitrogen-fixing apparatus.

Flavonoid co-inducers derived from host plants promote attachment of rhizobia to the young growing root hairs of legumes. More than 4000 different flavonoids have been identified in vascular plants, and a particular subset is involved in mediating host specificity in legumes (Perret et al. 2000). In the presence of flavonoids, regulatory genes such as *nodD, syrM, nolA, nolR* and *nodVW* regulate the activity of nodulation genes (Spaink et al. 1987; Perret et al. 2000).

4.1
Molecular Interactions Between Partners

Nod factors are the main molecular signals produced by rhizobia that advertise the presence of a suitable symbiotic microbe to the plant. Nod factors are synthesized through the action of enzymes encoded by the induced common *nodABC* and host specific nodulation genes (*nodEIJFUNS*) and other *nol* and *noe* genes.

The *nodABC* genes among others are the early nodulation genes, which are responsible for the core structure of Nod factors which usually consists of four or five β-1-4 linked N-acetylglucosamine residues. They are called common because they are structurally and functionally conserved and have been found in all rhizobia studied to date. Previously, these genes were regarded as interchangeable between species (Martinez 1990). Later, it was found that in

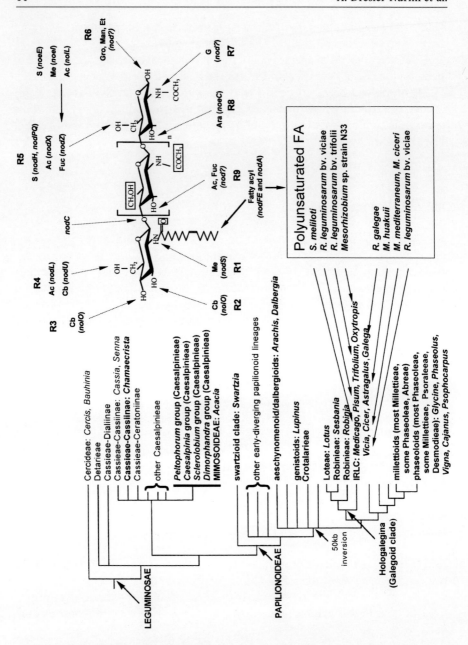

many cases mutation in these genes can not be complemented by genes from other species without changing host range (Roche et al. 1996).

In addition to common *nod* genes, the host-specific nodulation genes such as *nodEF*, *nodH*, *nodL*, *nodPQ* and *nodSU*, *nodX*, *nodZ* and *nol* and *noe* participate in modification of the core structure of Nod factor (Fig. 2). These substitutions

Fig. 2 Phylogenetic relationships in Leguminosae (*left*) and Nod factor structure (*right*). In the phylogenetic tree, taxa in bold face are dominated by nodulating species. The IRLC (inverted repeat-lacking clade) comprises among others galegoid legumes the symbionts of which have been described as producing Nod factors with polyunsaturated fatty acids (frame on the left). *Left* and *right* parts of figure are adapted and reprinted with the permission from the Plant Physiology, Doyle and Luckow 2003 and Annual Review of Microbiology, Spaink 2000, respectively. Abbreviations for substitutions in Nod factor: Ac, acetyl; Ara, arabinosyl; Et, ethyl; Cb, carbamoyl; Fuc, fucosyl; G, N-glycolyl; Gro, glyceryl; H, hydrogen; Man, mannosyl; Me, methyl; S sulphate

play more subtle roles in nodulation by enhancing rather than supporting the basic structure of Nod factors (Perret et al. 2000; Terefework et al. 2000).

4.2
Determinants of Host Specifity in Rhizobia

In rhizobia, host specificity plays a key role in the establishment of the effective symbiosis.

Although many host plants and rhizobia have the ability to enter into symbiosis with more than one partner, only a certain set of symbionts lead to the formation of nitrogen-fixing nodules. Exceptionally, tropical leguminous trees, such as *Acacia*, *Prosopis* or *Calliandra* can form nodulation symbioses with various rhizobia from different genera. However, the specificity between symbiotic partners minimizes the formation of ineffective, nonfixing nodules by the host plant (Perret et al. 2000). The construction of the nodules requires extra energy and nutrient sources from the host.

Rhizobia differ in their response to different signal molecules produced by legumes. Some rhizobia have a narrow host range and form nodules with a limited number of legumes. For example *A. caulinodans*, *S. saheli* and the sesbaniae biovar of *S. terangae* nodulate only *Sesbania rostrata* (Boivin et al. 1997) and *R. galegae* is the only symbiont of *Galega officinalis* and *Galega orientalis* (Lindström 1989). In contrast some rhizobia have a broad host range and are capable of nodulating a wide spectrum of legumes with various degrees of promiscuity. For example, *Sinorhizobium* sp. NGR234 and the closely related *S. fredii* USDA257 nodulates at least 112 and 77 legumes from two different tribes, respectively (Pueppke and Broughton 1999). Conversely, legumes may also be host to only one kind of symbiont (*Galega* spp.) or establish symbioses with a wide range of rhizobia (*Leucaena leucocephala*, *Calliandra calothyrsus*, *Phaseolus vulgaris*).

Distantly related rhizobia can nodulate the same host, e.g. *S. fredii*, *B. japonicum* and *B. elkanii* all nodulate *Glycine max*. Members of *Rhizobium*, *Sinorhizobium* and *Bradyrhizobium* are less related to each other than to non-rhizobial genera (Fig. 1). Stem- and root-nodulating *A. calidonans* and root-nodulating *S. fredii* and *S. terangae* bv. sesbaniae, both symbionts of *Sesbania rostrata*, also represent two taxonomically distant genera.

R. etli, R. gallicum, R. giardinii, R. leguminosarum and *S. terangae* are heterogenous from the symbiotic point of view and are subdivided into biovars (Table 1). It is important to note that whereas all the host plants for rhizobia from the three *R. leguminosarum* biovars belong to the Papilionoideae subfamily, some of host legumes that are nodulated by *R. etli* bv. *mimosae*, *R. gallicum* bv. *gallicum*, *R. giardinii* bv. *giardinii* and *S. terangae* bv. *acaciae* belong to the Mimosoideae subfamily (Table 1; Pueppke and Broughton 1999).

4.2.1
The NodD/Flavonoid Complex

The *nodD* gene is ubiquitous in rhizobia. Due to various copies of *nodD* and their different flavonoid preferences, symbiotic characteristics of *nodD* often vary among species (Sect. 5.1.2) (Perret et al. 2000). For example, in case of *R. leguminosarum* possessing only one *nodD*, its mutation leads to Nod⁻ phenotype. In contrast, *S. meliloti* carries three copies of *nodD* and only mutation of all three copies results in a Nod⁻ phenotype. Moreover, two out of three *nodD* copies in *S. meliloti* recognize different plant exudates and thus can play a role in host specificity (Honma et al. 1990). A different case is observed with the broad host- range *Sinorhizobium* sp. strain NGR234, which possesses two *nodD* loci (Perret et al. 1991). Mutation in *nodD1* results in a Nod⁻ phenotype on different host plants tested, but mutation of *nodD2* does not play an active role in the control of nodulation by NGR234 (Bassam et al. 1988). In comparison with other rhizobia, the *nodD1* gene of NGR234 responds to a large number of flavonoids. Thus, a major difference between the narrow host range rhizobia and the most promiscuous rhizobium NGR234 is that the NodD1 of NGR234 can interact with a broad spectrum of inducing flavonoids (Bassam et al. 1988).

However, for all correlations between types of flavonoids interacting with NodD protein and the breadth of the host range, only the type of NodD/flavonoid complex is insufficient to explain the phenomenon of host specificity (Perret et al. 2000).

4.2.2
Nod Factors

Depending on the nature of the fatty acids N-linked to the terminal non-reducing end of the Nod factor molecule (position "fatty acyl" in Fig. 2), two major Nod factor groups can be distinguished (Terefework et al. 2000; Yang et al. 1999).

The first one is composed of Nod factors acylated by common saturated or monounsaturated fatty acids coming from the general lipid metabolism. Moreover, Nod factor molecules of strains belonging to this group consist of at least one of the two the following substitutions: 1) N-methyl group at the terminal non-reducing glucosamine residue (position R1, Fig. 2), 2) a fucose or fucose-derived group at the terminal reducing residue (position R5, Fig. 2; Terefework et al. 2000).

Second, a more rare group consists of Nod factors with polyunsaturated α, β-fatty acids containing two to four double bounds. Synthesis of such Nod factors is directed by the products of *nodEF* genes. Polyunsaturated Nod factors produced by *M. huakuii*, *Mesorhizobium* sp. strain N33, *R. galegae*, *R. leguminosarum* bv. vicieae, *R. leguminosarum* bv. trifolii and *S. meliloti* are necessary for nodulation of legumes in the genera *Astragalus* (Galegae), *Oxytropis* (Galegae), *Galega* (Galegae), *Vicia* (Viciae), *Trifolium* (Trifolieae) and *Medicago* (Trifolieae), respectively.

Rhizobia associated with these legumes are highly specialized, having restricted host ranges (Table 1). Their *nodA* phylogeny reveals a common ancestor (Fig. 3). Moreover, the legume phylogeny also displays a common origin for the host plants of these rhizobia (Figs. 2 and 3; Doyle and Luckow 2003). These legumes are part of the galegoid clade (Hologalegina), which, together with the phaseoloid clade, constitute two well supported major subclasses of papilionoids (Fig. 2; Wojciechowski et al. 2000).

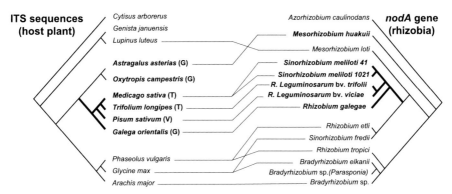

Fig. 3 Phylogenetic congruence between ITS (internal transcribed spacer) sequences obtained from host legumes (*left*) and rhizobial nodA (*right*). *Lines* between two trees indicate plant host-rhizobia associations. Galegoid host plants and rhizobial species associated with them are shown in **bold**. (G), (T) (V) are abbreviations for three tribes of the Leguminosae family: Galegae, Trifolieae and Viciae, respectively. Adapted and reprinted with permissions from Suominen et al. (2001) and Oxford University Press. See also Fig. 2

The presence of two different classes of Nod factors indicates two different recognition strategies by plant receptors. For strains from the first group the terminal reducing residue might be the most important one, whereas for strains producing Nod factors with polyunsaturated fatty acids, selection of symbiotic partner depends on fatty acid group (Terefework et al. 2000).

Thus, structures of Nod factors play important role in host specificity but they are not only determinants. Nod factors produced by *R. etli* and *M. loti* are identical but they have very distinct host ranges, *Phaseolus* sp. and *Lotus* sp., respectively (Cárdenas et al. 1995). In contrast, two different rhizobia (*R. tropici* and *R. etli*) that nodulate the same plant (*P. vulgaris*) may secrete different Nod factors. Thus, although Nod factors are essential for nodulation and their modifications often contribute to host-specificity, this family of signal molecules is probably just one of several elements specifying host range (Perret et al. 2000).

It seems that a combination of flavonoids/NodD proteins/Nod factors is insufficient to explain host specificity completely. Some additional recognition factors which act within the plant must control development of symbiosis (Perret et al. 2000; Radutoiu et al. 2003).

5
Diversity and Taxonomy of Rhizobia

Rhizobia belong to the α- and β- subdivisions of *Proteobacteria*. Until recently rhizobia were believed to all belong to just the α-subdivision of the *Proteobacteria*. Traditionally rhizobia were split into six different genera according to phenotypic and genetic characterizations: *Allorhizobium, Azorhizobium, Bradyrhizobium, Mesorhizobium, Rhizobium* and *Sinorhizobium* (Sawada et al. 2003). Recently, a new kind of root-nodule bacteria from the β-subdivision of *Proteobacteria* were isolated and identified: *Bulkholderia phymatum* and *B. tuberum* from *Aspalathus* and *Machaerium* in Africa and South America (Moulin et al. 2001); *Ralstonia taiwanensis* as a symbiont of *Mimosa pudica* from Taiwan (Chen et al. 2001) and *Herbaspirillum lusitanum* nodulating *Phaseolus vulgaris* in Portugal (Valverde et al. 2003). Later, *Ralstonia* species was transferred to the genus *Cupriavidus* (Vandamme and Coenye 2004).

Moreover, when new legumes have been investigated, nodulating isolates from α-*Proteobacteria* have been identified in genera not commonly referred to as a rhizobia, including *Methylobacterium* (Jourand et al. 2004; Sy et al. 2001), *Ochrobactrum* (Ngom et al. 2004; Trujillo et al. 2005), *Devosia* (Rivas et al. 2002) and *Blastobacter* (van Berkum and Eardly 2002; Fig. 1; Table 1).

5.1
Phenotypic and Genotypic Diversity

The formation of nodules on the legume host continues to be regarded as the most important phenotypic trait because of the practical agricultural importance of rhizobia.

In addition to symbiotic phenotype, other phenotypic features such as FAME (fatty acids methyl esters), SDS-PAGE (whole-cell protein analysis using sodium dodecyl sulfate-polyacrylamide gel electrophoresis) and MLEE (multilocus enzyme electrophoresis) followed by numerical taxonomy have successfully been used for grouping and characterization of unknown strains and for the description of novel species of rhizobia (Table 2; Vandamme et al. 1996).

Later, techniques based on the PCR have dominated studies of rhizobial diversity. PCR-RFLP analysis of the 16S rDNA has been the most popular one. This method has been regarded as an initial and rapid tool for identification and estimation of genetic relationship at species and higher level (Table 2).

PCR fingerprinting methods such as rep-PCR, PCR with random or arbitrary primers and AFLP (Amplified Fragment Length Polymorphism) are based on whole genome. Depending on the rhizobia, these tools have been reported sometimes to have a higher resolving power than results from DNA sequences of protein-coding genes or ITS regions, from DNA-DNA hybridization and from MLEE analyses (Rademaker et al. 2000; Vinuesa et al. 2005c).

Moreover, AFLP or rep-PCR together with population genetic analyses of sequence data for multiple loci (Sects. 5.1.1, 5.1.2 and 6) can help molecular ecologists to delineate genetic structures of rhizobial populations. The resolving power of such combinations is maximized by using a genetically and geographically diverse large collection of isolates (Vinuesa et al. 2005c).

Table 2 Techniques applied for studying the phenotypic and genotypic diversity of rhizobia

Method	Refs.
Phenotypic diversity	
Cross-nodulation tests	Fred et al. 1932
Morphological (shape, Gram strain, dimensions etc.); physiological and biochemical tests ex g.: pH, salt conc., carbon and amino acids sources) followed by numerical analysis	Vandamme et al. 1996
FAME (fatty acids methyl esters)	Jarvis et al. 1996; Jarvis and Tighe 1994
SDS-PAGE (profiles of whole cell proteins)	Jourand et al. 2004, 1993; de Lajudie et al. 1994, 1998; de Lajudie et al. 1998; Moreira et al. 1993, 1998; Nick et al. 1999
MLEE (multilocus enzyme electrophoresis)	Eardly and van Bercum 2005; Martinez-Romero et al. 1991; Nick et al. 1999); Silva et al. 2003; Vinuesa et al. 2005a,b; Wang et al. 1998
Genotypic diversity	
DNA-DNA hybridization	de Lajudie et al. 1994; de Lajudie et al. 1998; Laguerre et al. 1993; Vandamme et al. 1996
PFGE (Pulsed-field gel electophoresis)	Haukka and Lindstöm 1994

Table 2 (continued)

Method	Refs.
RFLP (Restriction fragment lenth polymorphism) coupled with DNA hybridization probes:	
symbiotic genes	Kaijalainen and Lindström 1989; Krishnan and Pueppke 1994
IS elements (IS-fingerprinting)	Andronow et al. 2003; Selbitschka et al. 1999
Ribosomal genes (ribotyping)	Martinez-Romero et al. 1991
PCR-RFLP (Restriction fragment polymorphism of PCR amplified fragments) of	
ribosomal genes	Laguerre et al. 1994, 1996; Gao et al. 2001; Terefework et al. 1998; Vinuesa et al. 2005b
rRNA ITS	Andronow et al. 2003; Doignon-Bourcier et al. 2000; Laguerre et al. 2003
symbiotic loci	Guo et al. 1999; Laguerre et al. 1996
16S rRNA sequencing	(Sect. 5.1.1.1, this chapter)
PCR based on repetitive elements (REP, ERIC, BOX and GTG$_5$)	Laguerre et al. 1996; Nick and Lindström 1994; Nick et al. 1999; Zhang et al. 1999; Vinuesa et al. 2005c
PCR with random or arbitrary primers (RAPD, AP-PCR)	Moschetti et al. 2005; Selenska-Pobell et al. 1996
AFLP	Doignon-Bourcier et al. 2000; Gao et al. 2001; Terefework et al. 2001; Wdowiak-Wrobel and Malek 2005; Willems et al. 2001; Wolde-Meskel et al. 2004

Variation in copy number of IS elements and their distribution between closely related species may provide high resolution fingerprints of rhizobial strains and can help to understand the role and dynamics of these elements in rhizobial populations (Martinez 1990).

5.1.1
Diversity Based on "Core" Chromosomal Genes

5.1.1.1
Ribosomal Genes

Genus affiliation of a new rhizobial species has been based on analysis of the complete *rrs* gene (Martinez-Romero et al. 2006). Additional analysis of the complete *rrl* may provide support of genus establishment (Terefework et al.

1998). It is now clear that single genomes may harbour divergent ribosomal gene copies (Coenye and Vandamme 2003; Haukka et al. 1996). Ribosomal genes can be also subject to gene conversion and recombination. In rhizobia, a high level of sequence mosaicism is found in this locus and may complicate the prediction of common ancestry (Eardly and van Bercum 2005; Silva et al. 2005; Vinuesa et al. 2005c).

Moreover, the phylogenies of *rrs* or the *rrl* are based on a single gene and such single-gene phylogeny may not reflect the evolution of the genome as a whole. These genes are not under the same selective pressures as the accessory gene set is.

In contrast to *rrs* and *rrl*, the genetic information contained in the variable internal transcribed spacer (ITS), is not appropriate for phylogenetic inference. However, for closely related rhizobial strains that are poorly distinguished by their *rrs* and *rrl* loci, the number of published ITS sequences has increased in recent years (van Bercum and Furmann 2000; Vinuesa et al. 2005a; Willems et al. 2001, 2003).

ITS is well suited to screening and strain characterization of rhizobial collections. For example, *R. galegae* strains can be clearly separated by their ITS-RFLP profiles into two distinct groups indicating two *R. galegae* biovars, officinalis and orientalis (Andronov et al. 2003).

5.1.1.2
Other "Core" Chromosomal Housekeeping Genes

Due to conservation and ubiquity, several unlinked and chromosomally encoded housekeeping genes have been used for phylogenetic studies in rhizobia:

1. The *atpD* gene essential for energy production, encoding the beta subunit of the membrane ATP synthase (Gaunt et al. 2001; Silva et al. 2005; Stepkowski et al. 2005; Vinuesa et al. 2005a,b,c; Weir et al. 2004),
2. The *recA* encoding part of the DNA recombination and repair system (Gaunt et al. 2001; Stepkowski et al. 2005; Vinuesa et al. 2005a,b,c; Weir et al. 2004),
3. The *dnaK* gene encoding the 70 kDa chaperone that prevents protein aggregation and supports the refolding of damaged proteins (Stepkowski et al. 2003, 2005) and
4. The *glnII* gene encoding glutamine synthetase II, a key enzyme in nitrogen assimilation (Silva et al. 2005; Stepkowski et al. 2005; Turner and Young 2000; Vinuesa et al. 2005a,c).

Phylogenetic trees reflect the genealogy of individual genes and not of the species, and phylogenetic trees based on different genes such as: *recA*, *atpD*, *GSII*, *dnaK* and *rrs* are often incongruent, making it difficult to determine the true evolutionary history of organisms (Nichols 2001). In many cases, however, phylogenetic inferences based on the protein-coding chromosomal housekeeping genes having the same evolutionary history are very similar to each other and can therefore be concatenated (combined) in the analyses, which maximize the phylogenetic signal by forming strongly supported clades. (Silva et al. 2005; Vinuesa et al. 2005b,c).

5.1.2
Diversity Based on "Accessory" Symbiotic Genes

5.1.2.1
Nitrogen Fixation (*nif* and *fix*) Genes

The genes which are responsible for conversion of atmospheric dinitrogen to ammonia are called *nif* and *fix*.

Different proteins encoded by *nifHDKENBQVSU* genes are involved in making Mo-dependent nitrogenase which is widely distributed among diazotrophs belonging to both aerobic and anaerobic bacteria and archaea (Young 1992). Until 2006, *Methanothermococcus thermolithothrophicus* has been known as the most thermophilic N_2-fixing organisms being able to fix N_2 at up to 64 °C (Belay et al. 1984). Later, Mehta and Baross (2006) described a hyperthermophilic methanogenic archaeon from nitrogen-limited deep-sea vent fluid environments that could fix nitrogen (N_2) at temperatures of up to 92 °C.

In the case of rhizobia, the most studied nitrogen fixation genes are *nifH* and *nifD*, both being components of a *nifHDK* operon:

1. The *nifH* coding for two identical subunits of component II dinitrogenase reductase is the most studied nitrogen fixation gene (Chen et al. 2003; Dobert et al. 1994; Haukka et al. 1998; Laguerre et al. 2001; Rivas et al. 2002; Trujillo et al. 2005; Vinuesa et al. 2005b,c). In rhizobia *nifH* can exist in multiple copies. However, sequencing of multiple copies has shown them to be practically identical.
2. The *nifD* encoding for the alpha subunits of component I in nitrogenase reductase (Parker et al. 2002; Ueda et al. 1995c).

The *nif* and *fix* genes are found within highly conserved operons and have remarkably congruent phylogenetic histories. The high homology of nitrogenases between different organisms can be due to conservation because of functional constraints (Ruvkun and Ausubel 1980). The age of symbiotic nitrogen fixation is unknown. However, Raymond et al. (2004) proposed that nitrogenase may have first arisen before the divergence of the three branches of the life (Archaea, Eukaryota and Bacteria). According to recent analyses, nitrogenase may first have arisen in thermophilic archaeon and subsequently horizontally transferred from anaerobic archaea to bacteria domain (Ciccarelli et al. 2006; Raymond et al. 2004; Mehta and Baross 2006).

It was proposed that *rrs* and *nifH* genes share a similar evolutionary history (Ueda et al. 1995c; Widmer et al. 1999). In contrast, Haukka et al. (1998) in her study describing broadly diversity and phylogeny of *Acacia* and *Prosopis* rhizobia, presented evidence that the *nifH* tree in many aspects is not consistent with the tree based on *rrs* but reveals the same clustering patterns as the *nodA* tree. This result agrees with the fact that the *nod* and *nif* genes in these species are linked to the same plasmid, so it is not surprising that they showed similar evolutionary patterns. Parker et al. (2002) also presented incongruent phylogeographic patterns in *rrs* and *nifD* within *Bradyrhizobium* populations distributed across North America, Central America, Asia and Australia. According to his theory, these geographic areas were initially colonized by several diverse *rrs* lineages, with subsequent horizontal gene transfer of similar *nifD* leading to increased *nifD* sequence homogeneity within each regional *Bradyrhizobium* population.

5.1.2.2
Nodulation Genes

Allelic variation in sequences of common and host specific as well as regulatory nodulation genes contribute to the control of host range via variation in Nod factor structure. The organization of the *nod* genes, their genome position, the presence or absence, and copy number vary considerably in different rhizobia (Fig. 4A,B).

Due to conservation and ubiquity, several nodulation genes have been used for phylogenetic studies:

1. The *nodD* gene and its alleles encoding the NodD transcriptional activator of other *nod* genes show clearly host specific features. Rhizobia typically contain one to five copies of *nodD* genes (Figs. 4B, 5). Complex effects on the breadth of host range by occurrence of several copies of *nodD* in rhizobia and by the fact that NodD homologues from the same strain may have different flavonoid preferences makes data derived from *nodD* genes difficult to interpret in phylogenetic studies. Because of the ability of NodD to recognise specific plant flavonoids, a correlation between *nodD* genes and host legumes has been sought, but has often led to misleading conclusions (Dobert et al. 1994; Rivas et al. 2002; Suominen et al. 2003; van Rhijn et al. 1993). In addition to analyses of ribosomal genes, sequences of *nodD* followed by an RFLP approach were used as markers to investigate symbiotic components of the genome in soils communities (Laguerre et al. 1996, 2003; Zézé et al. 2001).
2. The *nodC* gene encodes an N-acetylglucosaminyl transferase is the first step in Nod factor assembly, affecting the length of the chitin oligosacharide chain and determining the host plant range (Laguerre et al. 2001; Ueda et al. 1995b; Wernegreen and Riley 1999).

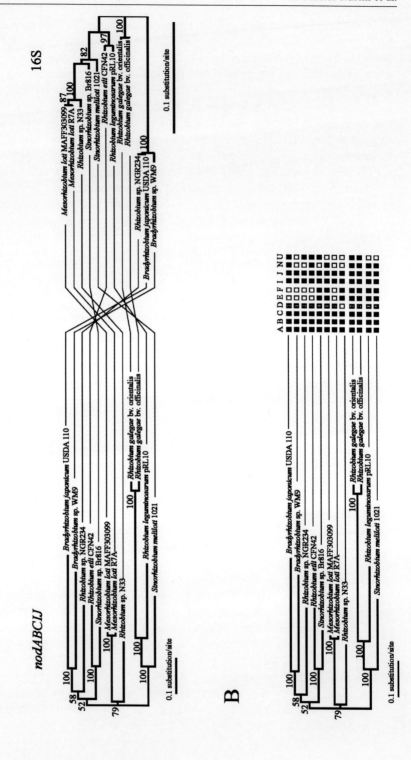

◄ **Fig. 4 A** Incongruence between *nod* (*right*) and *rrs* (*left*) sequences. Maximum likelihood *nod* and *rrs* trees are based on concatenated *nodA*, *nodB*, *nodC*, *nodI*, and *nodJ* and *rrs* sequences, respectively. Bootstrap values are given at the node and are based on 1000 replicates. **B** The presence or absence of nodulation genes mapped to a maximum likelihood tree based on concatenated sequences of *nodA*, *nodB*, *nodC*, *nodI* and *nodJ*. Bootstrap values from 1000 replicates are given at the node

3. The *nodB* gene encoding N-deacetylase, which removes the N-acetyl moiety from the non-reducing terminus of the N-acetylglucosamine oligosaccharides (Silva et al. 2005; Wernegreen and Riley 1999).
4. The *nodA* gene encoding an essential host-specific determinant, an acylotransferase enzyme that transfers fatty acids into the backbone of the Nod factor (Ba et al. 2002; Chen et al. 2003; Haukka et al. 1998; Moulin et al. 2004; Suominen et al. 2001; Sy et al. 2001).
5. The *nodZ*, *nolL*, and *noeI* genes involved in specific modifications of Nod factors common to bradyrhizobia, the transfer of a fucosyl group on the Nod factor core, fucose acetylation and fucose methylation, respectively (Moulin et al. 2004; Stepkowski et al. 2005).

Phylogenies of *nod* genes are well correlated with the host plant showing symbiotic similarity within or between nodulating species. In contrast, Phylogenies inferred from "core" chromosomal housekeeping and "accessory" *nod* genes have been reported to be incongruent (Ba et al. 2002; Haukka et al. 1998; Laguerre et al. 2001; Suominen et al. 2001; Ueda et al. 1995b; Wernegreen and Riley 1999).

6
Phylogenetic and Biogeographic Inferences Based on the Analysis of Rhizobial Core and Accessory Genes

Each gene has its own history and traces back to an individual molecular ancestor. Molecular ancestors were also likely to be present in different organisms at different times. On the other hand, genes are also indirectly records of the histories of the organisms possessing them (Zhaxybayeva and Gogarten 2004).

Similar symbiotic gene types can be found in different chromosomal backgrounds, and the same chromosomal background can harbour different symbiotic loci. Thus, chromosomal loci usually have an evolutionary history independent of nodulation (*nod*) and nitrogen fixation genes, which are marked by horizontal gene transfer of symbiotic loci across species boundaries (Fig. 4A). Trees based on *nodA* gene show some correlation with host plant range (Fig. 3) (Dobert et al. 1994; Haukka et al. 1998; Suominen et al. 2001; Ueda et al. 1995b; Wernegreen and Riley 1999).

Haukka et al. (1998) proposed five different taxonomic levels to evaluate factors influencing evolution of rhizobia: i) rhizobial level where *rrs* phy-

Fig. 5 A maximum likelihood *nodD* tree showing the relationship between *nodD* homologs from 13 taxa. The *nodD1* is marked with *black filled box*, *nodD2* with *black filled circle*, *nodD3* with *open circle* and *nodD4* with *black filled circle*

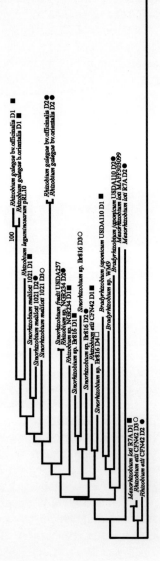

logeny follows nifH gene phylogeny; ii) the gene level where nod and nif phylogenies are congruent with host plant range; iii) symbiotic systems require the correct chromosomal background (at this level events of gene exchange can be possible); iv) geographical factors, like physical isolation and adaptation to new environmental conditions, can influence the range of legume hosts available; and finally v) the strain level where different combinations of nod, nif and rrs occur.

There are also some biogeographic implications of using core and accessory genes for rhizobial phylogeny. When a large number of isolates are used, genetic differentiation analyses based on the chromosomal protein-coding genes can also reveal a biogeographic pattern of isolates, unlike *rrs* sequences, for which such a correlation may not exist (Silva et al. 2005). Moreover, evidences for distribution of rhizobial isolates around the world, possibly by soil contaminated seeds (Pérez-Ramírez et al. 1998) has been discussed and confirmed with the help of housekeeping and nodulation genes phylogenies (Stepkowski et al. 2005).

7
Taxonomy and Phylogeny of Legumes

The angiosperms consist of 380 families of flowering plants. The family Leguminosae, or legumes, is the third largest family of dicotyledonous flowering plants, with approximately 650 genera and nearly 20 000 cosmopolitan species from tropical trees to alpine herbs (Doyle, 1994). Legumes dominate the lowland tropical rainforests of Africa and South America and are major components of dry and seasonally dry tropical forests as well. Legumes outnumbered all other plant families both in numbers of individuals and of species in a survey of various South American sites and in the lowland rainforests of Africa and Madagascar (Doyle 1994). Considered to be a tropical, the first ancestor of legumes with perhaps a late Cretaceous origin (65–70 Mya), the Leguminosae has an abundant and continuous fossil record since the Tertiary (65–1.8 Mya; Lavin et al. 2005). Many legume species have the ability to enter into a symbiotic relationship with rhizobia. The Leguminosae family is divided into three subfamilies, the probably monophyletic Mimosoideae (ca. 77 genera, 3000 spp.), the Papilionoideae (476 genera, 14 000 spp.) and the paraphyletic Caesalpinioideae (ca. 162 genera, 3000 spp.). The three subfamilies are broken into groups of genera called tribes (Doyle and Luckow 2003) (Fig. 2).

The subfamilies Caesalpinioideae and Mimosoideae contain mainly woody species growing in the tropics, whereas the species in the family Papilionoideae are largely herbaceous, often temperate, but some are tropical woody legumes (Sprent 1995). Fossil evidence suggests that the Caesalpinioideae is the oldest of the three subfamilies. The formation of nodules and presumably nitrogen-fixing ability are not 100% correlated within the legume family but although only a fraction of legumes have been studied, nodulation has been found in at least 95% of the Papilionoideae and 90% in Mimosoideae. In contrast, in the paraphyletic Caesalpinoideae subfamily only 30% of the members nodulate and this group includes many non-nodulating genera (Sprent 1995). Since the majority of Mimosoideae and Papilionoideae can nodulate and it is generally agreed that these two subfamilies evolved from the Caesalpinoideae, it is reasonable to suppose that their caesalpinoid ancestor also nodulated more commonly.

It was also suggested that the nodulation ability of legumes and the ability to fix atmospheric nitrogen in symbioses with rhizobia may have originated multiple times in the Leguminosae: once in *Chamaecrista*, the only caesalpinioid genus that is confirmed to be nodulated; once in the mimosoid line, and last in the papilinoid line (Doyle 1994; Sprent and Sprent 1990).

The morphologically based classification showed that plants from 10 families that are involved in nodulation symbioses with *Frankia* species and rhizobia are only distantly related. Revisions to the phylogeny of angiosperms show that these plant species are closely related (Benson and Silvester 1993; Soltis et al. 1995; Vessey et al. 2005) (see also chapters on *Frankia* and actinorhizal symbioses in this volume).

This fact (again) raised the hypothesis about independently evolving nitrogen-fixing capacity.

Phylogenetic analysis of DNA sequences for the chloroplast *rbcL gene* showed that representatives of all 10 families in which nitrogen-fixing symbiosis occur, could be placed in a single clade, "the nitrogen-fixing clade" together with several families lacking this association, suggesting that there was a single origin and a predisposition for nodulation in this lineage (Soltis et al. 1995).

However, the same study also suggested that within the nitrogen-fixing clade, the legume-rhizobia symbioses and actinorhizal-*Frankia* symbioses fall into distinct lineages. It suggests that in this group, some factor evolved enabling plants to form nitrogen-fixing symbioses or it arose but was afterwards lost. Another possibility is that nitrogen fixing and nodulation capacity has arisen quite independently several times among and even within-particular families (Doyle and Luckow 2003).

8
Evolution of Interaction Between Rhizobia and Their Legume Hosts

The fossil record indicates that legumes were probably not yet present among the flora of the mid Cretaceous, approximately 90 million years ago (Mya). The first distinct ancestor of legumes probably arose some 70 million years ago in the late Cretaceous in Africa. However, a clear fossil record of the legumes is obvious only from about 35 to 54 Mya (Eocene in mid-Tertiary), when papilionoid and mimosoid legumes become abundant and diverse in both North American and European fossil floras (Lavin et al. 2005). During the Eocene, Africa, Europe and North America were one continuous land mass. Thus, it is easy to understand the often pantropical distributions of the genera and major groups of legumes (Raven and Polhill 1981). Next, continental break-up and progressively cooler climate shaped the legume distribution patterns we observe today.

8.1
Divergence Events in Rhizobia–Legume Symbiosis

Deeper analyses of Eocene fossils also indicate the direction of legume migration in the Americas from north to south (boreotropical hypothesis), not like thought before from south to north (Gondwanan hypothesis). The three subfamilies in the Leguminosae family diverged from one another about 50 Mya (Fig. 6). After that, different legumes evolved and diversified in various regions of the world.

The origin of legumes and their distributions is best sought in the events during and after the Tertiary (6.5–1.8 Mya). The mostly leaf fossils date from Creaceous era, but there are no fossils with identifiable nodules associated with legumes. Thus, it is not known how long ago the fist legumes started to associate with rhizobia. However, according to one hypothesis, two separate nodulation events occurred in the humid tropics in late Cretaceous, one involved an ancestor of *Rhizobium* and the other a photosynthetic ancestor of *Bradyrhizobium* (Sprent 1994). However, according to Provorov (1998) rhizobia–legume symbioses may have arise from some earlier pathogenic interactions.

In the absence of a bacterial fossil record, it is difficult to date speciation within rhizobia. However, analysis of evolutionary changes in highly conserved genes can be used as a "molecular clock". Meta-analysis of both glutamine synthetase (GS) genes from the divergent genera of rhizobia and other reference organisms suggests that the divergence times of the different *Rhizobium* genera predate the existence of legumes, their host plants (Fig. 6; Turner and Young 2000).

Since the chromosomal backgrounds of rhizobia vary considerably, nodulation capacity is thought to have been acquired after bacterial divergence and horizontally spread among different genera within α- and β-*Proteobacteria* (Moulin et al. 2001) (Fig. 1, Table 1). Nodulation genes of α-rhizobia are clearly similar to those from β-rhizobia (Moulin et al. 2001).

This finding lead to a number of hypotheses to explain the sporadic distribution of rhizobia in the *Proteobacteria* lineage. It was proposed that the common ancestor of all rhizobia was a free-living organism that could live heterotrophically and simultaneously fix N_2 with rhizobial descendants having almost completely lost this feature during evolution and adapted to fix N_2 only within the nodule (Provorov 1998). Rhizobia are also proposed to have

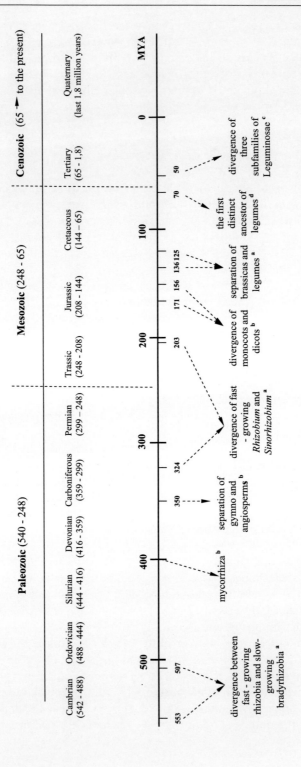

◄ **Fig. 6** Estimated divergence times in rhizobium-legume symbiosis. The drawing is based on date estimations from Turner and Young (2000)[a], Wang et al. (1999)[b], Wojciechowski et al. (2003)[c] and Wojciechowski et al. (2004)[d]. Dates in Mya (million years ago) are indicated in *brackets*

gained the ability to nodulate through recent genetic exchange (Young and Haukka 1996). *Bradyrhizobium* is believed to have diverged from the *Rhizobium* lineage before the leguminous plants evolved on earth. Thus, according to present knowledge, there is no biological reason for the early existence of any symbiotic genes (Fig. 6; Young and Haukka 1996).

8.2
Co-evolution of Rhizobia–Legume Symbiosis

Analysis of core and accessory genes may reveal rearrangements and horizontal genetic transfer events occurring within and across species in the course of evolution (Ba et al. 2002; Haukka et al. 1998; Silva et al. 2005; Ueda et al. 1995b,c). Analysis of symbiotic accessory genes suggests that either these genes have evolved from a common ancestral gene or have been acquired by different taxonomic and geographical lineages of rhizobial and non-rhizobial genera at different times by lateral transfer events (Moulin et al. 2001; Sawada et al. 2003). Subsequently, their evolution could have progressed via interaction with the hosts, under plant functional constraints with the Nod factors and other rhizobial determinants as targets for the selection pressure (Provorov 1998; Suominen et al. 2001; Terefework et al. 2000). Thus, there are some relationships between parts of the accessory gene pool and the host plant (Figs. 2 and 3; Dobert et al. 1994; Suominen et al. 2001; Terefework et al. 2000; Ueda et al. 1995a). Dobert et al. (1994) was among the first ones to claim that their data demonstrated co-evolution between *nod* genes and legumes. This case is an example not exactly for rhizobium-legume co-evolution, but precisely *nod* genes-legume co-evolution *sensu lato* (Dobert et al. 1994; Page and Charleston 1998; Ueda et al. 1995b).

If symbiotic partners in the course of evolution track one another with a greater or a lesser degree of fidelity, their phylogenies might be expected to be similar (Page and Charleston 1998, Young and Johnston 1989). In case of the rhizobia–legume symbiosis, any correspondence between phylogenies of rhizobial core genome parts and host plants provides evidence suggesting organism-organism association (co-evolution). The major reason for this is probably that rhizobia are not obligate symbionts and can successfully also live ex planta in the soil (Young and Johnston 1989).

Recent advances in the field of legume genomics and genetics have unraveled many plant molecules involved in Nod factor perception and signal transduction (Radutoiu et al. 2003). They open up new avenues for research into the co-evolution of rhizobia and legumes with opportunities to detect molecular changes in the plant as adaptations to the bacterial symbionts. In spite of incongruence between phylogenies based on accessory *nod* and core chromosomal housekeeping genes, phylogenetic classification of nodulation genes provide very important knowledge for our understanding of the rhizobia–legume symbiosis.

9
Conclusions

Phylogenetic analysis of several carefully selected housekeeping genes and accessory loci by multilocus sequence analysis (MLSA) in combination with population genetics would be the right way to faithfully delineate the process of evolution in rhizobial genome, genetic structures of populations and uncover bacterial species. The resolving power of such approaches is maximized when large rhizobial strain collections are used. Such sequence comparisons suggest that gene duplication, lateral transfer and recombination events play an important role in rapid adaptation to a changing environment and in an adaptive evolution of bacteria. In the case of rhizobia, changes in accessory gene pools in particular have shaped rhizobia–legume evolution.

Acknowledgements We thank Tadeusz Dresler for his excellent help with creating some of the figures. We also thank Jeff Doyle, Herman Spaink and Leena Suominen for providing us the permissions for publishing of the figures.

References

Andronov EE, Terefework Z, Roumiantseva ML, Dzyubenko NI, Onichtchouk OP, Kurchak ON, Dresler-Nurmi A, Young JP, Simarov BV, Lindström K (2003) Symbiotic and genetic diversity of *Rhizobium galegae* isolates collected from the *Galega orientalis* gene center in the Caucasus. Appl Environ Microbiol 69:1067–1074

Appleby CA, Tjepkema JD, Trinick MJ (1983) Haemoglobin in a non-leguminous plant. *Parasponia*: possible genetic origin and function in nitrogen fixation. Science 220:951–953

Ba S, Willems A, de Lajudie P, Roche P, Jeder H, Quatrini P, Neyra M, Ferro M, Prome JC, Gillis M, Boivin-Masson C, Lorquin J (2002) Symbiotic and taxonomic diversity of rhizobia isolated from *Acacia tortilis* ssp. *raddiana* in Africa. Syst Appl Microbiol 25:130–145

Barnett MJ, Fisher RF, Jones T, Komp C, Abola AP, Barloy-Hubler F, Bowser L, Capela D, Galibert F, Gouzy J, Gurjal M, Hong A, Huizar L, Hyman RW, Kahn D, Kahn ML, Kalman S, Keating DH, Palm C, Peck MC, Surzycki R, Wells DH, Yeh KC, Davis RW, Federspiel NA, Long SR (2001) Nucleotide sequence and predicted functions of the entire *Sinorhizobium meliloti* pSymA megaplasmid. Proc Natl Acad Sci USA 98:9883–9888

Bassam BJ, Djordjevic MA, Redmond JW, Batley M, Rolfe BG (1988) Identification of a nodD-dependent locus in the *Rhizobium* strain NGR234 activated by phenolic factors secreted by soybeans and other legumes. Mol Plant Microbe Interact 1:161–168

Beijerinck MW (1890) Künstliche Infection von *Vicia faba* mit *Bacillus radicola*. Ernährungsbedingungen dieser Bacterie. Bot Zeitung 52:837–843

Belay N, Sparling R, Daniels L (1984) Dinitrogen fixation by a thermophilic methanogenic bacterium. Nature 312:286–288

Benson DR, Silvester WB (1993) Biology of *Frankia* strains, actionomycete symbionts of actiorhizal plants. Microbiol Rev 57:293–319

Boivin C, Ndoye I, Lortet G, Ndiaye A, De Lajudie P, Dreyfus B (1997) The *Sesbania* root symbionts *Sinorhizobium saheli* and *S. teranga* bv. *sesbaniae* can form stem nodules on *Sesbania rostrata*, although they are less adapted to stem nodulation than *Azorhizobium caulinodans*. Appl Environ Microbiol 63:1040–1047

Campbell A (1981) Evolutionary significance of accessory DNA elements in bacteria. Annu Rev Microbiol 35:55–83

Cárdenas L, Dominguez J, Quinto C, Lopez-Lara IM, Lugtenberg BJ, Spaink HP, Rademaker GJ, Haverkamp J, Thomas-Oates JE (1995) Isolation, chemical structures and biological activity of the lipo-chitin oligosaccharide nodulation signals from *Rhizobium etli*. Plant Mol Biol 29:453–464

Chen WM, Moulin L, Bontemps C, Vandamme P, Bena G, Boivin-Masson C (2003) Legume symbiotic nitrogen fixation by beta-*Poteobacteria* is widespread in nature. J Bacteriol 185:7266–7272

Chen WM, Laevens S, Lee TM, Coenye T, De Vos P, Mergeay M, Vandamme P (2001) *Ralstonia taiwanensis* sp. nov., isolated from root nodules of *Mimosa* species and sputum of a cystic fibrosis patient. Int J Syst Evol Microbiol 51:1729–1735

Ciccarelli FD, Doerks T, von Mering C, Creevey CJ, Snel B, Bork P (2006) Toward automatic reconstruction of a highly resolved tree of life. Science 311(5765):1283–1287

Cohan FM (2002) What are bacterial species. Annu Rev Microbiol 56:457–487

Coenye T, Vandamme P (2003) Intragenomic heterogeneity between multiple 16S ribosomal RNA operons in sequenced bacterial genomes. FEMS Microbiol Lett 228:45–49

Davey RB, Reanney DC (1980) Extrachromosomal genetic elements and the adaptive evolution of bacteria. Evol Biol 13:113–147

de Lajudie P, Willems A, Nick G, Moreira F, Molouba F, Hoste B, Torck U, Neyra M, Collins MD, Lindström K, Dreyfus B, Gillis M (1998) Characterization of tropical tree rhizobia and description of *Mesorhizobium plurifarium* sp. nov. Int J Syst Bacteriol 48:369–382

de Lajudie P, Willems A, Pot B, Dewettinck D, Maestrojuan G, Neyra M, Collins MD, Dreyfus B, Kersters K, Gillis M (1994) Polyphasic taxonomy of rhizobia: emendation of the genus *Sinorhizobium* and description of *Sinorhizobium meliloti* comb. nov., *Sinorhizobium saheli* sp. nov., and *Sinorhizobium teranga* sp. nov. Int J Syst Bacteriol 44:715–733

Dobert RC, Breil BT, Triplett EW (1994) DNA sequence of the common nodulation genes of *Bradyrhizobium elkanii* and their phylogenetic relationship to those of other nodulating bacteria. Mol Plant Microbe Interact 7:564–572

Doignon-Bourcier F, Willems A, Coopman R, Laguerre G, Gillis M, de Lajudie P (2000) Genotypic characterization of *Bradyrhizobium* strains nodulating small Senegalese legumes by 16S-23S rRNA intergenic gene spacers and Amplified Fragment Length Polymorphism fingerprint analyses. Appl Environ Microbiol 66:3987–3997

Doyle JJ (1994) Phylogeny of the legume family: An approach to understanding the origin of nodulation. Ann Rev Ecol Syst 25:325–349

Doyle JJ, Luckow MA (2003) The rest of the iceberg. Legume diversity and evolution in a phylogenetic context. Plant Physiol 131:900–910

Eardly B, van Berkum P (2005) Use of population genetic structure to define species limits in Rhizobiaceae. Symbiosis 38:109–122

Finan TM, Weidner S, Wong K, Buhrmester J, Chain P, Vorholter FJ, Hernández-Lucas I, Becker A, Cowie A, Gouzy J, Golding B, Puhler A (2001) The complete sequence of the 1,683-kb pSymB megaplasmid from the N_2-fixing endosymbiont *Sinorhizobium meliloti*. Proc Natl Acad Sci USA 98:9889–9894

Fred EB, Baldwin IL, McCoy E (1932) Root nodule bacteria and leguminous plants. University of Wisconsin Press, Madison, WI, USA

Freiberg C, Fellay R, Bairoch A, Broughton WJ, Rosenthal A, Perret X (1997) Molecular basis of symbiosis between *Rhizobium* and legumes. Nature 387:394–401

Galibert F, Finan TM, Long SR, Pühler A, Abola P, Ampe F, Barloy-Hubler F, Barnett MJ, Becker A, Boistard P, Bothe G, Boutry M, Bowser L, Buhrmester J, Cadieu E, Capela D, Chain P, Cowie A, Davis RW, Dreano S, Federspiel NA, Fisher RF, Gloux S, Godrie T, Goffeau A, Golding B, Gouzy J, Gurjal M, Hernandez-Lucas I, Hong A, Huizar L, Hyman RW, Jones T, Kahn D, Kahn ML, Kalman S, Keating DH, Kiss E, Komp C, Lelaure V, Masuy D, Palm C, Peck MC, Pohl TM, Portetelle D, Purnelle B, Ramsperger U, Surzycki R, Thebault P, Vandenbol M, Vorholter FJ, Weidner S, Wells DH, Wong K, Yeh KC, Batut J (2001) The composite genome of the legume symbiont *Sinorhizobium meliloti*. Science 293:668–672

Gao J, Terefework Z, Chen W, Lindström K (2001) Genetic diversity of rhizobia isolated from *Astragalus adsurgens* growing in different geographical regions of China. J Biotech 91:155–168

Gaunt MW, Turner SL, Rigottier-Gois L, Lloyd-Macgilp SA, Young JP (2001) Phylogenies of *atpD* and *recA* support the small subunit rRNA-based classification of rhizobia. Int J Syst Evol Microbiol 51:2037–2048

González V, Santamaria RI, Bustos P, Hernandez-Gonzalez I, Medrano-Soto A, Moreno-Hagelsieb G, Janga SC, Ramirez MA, Jimenez-Jacinto V, Collado-Vides J, Davila G (2006) The partitioned *Rhizobium etli* genome: genetic and metabolic redundancy in seven interacting replicons. Proc Natl Acad Sci USA 103:3834–3839

Guo XW, Zhang XX, Zhang ZM, Li FD (1999) Characterization of *Astragalus sinicus* rhizobia by restriction fragment length polymorphism analysis of chromosomal and nodulation genes regions. Curr Microbiol 39:358–364

Haukka K, Lindström K (1994) Pulsed-field gel electrophoresis for genotypic comparison of *Rhizobium* bacteria that nodulate leguminous trees. FEMS Microbiol Lett 119:215–2201

Haukka K, Lindström K, Young JP (1998) Three phylogenetic groups of *nodA* and *nifH* genes in *Sinorhizobium* and *Mesorhizobium* isolates from leguminous trees growing in Africa and Latin America. Appl Environ Microbiol 64:419–426

Hirsch AM, Lum MR, Downie JA (2001) What makes the rhizobia–legume symbiosis so special? Plant Physiol 127:1484–1492

Honma MA, Asomaning M, Ausubel FM (1990) *Rhizobium meliloti nodD* genes mediate host-specific activation of *nodABC*. J Bacteriol 172:901–911

Jarvis BDW, Sivacumaran S, Tighe SW, Gillis M (1996) Identification of *Agrobacterium* and *Rhizobium* species based on cellular fatty acids composition. Plant Soil 184:143–158

Jarvis BDW, Tighe SW (1994) Rapid identification of *Rhizobium* species based on cellular fatty acid analysis. Plant Soil 161:31–41

Jourand P, Giraud E, Bena G, Sy A, Willems A, Gillis M, Dreyfus B, de Lajudie P (2004) *Methylobacterium nodulans* sp. nov., for a group of aerobic, facultatively methylotrophic, legume root-nodule forming and nitrogen fixuing bacteria. Int J Syst Evol Microbiol 54:2269–2273

Kaijalainen S, Lindstrom K (1989) Restriction fragment length polymorphism analysis of *Rhizobium galegae* strains. J Bacteriol 171:5561–5566

Kaneko T, Nakamura Y, Sato S, Asamizu E, Kato T, Sasamoto S, Watanabe A, Idesawa K, Ishikawa A, Kawashima K, Kimura T, Kishida Y, Kiyokawa C, Kohara M, Matsumoto M, Matsuno A, Mochizuki Y, Nakayama S, Nakazaki N, Shimpo S, Sugimoto M, Takeuchi C, Yamada M, Tabata S (2000) Complete genome structure of the nitrogen-fixing symbiotic bacterium *Mesorhizobium loti* (supplement). DNA Res 31:381–406

Kaneko T, Nakamura Y, Sato S, Minamisawa K, Uchiumi T, Sasamoto S, Watanabe A, Idesawa K, Iriguchi M, Kawashima K, Kohara M, Matsumoto M, Shimpo S, Tsuruoka H, Wada T, Yamada M, Tabata S (2002) Complete genomic sequence of nitrogen-fixing symbiotic bacterium *Bradyrhizobium japonicum* USDA110. DNA Res 9:189–197

Kondorosi E, Banfalvi Z, Kondorosi A (1984) Physical and genetic analysis of a symbiotic region of Rhizobium meliloti: identification of nodulation genes. Mol Gen Genet 193:445–452

Krishnan HB, Pueppke SG (1994) Host range, RFLP, and antigenic relationships between *Rhizobium fredii* strains and *Rhizobium* sp. NGR234. Plant Soil 161:21–29

Laguerre G, Allard MR, Revoy F, Amarger N (1994) Rapid identification of rhizobia by restriction fragment polymorpfism analysis of PCR-amplified 16S rRNA genes. Appl Environ Microbiol 60:56–63

Laguerre G, Fernandez MP, Edel V, Normand P, Amarger N (1993) Genomic heterogeneity among french *Rhizobium* strains isolated from *Phaseolus vulgaris* L. Int J Syst Bacteriol 43:761–767

Laguerre G, Louvrier P, Allard MR, Amarger N (2003) Compatibility of rhizobial genotypes within natural populations of *Rhizobium leguminosarum* biovar *viciae* for nodulation of host legumes. Appl Environ Microbiol 69:2276–2283

Laguerre G, Mavingui P, Allard MR, Charnay MP, Louvrier P, Mazurier SI, Rigottier-Gois L Amarger N (1996) Typing of rhizobia by PCR DNA fingerprinting and PCR-restriction fragment length polymorphism analysis of chromosomal and symbiotic gene regions: application to *Rhizobium leguminosarum* and its different biovars. Appl Environ Microbiol 62:2029–2036

Laguerre G, Nour SM, Macheret V, Sanjuan J, Drouin P, Amarger N (2001) Classification of rhizobia based on *nodC* and *nifH* gene analysis reveals a close phylogenetic relationship among *Phaseolus vulgaris* symbionts. Microbiology 147:981–993

Lan RT, Reevers PR (2000) Intraspecies variation in bacterial genomes: the need for a species genome concept. Trends Microbiol 8:396–401

Lan RT, Reevers PR (2001) When does a clone deserve a name? A perspective on bacterial species based on population genetics. Trends Microbiol 9:419–324

Lavin M, Herendeen PS, Wojciechowski MF (2005) Evolutionary rates analysis of Leguminosae implicates a rapid diversification of lineages during the tertiary. Syst Biol 54:575–594

Leigh JA (2000) Nitrogen fixation in methanogens: the archaeal perspective. Curr Issues Mol Biol 2:125–131

Lindstöm K (1989) *Rhizobium galegae*, a new species of root nodule bacteria. Int J Syst Bacteriol 39:365–367

Lindström K, Gyllenberg HG (2006) The species paradigm in bacteriology: proposal for cross-disciplinary species concept. World federation of culture collections Newsletter-July 2006: 4–13 http://wdcm.nig.ac.jp/wfcc/

Louvrier P, Laguerre G, Amarger N (1996) Distribution of symbiotic genotypes in *Rhizobium leguminosarum* biovar viciae populations isolated directly from soils. Appl Environ Microbiol 62:4202–4205

Martinez E, Romero D, Palacios R (1990) The *Rhizobium* genome. Crit Rev Plant Sci 9:59–93

Martinez-Romero E, Lindström K, van Bercum P, Eardly B, Chen WX, de Lajudie P, Graham PH, Jarvis BDW, Laguerre G, Nesme X, Young JPW, Vinuesa P, Willems A (2006) ICSP Subcommittee on the taxonomy of *Rhizobium* and *Agrobacterium*-diversity, phylogeny and systematics. http://edzna.ccg.unam.mx/csb-sra/. Last full update 16th of March 2007

Martinez-Romero E, Segovia L, Martins Mercante F, Franco AA, Graham P, Pardo MA (1991) *Rhizobium tropici*, a novel species nodulating *Phaseolus vulgaris* L. bean and *Leucaena* sp. trees. Int J Syst Bacteriol 41:417–426

Martinez-Romero E, Vinuesa P, Young PJW, de Lajudie P, Eardly B, Laguerre G, van Bercum P, Willems A, Javis BDW, Lindstöm K (2007) Guidelines to propose new rhizobial species. Manuscript submitted

Mayr E (1957) Species concept and definitions: In: Mayr E (ed) The species problem. The American Association for the Advancement of Science, Washinton D.C., USA, pp. 1–22

Mehta MP, Baross JA (2006) Nitrogen fixation at 92 degrees C by a hydrothermalvent archaeon. Science 314:1783–1786

Moreira FMS, Gillis M, Pot B, Kersters K, Franco AA (1993) Characterization of rhizobia isolated from different divergence groups of tropical Leguminosae by comparative polyacrylamide gel electrophoresis of their total proteins. Syst Appl Microbiol 16:135–146

Moschetti G, Peluso A, Protopapa A, Anastasio M, Pepe O, Defez R (2005) Use of nodulation pattern, stress tolerance, nodC gene amplification, RAPD-PCR and RFLP-16S rDNA analysis to discriminate genotypes of *Rhizobium leguminosarum* biovar *viciae*. Syst Appl Microbiol 28:619–631

Moulin L, Bena G, Boivin-Masson C, Stepkowski T (2004) Phylogenetic analyses of symbiotic nodulation genes support vertical and lateral gene co-transfer within the *Bradyrhizobium* genus. Mol Phylogenet Evol 30:720–732

Moulin L, Munive A, Dreyfus B, Bovin-Masson C (2001) Nodulation of legumes by members of the beta-subclass of *Poteobacteria*. Nature 411:948–950

Ngom A, Nakagawa Y, Sawada H, Tsukahara J, Wakabayashi S, Uchiumi T, Nuntagij A, Kotepong S, Suzuki A, Higashi S, Abe M (2004) A novel symbiotic nitrogen-fixing member of the *Ochrobactrum* clade isolated from root nodules of *Acacia mangium*. J Gen Appl Microbiol 50:17–27

Nichols R (2001) Gene trees and species trees are not the same. Trends Ecol Evol 16:358–364

Nick G, Lindström K (1994) Use of repetitive sequences and the polymerase chain reaction to fingerprint the genomic DNA of *Rhizobium galegae* strains and to identify the DNA obtained by sonicating the liquid cultures and root nodules. Syst Appl Microbiol 17:265–273

Nick G, de Lajudie P, Eardly BD, Suominen S, Paulin L, Zhang X, Gillis M, Lindström K (1999) *Sinorhizobium arboris* sp. Nov., and *Sinorhizobium kostiense* sp. nov., isolated from leguminous trees in Sudan and Kenya. Int J Syst Bacteriol 49:1359–1368

Page RDM, Charleston MA (1998) Trees within trees: Phylogeny and historical associations. Trends Ecol Evol 13:356–359

Palys T, Nakamura LK, Cohan FM (1997) Discovery and classification of ecological diversity in the bacterial world: the role of DNA sequence data. Int J Syst Bacteriol 47:1145–1156

Parker MA, Lafay B, Burdon J, van Berkum P (2002) Conflicting phylogeographic patterns in rRNA and *nifD* indicate regionally restricted gene transfer in *Bradyrhizobium*. Microbiol 148:2557–2565

Pérez-Ramírez NO, Rogel MA, Wang E, Castellanos JZ, Martínez-Romero E (1998) Seeds of *Phaseolus vulgaris* bean carry *Rhizobium etli*. FEMS Microbiol Ecol 26:289–296

Perret X, Broughton WJ, Brenner S (1991) Canonical ordered cosmid library of the symbiotic plasmid of *Rhizobium* species NGR234. Proc Natl Acad Sci USA 88:1923–1927

Perret X, Staehelin C, Broughton WJ (2000) Molecular basis of symbiotic promiscuity. Microbiol Mol Biol Rev 64:180–201

Postgate J (1998) Nitrogen fixation, 3rd edn. Cambridge University Press, Cambridge UK

Provorov NA (1998) Coevolution of rhizobia with legumes: Facts and Hypotheses. Symbiosis 24:337–368

Pueppke SG, Broughton WJ (1999) *Rhizobium* sp. strain NGR234 and *R. fredii* USDA257 share exceptionally broad, nested host ranges. Mol Plant Microbe Interact 12:293–318

Rademaker JL, Hoste B, Louws FJ, Kersters K, Swings J, Vauterin L, Vauterin P, de Bruijn FJ (2000) Comparison of AFLP and rep-PCR genomic fingerprinting with DNA-DNA homology studies: *Xanthomonas* as a model system. Int J Syst Evol Microbiol 50:665–677

Radutoiu S, Madsen LH, Madsen EB, Felle HH, Umehara Y, Gronlund M, Sato S, Nakamura Y, Tabata S, Sandal N, Stougaard J (2003) Plant recognition of symbiotic bacteria requires two LysM receptor-like kinases. Nature 425:585–592

Raven PH, Polhill RM (1981) Biogeography of the Leguminosae. In: Polhill RM, Raven PH (eds) Advances in legume systematics, part 2. Royal Botanic Gardens, Kew, pp. 27–34

Raymond J, Siefert JL, Staples CR, Blankenship RE (2004) The natural history of nitrogen fixation. Mol Biol Evol 21:541–554

Rivas R, Velazquez E, Willems A, Vizcaino N, Subba-Rao NS, Mateos PF, Gillis M, Dazzo FB, Martinez-Molina E (2002) A new species of *Devosia* that forms a unique nitrogen-fixing root-nodule symbiosis with the aquatic legume *Neptunia natans* (L.f.) druce. Appl Environ Microbiol 68:5217–5222

Roche P, Maillet F, Plazanet C, Debelle F, Ferro M, Truchet G, Prome JC, Denarie J (1996) The common *nodABC* genes of *Rhizobium meliloti* are host-range determinants. Proc Natl Acad Sci USA 93:15305–15310

Rodriguez-Quiñones F, Maguire M, Wallington EJ, Gould PS, Yerko V, Downie JA, Lund PA (2005) Two of the three *groEL* homologues in *Rhizobium leguminosarum* are dispensable for normal growth. Arch Microbiol 183:253–265

Ruvkun GB, Ausubel FM (1980) Interspecies homology of nitrogenase genes. Proc Natl Acad Sci USA 77:191–195

Sawada H, Kuykendall LD, Young JM (2003) Changing concepts in the systematics of bacterial nitrogen-fixing legume symbionts. J Gen Appl Microbiol 49:155–179

Selbitschka W, Zekri S, Schröder G, Pühler A, Toro N (1999) The *Sinorhizobium meliloti* insertion sequence (IS) elements IS*Rm*102F34-1/IS*Rm*7 and IS*Rm*220-13-5 belong to a new family of insertion sequence elements. FEMS Microbiol Lett 172:1–7

Selenska-Pobell S, Evguenieva-Hackenberg E, Radeva G, Squartini A (1996) Characterization of *Rhizobium "hedysari"* by RFLP analysis of PCR amplified rDNA and by genomic PCR fingerprinting. J Appl Bacteriol 5:517–528

Silva C, Vinuesa P, Eguiarte LE, Martinez-Romero E, Sousa V (2003) *Rhizobium etli* and *Rhizobium gallicum* nodulate common bean (*Phaseolus vulgaris*) in a traditionally managed milpa plot in Mexico: population genetics and biogeographic implications. Appl Environ Microbiol 69:884–893

Silva C, Vinuesa P, Eguiarte LE, Souza V, Martnez-Romero E (2005) Evolutionary genetics and biogeographic structure of *Rhizobium gallicum sensu lato*, a widely distributed bacterial symbiont of diverse legumes. Mol Ecol 14:4033–4050

Soltis DE, Soltis PS, Morgan DR, Swensen SM, Mullin BC, Dowd JM, Martin PG (1995) Chloroplast gene sequence data suggest a single origin of the predisposition for symbiotic nitrogen fixation in angiosperms. Proc Natl Acad Sci USA 92:2647–2651

Spaink HP, Wijffelman CA, Pees E, Okker RH, Lugtenberg BJJ (1987) *Rhizobium* nodulation gene *nodD* as a determinant of host specificity. Nature 328:337–340

Sprent JI (1994) evolution and diversity in the legume-rhizobium symbiosis: chaos theory? Plant Soil 161:1–10

Sprent JI (1995) Legume trees and shrubs in the tropics: N_2 fixation in perspective. Soil Biol Biochem 27:401–407

Sprent JI, Sprent P (1990) Nitrogen fixing organisms. Chapman and Hall, New York

Stackebrandt E, Frederiksen W, Garrity GM, Grimont PA, Kampfer P, Maiden MC, Nesme X, Rossello-Mora R, Swings J, Truper HG, Vauterin L, Ward AC, Whitman WB (2002) Report of the ad hoc committee for the re-evaluation of the species definition in bacteriology. Int J Syst Evol Microbiol 52:1043–1047

Stepkowski T, Czaplinska M, Miedzinska K, Moulin L (2003) The variable part of the *dnaK* gene as an alternative marker for phylogenetic studies of rhizobia and related alpha *Poteobacteria*. Syst Appl Microbiol 26:483–494

Stepkowski T, Moulin L, Krzyzanska A, McInnes A, Law IJ, Howieson J (2005) European origin of *Bradyrhizobium* populations infecting lupins and serradella in soils of Western Australia and South Africa. Appl Environ Microbiol 71:7041–7052

Suominen L, Roos C, Lortet G, Paulin L, Lindström K (2001) Identification and structure of the *Rhizobium galegae* common nodulation genes: evidence for horizontal gene transfer. Mol Biol Evol 18:907–916

Suominen L, Luukkainen R, Roos C, Lindström K (2003) Activation of the nodA promoter by the *nodD* genes of *Rhizobium galegae* induced by synthetic flavonoids or *Galega orientalis* root exudate. FEMS Microbiol Lett 219:225–232

Sullivan JT, Eardly BD, van Berkum P, Ronson CW (1996) Four unnamed species of nonsymbiotic rhizobia isolated from the rhizosphere of *Lotus corniculatus*. Appl Environ Microbiol 62:2818–2825

Sy A, Giraud E, Jourand P, Garcia N, Willems A, de Lajudie P, Prin Y, Neyra M, Gillis M, Boivin-Masson C, Dreyfus B (2001) Methylotrophic *Methylobacterium* bacteria nodulate and fix nitrogen in symbiosis with legumes. J Bacteriol 183:214–220

Terefework Z, Kaijalainen S, Lindström K (2001) AFLP fingerprinting as a tool to study the genetic diversity of *Rhizobium galegae* isolated from *Galega orientalis* and *G. officinalis*. J Biotechnol 91:169–180

Terefework Z, Lortet G, Suominen L, Lindström K (2000) Molecular evolution of interactions between rhizobia and their legume hosts. In: Triplett E (ed) Prokaryotic nitrogen fixation: a model system for the analysis of a biological process. Horizon Scientific Press, Wymondham, UK, pp 187–206

Terefework Z, Nick G, Suomalainen S, Paulin L, Lindström K (1998) Phylogeny of *Rhizobium galegae* with respect to other rhizobia and agrobacteria. Int J Syst Bacteriol 48:349–356

Trujillo ME, Willems A, Abril A, Planchuelo AM, Rivas R, Ludena D, Mateos PF, Martinez-Molina E, Velazquez E (2005) Nodulation of *Lupinus albus* by strains of *Ochrobactrum lupini* sp. nov. Appl Environ Microbiol 71:1318–1327

Turner SL, Young JP (2000) The glutamine synthetases of rhizobia: phylogenetics and evolutionary implications. Mol Biol Evol 17:309–319

Turner SL, Zhang XX, Li FD, Young JP (2002) What does a bacterial genome sequence represent? Mis-assignment of MAFF 303099 to the genospecies *Mesorhizobium loti*. Microbiol 148:3330–3331

Ueda T, Suga Y, Yahiro N, Matsuguchi T (1995a) Genetic diversity of N_2-fixing bacteria associated with rice roots by molecular evolutionary analysis of *nifD* library. Can J Microbiol 41:235–240

Ueda T, Suga Y, Yahiro N, Matsuguchi T (1995b) Phylogeny of Sym plasmids of rhizobia by PCR-based sequencing of a *nodC* segment. J Bacteriol 177:468–472

Ueda T, Suga Y, Yahiro N, Matsuguchi T (1995c) Remarkable N_2-fixing bacterial diversity detected in rice roots by molecular evolutionary analysis of *nifH* gene sequences. J Bacteriol 177:1414–1417

Valverde A, Velazquez E, Gutiérrez C, Cervantes E, Ventosa A, Igual J-M (2003) *Herbaspirillum lusitanum* sp. nov., a novel nitrogen-fixing bacterium associated with root nodules of *Phaseolus vulgaris*. Int J Syst Evol Microbiol 53:1979–1983

van Bercum P, Eardly BD (1998) I: Spaink HP, Kondorosi A, Hooykaas PJ (eds) JMolecular evolutionary systematics of the Rhizobiaceae. Kluwer Academic Publishers, Dordrecht, The Netherlands, pp 1–24

van Berkum P, Eardly BD (2002) The aquatic budding bacterium *Blastobacter denitrificans* is a nitrogen-fixing symbiont of *Aeschynomene indica*. Appl Environ Microbiol 68:1132–1136

van Berkum P, Fuhrmann JJ (2000) Evolutionary relationships among the soybean bradyrhizobia reconstructed from 16S rRNA gene and internally transcribed spacer region sequence divergence. Int J Syst Evol Microbiol 50:2165–2172

van Berkum P, Leibold JM, Eardly BD (2006) Proposal for combining *Bradyrhizobium* spp. (*Aeschynomene indica*) with *Blastobacter denitrificans* and to transfer *Blastobacter denitrificans* (Hirsch and Muller, 1985) to the genus *Bradyrhizobium* as *Bradyrhizobium denitrificans* (comb. nov.). Syst Appl Microbiol 29:207–215

Vandamme P, Coenye T (2004) Taxonomy of the genus *Cupravidus*: a tale of lost and found. Int J Syst Evol Microbiol 54:2285–2289

Vandamme P, Pot B, Gillis M, De Vos PK, Kersters K, Swings J (1996) Polyphasic taxonomy, a consensus pproach to bacterial systematics. Microbiol Rev 60:407–438

van Rhijn PJ, Feys B, Verreth C, Vanderleyden J (1993) Multiple copies of *nodD* in *Rhizobium tropici* CIAT899 and BR816. J Bacteriol 175:438–447 (Erratum in: J Bacteriol (1993) 175:3692)

Vessey JK, Pawlowski K, Bergman B (2005) Root-based N_2-fixing symbioses: Legumes, actinorhizal plants, *Parasponia* sp. and cycads. Plant Soil 274:51–78

Vinuesa P, Leon-Barrios M, Silva C, Willems A, Jarabo-Lorenzo A, Perez-Galdona R, Werner D, Martinez-Romero E (2005a) *Bradyrhizobium canariense* sp. nov., an acid-tolerant endosymbiont that nodulates endemic genistoid legumes (Papilionoideae: Genisteae) from the Canary Islands, along with *Bradyrhizobium japonicum* bv. *genistearum*, *Bradyrhizobium* genospecies alpha and *Bradyrhizobium* genospecies beta. Int J Syst Evol Microbiol 55:569–575

Vinuesa P, Silva C (2004) Species delineation and biogeography of symbiotic bacteria associated with cultivated and wild legumes. In: Werner D (ed) Biological resources and migration. Springer Verlag, Berlin, pp 143–155

Vinuesa P, Silva C, Lorite MJ, Izaguirre-Mayoral ML, Bedmar EJ, Martinez-Romero E (2005b) Molecular systematics of rhizobia based on maximum likelihood and Bayesian phylogenies inferred from *rrs*, *atpD*, *recA* and *nifH* sequences, and their use in the classification of *Sesbania* microsymbionts from Venezuelan wetlands. Syst Appl Microbiol 28:702–716

Vinuesa P, Silva C, Werner D, Martinez-Romero E (2005c) Population genetics and phylogenetic inference in bacterial molecular systematics: the roles of migration and recombination in *Bradyrhizobium* species cohesion and delineation. Mol Phylogenet Evol 34:29–54

Wang DY, Kumar S, Hedges SB (1999) Divergence time estimates for the early history of animal phyla and the origin of plants, animals and fungi. Proc Biol Sci 266:163–171

Wdowiak-Wrobel S, Malek W (2005) Genomic diversity of *Astragalus cicer* microsymbionts revealed by AFLP fingerprinting. J Gen Appl Microbiol 51:369–378

Weir BS, Turner SJ, Silvester WB, Park DC, Young JM (2004) Unexpectedly diverse *Mesorhizobium* strains and *Rhizobium leguminosarum* nodulate native legume genera of New Zealand, while introduced legume weeds are nodulated by *Bradyrhizobium* species. Appl Environ Microbiol 70:5980–5987

Wernegreen JJ, Riley MA (1999) Comparison of the evolutionary dynamics of symbiotic and housekeeping loci: a case for the genetic coherence of rhizobial lineages. Mol Biol Evol 16:98–113

Widmer R, Shaffer BT, Porteous LA, Seidler J (1999) Analysis of *nifH* gene pool complexity in soil and litter at a Douglas fir forest site in the Oregon cascade mountain range. Appl Environ Microbiol 65:374–380

Willems A, Coopman R, Gillis M (2001) Comparison of sequence analysis of 16S–23S rDNA spacer regions, AFLP analysis and DNA-DNA hybridizations in *Bradyrhizobium*. Int J Syst Evol Microbiol 51:623–632

Willems A, Collins MD (1993) Phylogenetic analysis of rhizobia and agrobacteria based on 16S rRNA gene sequences. Int J Syst Bacteriol 43:305–313

Willems A, Munive A, de Lajudie P, Gillis M (2003) In most *Bradyrhizobium* groups sequence comparison of 16S–23S rDNA internal transcribed spacer regions corroborates DNA-DNA hybridizations. Syst Appl Microbiol 26:203–210

Wojciechowski MF (2003) Reconstructing the phylogeny of legumes (Leguminosae): an early 21st century perspective. In: Klitgaard BB, Bruneau A (eds) Advances in legume systematics, part 10. Royal Botanic Gardens, Kew, UK

Wojciechowski MF Lavin M, Sanderson MJ (2004) A phylogeny of legumes (Leguminosae) based on analysis of the plastid *matK* gene resolves many well-supported subclades within the family. Am J Bot 91:1846–1862

Wojciechowski M, Sanderson MJ, Steele KP, Liston A (2000) Molecular phylogeny of the "temperate herbaceous tribes" of papilionoid legumes: a supertree approach. In: Herendeen PS, Bruneau A (eds) Advances in legume systematics, part 9. Royal Botanic Gardens, Kew, pp 277–298

Wolde-Meskel E, Terefework Z, Lindstrom K, Frostegard A (2004) Metabolic and genomic diversity of rhizobia isolated from field standing native and exotic woody legumes in southern Ethiopia. Syst Appl Microbiol 27:603–611

Yanagi M, Yamasato K (1993) Phylogenetic analysis of the family Rhizobiaceae and related bacteria by sequencing of 16S rRNA gene using PCR and DNA sequencer. FEMS Microbiol Lett 107:115–120

Yang G-P, Debelle F, Savgnac A, Ferro M, Schiltz O, Maillet F, Prome D, Treilhou M, Vialas C, Lindström K, Denarie J, Prome J-C (1999) Structure of the *Mesorhizobium huakuii* and *Rhizobium galegae* Nod factors: a cluster of phylogenetically related legumes are nodulated by rhizobia producing Nod factors with α,β-unsaturated N-acyl substitutions. Mol Microbiol 34:227–237

Young JPW (1992) Phylogenetic classification of nitrogen-fixing organisms. In Stacey G, Burris RH, Evans HJ (eds) Biological nitrogen fixation. Chapman and Hall, New York, NY, pp 43–86

Young JPY, Crossman LC, Johnston AWB, Thomson NR, Ghazoui ZF, Hull KH, Wexler M, Curson AR, Todd JD, Poole PS, Mauchline TH, East AK, Quail MA, Churcher C, Arrowsmith C, Cherevach I, Chillingworth T, Clarke K, Cronin A, Davis P, Fraser A, Hance Z, Hauser H, Jagels K, Moule S, Mungall K, Norbertczak H, Rabbinowitsch E, Sanders M, Simmonds M, Whitehead S, Parkhill J (2006) The genome of *Rhizobium leguminosarum* has recognizable core and accessory components. Genome Biol 7:R34

Young JPW, Downer HL, Eardly BD (1991) Phylogeny of the phototrophic *Rhizobium* strain BTAi1 by polymerase chain reaction-based sequencing of a 16S rRNA gene segment. J Bacteriol 173:2271–2277

Young JPW, Haukka KE (1996) Diversity and phylogeny of rhizobia. New Phytol 133:87–94

Young JPW, Johnston AWB (1989) The evolution of specificity in the legume-*rhizobium* symbiosis. Trends Ecol Evol 4:341–349

Zézé A, Mutch LA, Young JP (2001) Direct amplification of *nodD* from community DNA reveals the genetic diversity of *Rhizobium leguminosarum* in soil. Environ Microbiol 3:363–370

Zhang X-X, Guo X-W, Terefework Z, Cao Y-Z, Hu FR, Lindström K, Li F-D (1999) Genetic diversity among rhizobial isolates from field-grown *Astragalus sinicus* of Southern China. Syst Appl Microbiol 22:312–320

Zhaxybayeva O, Gogarten JP (2004) Cladogenesis, coalescence and the evolution of the three domains of life. Trends Genet 20:182–177 (Erratum in: Trends Genet 2004 20:291)

Erratum to
The Diversity and Evolution of Rhizobia

Aneta Dresler-Nurmi (✉) · David P. Fewer · Leena A. Räsänen · Kristina Lindström

Department of Applied Chemistry and Microbiology, Viikki Biocenter, University of Helsinki, PO Box 56 (Viikinkaari 9), 00014 Helsinki, Finland
aneta.dresler@helsinki.fi

Due to an oversight of the author, Fig. 1 of the above-mentioned article was published online with a mistake and have to be corrected as follows:

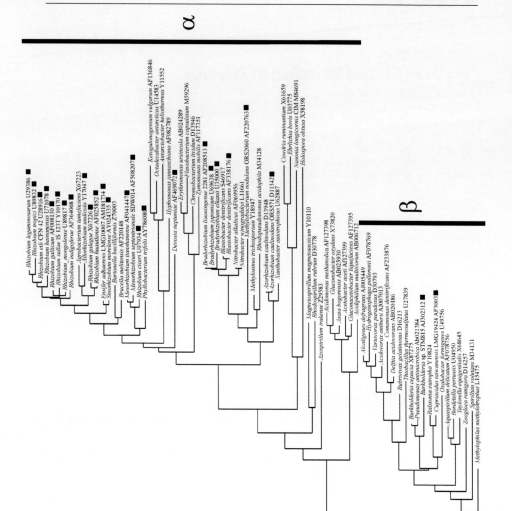

Fig. 1 A maximum likelihood tree based on the *rrs* gene from 75 taxa from α- and β- subdivisions of the proteobacteria. Representatives of species capable of forming nodules are marked with a *black box*

Making Rhizobium-Infected Root Nodules

Andreas Untergasser · Ton Bisseling · René Geurts (✉)

Laboratory of Molecular Biology, Department of Plant Science, Wageningen University, Dreijenlaan 3, 6703HA Wageningen, The Netherlands
rene.geurts@wur.nl

1	Introduction	45
2	The Species Involved	46
3	Making Nodules	48
4	Infection	50
4.1	The Molecular Dialogue	52
4.2	Flavonoid Perception by the Bacterium	52
4.3	Nod Factors	54
4.3.1	Nod Factor Perception by the Host Plant	55
4.3.2	Nod Factor-Controlled Bacterial Infection	59
4.4	Signaling during Crack Entry	61
5	Perspectives	62
	References	63

Abstract Rhizobium bacteria have the unique capability to establish a symbiosis with higher plants of the taxonomic family of *Fabaceae* (also named *Leguminosae*) in which a new root organ, the nodule, is formed. In this nodule atmospheric nitrogen (N_2) is fixed into ammonia and supplied to the plant. It is this symbiosis that will be central in this chapter. We will focus on the underlying molecular networks that are essential to make this interaction happen.

1
Introduction

The taxonomic order *Rhizobiales* contains a diverse group of bacterial families of which several species share the ability to life in close relation with eukaryotic species (Gupta 2005; Lee et al. 2005). Depending on the species, such interactions can result in pathogenicity, e.g., crown gall diseases in higher plants caused by *Agrobacterium tumefaciens* or human/animal brucellosis caused by *Brucella* species (Escobar and Dandekar 2003). In other cases several soil bacteria are able to live in symbiosis with higher plants, an interaction driven by the fact that these bacteria are able to fix atmospheric nitrogen (N_2) into ammonia. The plant species involved belong almost exclusively to the taxonomic family of *Fabaceae* (also named *Leguminosae*), giving them a unique ability to grow independent of organic nitrogen sources in soil.

Legumes form a new organ on the root of the plant to host rhizobium; the so-called root nodule. Within this nodule the bacteria find an environment low in oxygen, which is a prerequisite for the nitrogenase enzyme complex to be able to convert atmospheric nitrogen gas into ammonia (Peters and Szilagyi 2006). Inside the root nodule the bacteria find a protective niche and are provided with nutrients, especially carbohydrates. Upon nodule senescence a substantial fraction of bacteria captured in the nodule are released back into the environment, providing opportunities to populate surrounding soils.

Root nodule formation is set in motion upon perception of a bacterial signal molecule named the Nodulation (Nod) factor. Nod factors produced by different rhizobial species have a similar basic structure consisting of a *N*-acetyl-glucosamine backbone with four to six residues and additional chemical groups attached to both terminal glucosamine residues. The non-reducing terminal residue is substituted with an acyl chain of variable length and structure. Under nitrogen limiting growth conditions, legume plants are susceptible for this signal, resulting in the onset of a developmental program leading to root nodule formation. Simultaneously, guided penetration of the rhizobia into the plant root occurs, facilitating the transport of the prokaryotic partner to the newly formed organ. This infection process is set in motion by Nod factors as well.

2
The Species Involved

Legume root-nodule bacteria form a diverse group of 40–50 species divided into twelve genera within the alpha, beta and gamma proteobacteria (Moulin et al. 2001; Sawada et al. 2003; Benhizia et al. 2004). The group of bacteria that is most studied is collectively known as rhizobium (or rhizobia) and belongs to two taxonomic families within the alpha protobacteria, namely the *Rhizobiaceae* (the genera *Rhizobium*, *Allorhizobium*, *Sinorhizoibum* and *Mesorhizobium*) and *Bradyrhizobiaceae* (the genus *Bradyrhizobium* and *Azorhizobium*) (Sawada et al. 2003; Gupta 2005). Through comparing the bacterial genes involved in the symbiosis among all legume root-nodule bacteria, a high level of conservation is observed, which suggests horizontal gene transfer. Since the bacterial symbiosis genes are generally located on an extrachromosomal plasmid (sym plasmid; pSym), conjugational transfer in the rhizosphere seems to be the most important evolutionary force spreading key genetic information essential for symbiotic nitrogen fixation among different bacterial species. This hypothesis is supported by the finding that a sym plasmid is present in some *Agrobacterium* species, which now combine their pathogenic characteristics with the possibility of establishing nitrogen fixing root symbiosis with legumes (Velázquez et al. 2005).

The rhizobial root nodular nitrogen fixing symbiosis is essentially restricted to a single plant family, the *Fabaceae*. *Fabaceae* is the third largest taxonomic plant family encompassing ~ 18 000 species divided over three sub-families; *Caesalpinioideae*, *Mimosoideae* and *Papilionoideae*, respectively (Fig. 1). Rhizobial symbiosis can occur with most genera belonging to the latter two sub-families, whereas only ~ 5% of the genera of the most basal subfamily *Caesalpinioideae* contain species that can form nodules (Sprent 2001). Since these nodulating genera are rather scattered within the current phylogentic tree (Fig. 1), it is assumed that the interaction has evolved multiple times in evolution. A multiple occurrence of the symbiosis is supported by the discovery of a rhizobial interaction outside the *Fabaceae*. The genus *Parasponia* (encompassing ~ 4 of species) belonging to the *Ulmaceae* family can have a root nodule symbiosis with rhizobial species similar to legumes (Trinick 1973; Scott 1986; Bender et al. 1989; Lafay et al. 2006). The *Ulmaceae* and *Fabaceae* are phylogenetically related, but form two distinct lin-

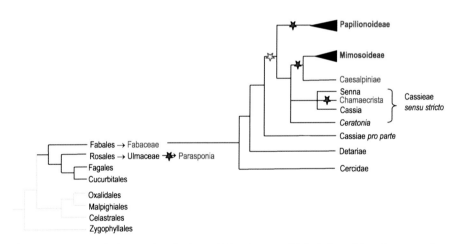

Fig. 1 Consensus phylogenetic relation within the eurosid I clade, with a focus on the basal *Fabaceae* subfamily *Caesalpinioideae*. The eurosid I clade contains eight orders of which four belong to the so-called Nitrogen fixation clade (given in *black lines*), which harbors all plant species that can establish N_2-fixing root nodules with either rhizobia (given in *red*) or *Frankia*. Nodulation is common in the *Fabaceae* within the sub-families *Mimosoideae* and *Papilionoideae*, whereas nodulation is rare in the basal sub-family *Caesalpinioideae*, though not absent (e.g. nodulation can occur in some genera of the tribes *Cassieae* and *Ceasalpinieae*). Possible origins of nodulation are given by stars, having either a single evolutionary event (*white star*) or multiple events (*red stars*) in the *Fabaceae* family (the figure is based on Soltis et al. (2000), Lewis et al. (2005) and Tree of Life web project (http://www.tolweb.org/tree))

eages (Fig. 1), which is evidence of a multiple occurrence of the rhizobium symbiosis in evolution (Soltis et al. 1995).

3
Making Nodules

The formation of a special nodular organ is not a unique feature exclusively associated with rhizobium symbiosis. Another group of plant species – collectively known as Actinorhizal plants – can make similar structures to host symbiotic *Frankia* bacteria (Pawlowski and Sirrenberg 2003). Actinorhizal plants, legumes and *Parasponia* species are phylogenetically related and belong to four orders within the Eurosid I clade (Fig. 1). These four orders, the Fabales, Fagales, Rosales and Cucerbitales, are also known as the nitrogen fixation clade, and it has been hypothesized that species within this clade could have a certain predisposition for root nodule formation (Fig. 1; Soltis et al. 1995).

Although species outside the *Fabaceae* family can make root nodules to host nitrogen fixing bacteria, the morphology of a legume nodule is significantly different when compared to nodules formed by *Parasponia* and actinorhizal plants. Legumes nodules have a large central zone containing cells that host the bacteria. This central zone is surrounded by an endodermis and peripheral vascular systems. In contrast, actinorhizal and *Parasponia* nodules have a lateral root-like architecture containing a central vascular bundle, and the bacteria are hosted in the cells of the cortical regions. The similarity with lateral roots points out the evolutionary origin of these symbiotic organs. There are several lines of evidence supporting the hypothesis that the developmental program underlying lateral root formation has formed the starting point in nodules evolution. In legumes for example, Nod factor perception triggers the initiation of nodule development, which is first visible upon activation of cells in the inner layers of the root. In general, nodules originate from re-differentiated cortical cells, though the first cells that respond upon Nod factor perception are located in the pericycle (the same cell layer lateral roots originate from) (Timmers et al. 1999; Compaan et al. 2001). Further, some legume plants are able to switch between nodule and lateral root development. In case rhizobium bacteria have activated nodule development by Nod factor secretion, but are unable to penetrate the plant root due to incompatibilities, nodule formation is stopped and interchanged by the onset of the lateral root developmental program (Ferraioli et al. 2004). In legumes, due to molecular genetic techniques and the establishment of *Medicago truncatula* and *Lotus japonicus* as model species, insights into the overlap of the nodule and root developmental program start to emerge. Forward genetic screens have revealed two types of plant mutants: (1) mutants that are effected in root as well as nodule development, and (2) mutants that make spontaneous nodules in the absence of bacteria under nitrogen limiting growth conditions. Both will be discussed here.

In *M. truncatula*, the *lateral root organ defective* (*latd*) mutant is affected in the functioning of three meristems: the primary root, lateral roots and nodules. The primary root growth eventually gets arrested resulting in a disorganized root meristem and root cap cells. Furthermore, lateral roots and nodules become arrested immediately upon emerging from the primary root and reveal a lack of organization (Bright et al. 2005). To date the *LATD* gene is not yet cloned, though studies with plant hormones showed that the *latd* mutant exhibits a reduced sensitivity to abscisic acid (ABA) and that the *latd* phenotpye can be – at least partially – rescued by adding an excess of ABA (Liang et al. 2007). This suggests an essential role of ABA signaling in nodule and root meristems. A second gene that underlines parallels between root and nodule formation encodes a CLAVATA1-type Leucine rich repeat receptor kinase that is expressed in roots and shoots (Krussel et al. 2002; Nashimura et al. 2002; Searle et al. 2003; Schnabel et al. 2005). This is especially true in the case of the *L. japonicus* mutant in this gene (named *har1*, hypernodulation aberrant root), which displays a phenotype that shows a functional overlap of the encoded protein in root and nodule formation. A knockout mutation in the *har1* locus alters root architecture by inhibiting root elongation and stimulating lateral root initiation. At the cellular level these developmental alterations are associated with changes in the position and duration of root cell growth and result in a premature differentiation of root cells (Wopereis et al. 2000). Upon inoculation with rhizobium the *har1* mutant displays an unrestricted nodulation (hypernodulation) phenotype accompanied by a drastic inhibition of root growth. This phenotype suggests an integrated regulatory processes controlling nodule organogenesis and nodule number (Wopereis et al. 2000). A second class of mutants that give insight in the underlying molecular mechanism of nodule formation are the so-called spontaneous nodulators. These mutants form nodules in the absence of rhizobium. To date two genes have been characterized; a calcium calmodulin dependent kinase (CCaMK) and cytokinin histidine receptor kinase (Gleason et al. 2006; Tirichine et al. 2006, 2007). In both cases the mutations result in an active form of the protein. The CCaMK is a key component in the Nod factor signaling pathway, demonstrating the importance of Nod factor signaling to trigger root nodule formation (see also*the molecular dialogue* subsection). The cloning of cytokinin histidine receptor kinase in *L. japonicus* (there named *LOTUS HISTIDINE KINASE 1* (*LHK1*) revealed that cytokinin signaling is a key factor in nodule organogenesis. A mutation in the putative cytokinin binding domain, resulting in cytokinin-independent activity, is causing the spontaneous nodulation phenotype, whereas a *lhk1* knockout mutant is unable to form nodule primordia upon Nod factor perception (Murray et al. 2006; Tirichine et al. 2007). In higher plants cytokinin histidine receptor kinases form a small gene family. Currently, it is challenging to investigate whether diversification of these genes during evolution have formed an essential evolutionary step to root nodule formation in legumes and possibly actinorhizal plants and *Parasponia*.

Evolution has shaped the nodule developmental program in legumes considerably. This diversity holds for nodule morphology as well as infection mechanisms. The most obvious difference is in whether or not a persistent meristem is present at the apical side of the nodule. Such meristem enables the nodule to grow continuously, and, therefore, this nodule type is named indeterminate. Indeterminate nodules have a clear developmental axis in apical-basal orientation. The apical meristem seemed to be lost several times in evolution. In such cases the nodule is named determinate and lacks a clear apical-basal organization. Determinate nodules can be found, for example, on *L. japonicus*, whereas the other model, *M. truncatula*, has nodules of the indeterminate type. Most basal indeterminate nodules are found on several *Ceasealpilonoid* (e.g., in the genus *Chamaecrista*) and in basal *Papilionoid* species (e.g., the genus *Andira*) (see Fig. 1) (De Faria et al. 1986; Naisbitt et al. 1992). Within these nodules, bacteria are hosted in a more primitive fashion, since they remain associated in persistent infection structures called fixation threads. The bacteria are surrounded by host cell wall material. Besides in more basal legumes, such infection structures have also been observed in *Parasponia*. In contrast, most legume species endocytose the rhizobia into the cytoplasm of nodule cells. The bacteria become surrounded by a plant membrane and form organelle-like structures named symbiosomes. Subsequent divisions of these symbiosomes ultimately result in nodule cells that are packed completely with N_2-fixing organelles. The formation of symbiosomes is presumed to represent a major step in the evolution of legume–nodule symbiosis, because they facilitate the exchange of metabolites between both partners.

4
Infection

The rhizobial infection process is essentially the same for most legumes; the rhizobia enter via root hairs. Nod factor secreting rhizobia get attached to the root hairs and trigger growth responses in these epidermal cells. The new direction of growth mostly differs from the original growth direction, by which the hairs obtain a deformed appearance (Heidstra et al. 1994). A few hairs form a tight curl around the Nod factor producing rhizobia. Curling is caused by a continuous redirection of growth towards the side where a bacterium is attached to the root hair. The plant cells most likely "sense" the rhizobia by secreted Nod factors, since a droplet containing Nod factor is sufficient to redirect growth of the hair towards the side where the droplet is applied (Esseling et al. 2004). Ultimately, the bacteria become entrapped in the pocket of the root hair curl. The function for this isolation of the rhizobia is not well understood, but could be required to resist the turgor pressure of the root hair cell during local cell wall degradation needed to penetrate the root hair cell. The small volume of the pocket would allow accumulation of Nod factors to relatively high concentrations. The rhizobia invaginates the plasmamembrane of the plant cell within the pocket of the curl. In this way, the bacteria enter the plant cell, where they trigger the formation of a thread-like structure (the so-called "infection thread"). Simultaneous with root hair curling, root cortical cells divide and form a nodule primordium. The infection thread will grow through the epidermal cell in the direction of this primordium that will differentiate. Upon arrival the rhizobium infection threads will infect cells from the central zone in the nodule and release the bacteria – surrounded by a plant derived membrane – into the nodule cells. The bacteria will differentiate in their symbiotic form and start to fix atmospheric nitrogen, leading to a genuine functional nodule. In indeterminate nodules bacteria remain present in the intercellular spaces directly proximal to the meristem. There they infect cells that are derived from the dividing meristem. The region in the nodule where this occurs is named the infection zone and consists of several cell layers. Since indeterminate nodules will continue to grow, such nodules will ultimately develop a zone where nitrogen fixation ceases due to age. In this zone cell senescence is observed. Symbiosomes will fuse with lytic vesicles and, ultimately, the bacteria and plant cells will die.

Some legumes gained an alternative strategy to root hair based infection (e.g., species of the genera *Arachis, Sesbania, Stylosanthes, Neptunia, Aeschynomene*). In these cases, rhizobia enter the root via epidermal breaches, by a mechanism called crack entry. Such breaches occur at sites where lateral or adventitious roots protrude. Upon entry, rhizobium occupy spaces between epidermal and cortical cells. Once the bacteria cross the epidermis, they can continue infection in different ways, e.g., by the formation of infection threads. Alternatively, infection threads are not formed, but instead the bacteria spread, intercellularly imbedded in an intercellular polysaccharide matrix (e.g., peanut (*Arachis hypogaea*)). Plant cells that become infected by bacteria repeatedly divide and become incorporated in the developing nodule arising at the axial branch of the newly formed lateral (or adventitious) root.

Crack entry is also the infection mechanism by which *Parasponia* gets infected by rhizobium. Rhizobia that have colonized the root surface trigger the formation of a so-called pre-nodule in the root outer cortex that ruptures the root epidermis creating openings for the bacteria to enter the root. Infection threads formed in the cortex grow from the pre-nodule toward the developing lateral root-type nodule (Lancelle and Torrey 1984). Occasionally a pre-nodule can get infected by bacteria via a root hair based infection mechanism, though it does not develop mature N_2-fixing nodules. Therefore it is unclear what function they have in the symbiosis.

4.1
The Molecular Dialogue

The molecular dialogue between legumes and rhizobial partners has been characterized in several species. Here we will focus on two related bacterial strains – *Sinorhizobium meliloti* and *Sinorhizobium* sp. NGR234 – and their subsequent host plants. Both species are distinct in their host range. Whereas *S. meliloti* almost exclusively nodulates plant species of the genus *Medicago*, NGR234 has an extremely broad host range containing plants of over 112 genera including *Lotus* species and the non-legume *Parasponia andersonii* though it is unable to nodulate *Medicago* species (Pueppke and Broughton 1999).

4.2
Flavonoid Perception by the Bacterium

The molecular dialogue between legume and rhizobium starts upon activation of the bacterial transcriptional network that is under primary control of NodD protein(s). Rhizobium NodD proteins belong to the class of LysR-type of transcriptional regulators that both become activated upon the binding of external signals. In case of NodD this signal is generally a plant secreted flavonoid, though it has to be mentioned that NodD variants have also been found that are less dependent on activation by flavonoids (Mulligan and Long 1989; Honma et al. 1990). Flavonoids bind directly to NodD, thereby causing a conformational change, which results in an increased binding affinity for nodulation gene promoters that contain a so-called nod box (Chen et al. 2005). Upon binding of an inducible flavonoid and in complex with the GroEL chaperonin protein, a tetrameric NodD protein complex activates a downstream network of genes (Ogawa and Long 1995; Feng et al. 2003). The way the NodD regulatory network function differs among rhizobium species. However, for several species – including *S. meliloti* and NGR234 – autoactivation loops within the network have been identified (Fig. 2). In *S. meliloti* two NodD proteins, NodD1 and NodD2, can be activated upon binding flavonoid-like compounds (methoxychalcone for both NodDs, luteolin specifically for NodD1 and the non-flavonoids trigonelline and stachydrine specifically for NodD2) (Peters et al. 1986; Maxwell et al. 1989; Philips et al. 1992). Upon binding both NodD proteins can activate the expression of genes containing a nod box in their promoter region. In *S. meliloti* at least seven such operons have been identified. Among these a second LysR-type regulator, named *Symbiotic regulator M* (*SyrM*), is present. SyrM subsequently regulates expression of NodD2 and NodD3 thereby creating an auto regulatory loop (Kondorosi et al. 1991; Barnett et al. 1996; Fig. 2A). In comparison, *Sinorhizobium* NGR234 harbors two NodD proteins, NodD1 and NodD2, of which NodD1 is the key activator and flavonoid-dependent, whereas NodD2 is in-

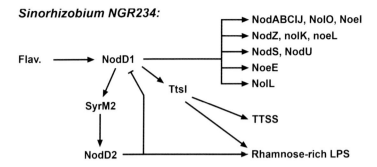

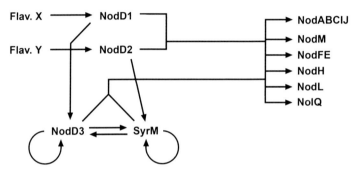

Fig. 2 The NodD regulatory network in *Sinorhizobium meliloti* and *Sinorhizobium* NGR234. Flavonoids bind directly to NodD, thereby causing a conformational change resulting in an increased binding affinity for nodulation gene promoters containing a so-called *nod* box. Among the activated genes are the nodulation genes essential for Nod factor production and secretion (the figure is based on Spaink et al. (1998) and Kobayashi et al. (2004))

volved in a delayed repression of several Nod-factor biosynthetic genes (Fellay et al. 1998). Both NodDs together with the SyrM2 protein form a regulatory network controlling expression of 18 *nod* box-containing operons (Fig. 2B; Kobayashi et al. 2004). Among the activated genes a two-component response regulator, named *TtsI*, is present which is essential for induction of a type III secretion system (TTSS) and biosynthesis of the rhamnan component in the lipopolysaccharides (LPS) (Marie et al. 2004). The TTSS is involved in secretion of several proteins that modulate the ability of *Sinorhizobium* NGR234 to nodulate many of its hosts (Viprey et al. 1998; Marie et al. 2003).

Legume plants secrete a diverse range of (iso)flavonoids that can differ between plant species. For example, *M. truncatula* produces a complex mixture of (iso)flavonoids that includes daidzein, formononetin, medicarpin and, probably, luteolin in greatest abundance. Though in *M. truncatula* the presence of the latter component has only been investigated in aerial parts, it is abundantly produced by roots of its close relative *M. sativa* (Peters et al. 1986; Wasson et al. 2006; Kowalska et al. 2007). Binding experiments with *S. meliloti* NodD1 showed that several different flavonoids bind to this protein and thereby increase the affinity to the nod box DNA element. However, not all (iso)flavonoids that increase DNA binding ultimately trigger transcriptional activation, suggesting a competition between inducing and noninducing flavonoids that are secreted by a potential host plant, and thereby narrowing down a potential host range of bacterium (Peck et al. 2006). Similar studies have not been done for the broad host range bacterium NGR234. However, in contrast to *S. meliloti*, expression of nodulation genes can be achieved by a series of isoflavonoids (including daidzein, apigenin, chrysin, genistein, kaempferol, luteolin and naringenin), suggesting a less stringent control of NodD1 activation in NGR234 than found for *S. meliloti* (Kobayashi et al. 2004).

4.3
Nod Factors

Among the genes activated by the NodD regulatory network are the nodulation (*nod*, *nol* and *noe*) genes that are responsible for the production and secretion of Nod factors. The basic structure of Nod factors produced by different rhizobial species is very similar (Fig. 3). Depending on the rhizobial species, the structure of the acyl chain (Fig. 3) can vary and substitutions at the reducing (position R4; Fig. 3) and nonreducing terminal glucosamine residues (positions R2 and R3; Fig. 3) can be present. These differences are due to the presence of species-specific nodulation genes or are the result of (allelic) variation causing a slightly different specificity of the encoded enzymes. The resulting variation in Nod factor structure plays a role in the ability of the bacterium to interact with its host plant.

In general, a certain rhizobium produces a mixture of Nod factors, and the variation within such mixture correlates with the size of the host range. For example, *S. meliloti* produces one major Nod factor, which is essential to nodulate *Medicago* species, including *M. truncatula*. This Nod factor is tetrameric and contains an acyl chain of 16 carbon atoms in length with two unsaturated bonds (C16:2). Furthermore, the terminal reducing glucosamine residue of this Nod factor is *O*-sulfated, whereas the other terminal glucosamine contains an *O*-acetyl group (Lerouge et al. 1990; Schultze et al. 1992) (Fig. 3). Besides this major compound, minor quantities of Nod factors containing C18:1 and C18-C26 (ω-1)-hydroxylated acyl chains and molecules that lack the *O*-acetyl group are also formed, though the biological function of those remains unclear (Demont et al. 1993). In contrast to a single major Nod factor produced by *S. meliloti*, *Sinorhizobium* NGR234 produces a number of Nod factors in higher quantities that have modifications to the core

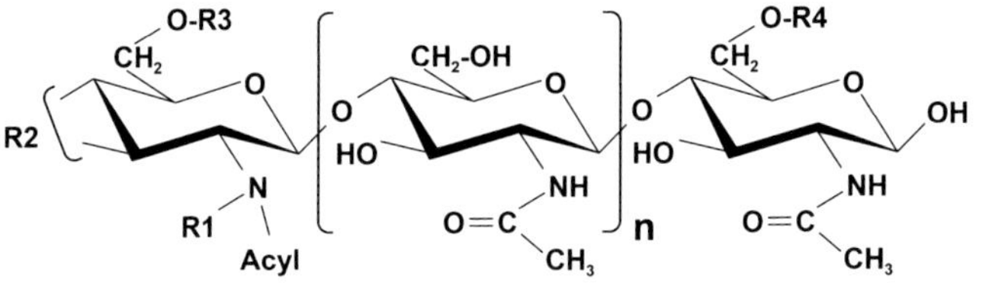

Fig. 3 Basal N-acetyl glucosamine backbone of Nod factors. An acyl chain is always attached to the terminal non-reducing glucosamine. Also, substitutions present on position R1 to R4 are given for main Nod factors produced by *Sinorhizobium meliloti* and *Sinorhizobium NGR234*

	Acyl	R1	R2	R3	R4	n
S. meliloti	$C_{16:2}$	H	OH	H Ac	Sulfate	2
S. NGR234	$C_{18:1}$ $C_{16:1}$ $C_{18:0}$	Met	Carb H	Carb H	MeFuc AcMeFuc SMeFuc	3

N-acyl chitomer structure, including O-carbamoylation and N-methylation of the acylated nonreducing sugar (position R2 and R3, Fig. 3), and 6-O-substitution of the reducing N-acetylglucosamine with either 4-O-acetylated or 3-sulfated 2-O-methylfucose (position R4, Fig. 3) (Price et al. 1996). This enables NGR234 to trigger symbiotic responses at a wide variety of legume plant species, and forms the basis of its broad host range. There are, however, some striking omissions in the structural repertoire of Nod factors produced by NGR234. For example, none of the Nod factors is O-sulfated directly at position R4, a substitution that is absolutely essential to trigger symbiotic responses in *Medicago* species. Furthermore, poly-unsaturated acyl chains are not present, but Nod factors are rather acylated with fatty acids from the general lipid metabolism. The presence of poly-unsaturated acyl chains is found exclusively in rhizobium species that nodulate legumes in the tribes Trifolieae (including *Medicago* and *Trifolium* (clover) species), Fabeae (including pea (*Pisum sativum*), broad bean (*Vicia faba*), and Galegeae (*Astragalus* and *Galega* sp.)). Again, this structural variation plays a role in the host specific nature of the symbiosis and over the course of evolution has resulted in an additional mechanism controlling the interaction, thereby restricting the number of potential host bacteria in this phylum (Yang et al. 1999).

4.3.1
Nod Factor Perception by the Host Plant

Perception of Nod factors by a potential host plant triggers a range of responses that are essential for nodule formation and bacterial infection. In some legume species application of purified Nod factors results in the formation of complete nodules (lacking bacteria). This underlines the importance of Nod factor signaling in this symbiosis. Cloning of plant genes that are essential for Nod factor perception and signaling (the so-called Nod factor signaling (NFS) genes) became possible through use of genetic approaches and provided insight in the Nod factor signaling network. Positional cloning of NFS genes has been achieved mainly in *L. japonicus* and *M. truncatula*. Subsequently, the obtained information was used to clone the corresponding pea mutants. Since for all three species a similar set of NFS genes has been identified, the screenings for NFS genes are close to saturation and has resulted in the identification of nine key genes that range from specific Nod factor receptors down to two transcription factors (Table 1). These genes are transcribed prior to Nod factor perception and regulate all symbiotic responses, including Nod factor induced gene expression.

Two independent genetic approaches have been used to clone putative Nod factor receptors. One method is to search for legume mutants that completely lack Nod factor induced responses, as it can be anticipated that some of these mutants might have a mutation in a Nod factor receptor gene. Alternatively, naturally occurring variation in stringency to Nod factor structure was used. Nod factor structure–function relationship studies showed that the rhizobium infection process has an especially more stringent demand on Nod factor structure and that natural variation occurs in this character (Ardourel et al. 1994; Geurts et al. 1997; Limpens et al. 2003). Therefore, it was hy-

Table 1 Nod factor signaling (NFS) genes as identified in *L. japonicus* and *M. truncatula*. NFS genes are essential for Nod factor perception and signaling. In *M. truncatula* knockout mutants of *NUP85* and *NUP133* have not been identified

	L. japonicus	M. truncatula
LysM-RKs	NFR1 [1]	LYK3, LYK4 [2]
	NFR5 [1,3]	NFP [4]
LRR-RK	SymRK [5]	DMI2 [6]
Cation channels	POLLUX, CASTOR [7]	DMI1 [8]
Nucleoporins	NUP85 [9]	–
	NUP133 [10]	
CCaMK	CCaMK [11]	DMI3 [12,13]
GRAS-type TFs	NSP1 [14]	NSP1 [15]
	NSP2 [14]	NSP2 [16]

References are indicated in superscript: (1) Radutoiu et al. (2003); (2) Limpens et al. (2003); (3) Madsen et al. (2003); (4) Arrighi et al. (2006); (5) Stracke et al. (2002); (6) Endre et al. (2002); (7) Imaizumi-Anraku et al. (2005); (8) Ane et al. (2004); (9) Saito et al. (2007); (10) Kanamori et al. (2006); (11) Tirichine et al. (2006); (12) Levy et al. (2004); (13) Mitra et al. (2004a); (14) Heckmann et al. (2007); (15) Smit et al. (2005); (16) Kalo et al. (2005)

pothesized that for infection a specific Nod factor receptor is involved, the "so-called Nod factor entry receptor" (see: Nod Factor Controlled Bacterial Infection).

Cloning of loss of function mutants in *L. japonicus* has resulted in the identification of two (putative) Nod factor receptors, LjNFR1 and LjNFR5, that have characteristic lysine motives (LysM) in the putative extracellular domain, which were previously found in peptidoglycan binding proteins (Madsen et al. 2003; Radutoiu et al. 2003). Therefore these LysM receptor kinases are good candidates to bind Nod factors as they contain an N-acetylglucosamine backbone. However, direct binding of Nod factors remains to be demonstrated. Corresponding mutants have been found for LjNFR5 in pea and *M. truncatula* (named *Pssym10* and *Mtnfp*). The encoded LysM receptor kinases share that the intra-cellular C-terminal domain in both is an atypical serine/threonine kinase, which lacks an activation loop that usually regulates kinase activity (Madsen et al. 2003; Arrighi et al. 2006). Therefore, activation of LjNFR5-type kinases will likely occur in a different fashion. For example it could be activated upon phosphorylation by a different interacting kinase. Also *LjNFR1* was shown to encode a LysM receptor kinase containing two LysM domains in its extracellular domain. Though, in contrast to the LjNRF5-type receptor kinases it has a more common serine/threonine kinase domain (Radutoiu et al. 2003). Since LjNFR1 and LjNFR5 are both essential to initiate all Nod factor induced responses, it is speculated that both receptors form a heterodimer protein complex (Radutoiu et al. 2003; Arrighi et al. 2006).

Seven other components downstream of the LysM receptor kinases, essential for early steps in Nod factor signaling, have been identified. These include an LRR-type receptor kinase, putative cation channel(s), a calcium calmodulin dependent kinase (CCaMK), two nucleoporins and two GRAS-type transcriptional regulators (Table 1; Fig. 4). These genes are all essential for Nod factor induced changes in gene expression, including the key regulatory genes *NIN* and the ethylene response factor/AP2 transcription factor *ERN* (Schauser et al. 1999; Mitra et al. 2004b; Marsh et al. 2007; Middleton et al. 2007). The LRR-receptor kinase is located in the plasma membrane, whereas in *M. truncatula* DMI1, which shows a low global similarity to ligand-gated cation channels, localizes to the nuclear envelope (Limpens et al. 2005; Riely et al. 2007). Both proteins, together with the LysM receptor kinases and the two nucleoporins, are essential for Nod factor induced oscillation of the cytosolic calcium concentration. These oscillations occur in and around the nucleus within minutes after Nod factor application, and are essential for Nod factor induced gene expression (Ehrhardt et al. 1996; Engstrom et al. 2002). Since the putative cation channel localizes in the vicinity of the place were Ca^{2+} spiking occurs, it is likely that it is a Ca^{2+} channel that is regulated upon Nod factor induced signaling, though experimental evidence for this hypothesis is not yet provided. Also, the nature of the secondary messenger that is essential for Ca^{2+} channel activation remains unknown. Pharmacological

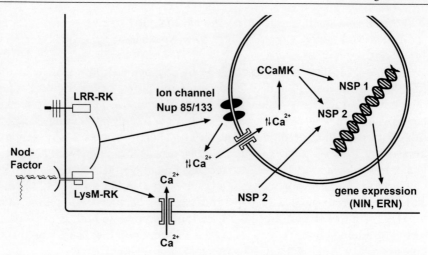

Fig. 4 A model of the Nod factor signaling Network. Nod factors are perceived by specific LysM receptor kinases that form a heterodimer (LjNFR1/LjNFR5, MtNFP/MtLYK). This complex is essential to trigger two calcium signals: a fast influx of external calcium and calcium oscillations in and around the nucleus. The latter response is triggered in conjunction with components also essential for mycorrhizae formation with symbiotic Glomales fungi; LRR-receptor kinase (LjSymRK, MtDMI2), a putative cation channel (LjCastor/LjPollux, MtDMI1) and two nucleoporins (NUP85 and NUP133). This calcium signal is interpreted by a calcium calmodulin dependent kinase (CCaMK), resulting in NSP1 and NSP2 dependent transcriptional activation of symbiotic genes

studies suggest that upon Nod factor perception a lipid signaling pathway is activated that links the plasma membrane to responses manifested in the nuclear region (Charron et al. 2004; Den Hartog et al. 2001, 2003). Phospholipase and inositol triphosphate (IP3) receptor inhibitors block Nod factor induced gene expression, implicating phospholipases and inositol tri-phosphates as potential second messengers.

The CCaMK is active downstream of Ca^{2+} oscillation. CCaMK can interpret different calcium signatures, and therefore it is likely that this protein percieves the Ca^{2+} oscillations induced upon Nod factor perception. CCaMK is localized in the nucleus, where it possibly regulates the activity of transcriptional regulators (Kaló et al. 2005; Smit et al. 2005). Since the GRAS-type transcription factor NSP1 is also localized in the nucleus, it could be a possible target of CCaMK (Smit et al. 2005). However, there is no evidence that there is a direct interaction between both proteins. In contrast to NSP1, NSP2 localizes in the nuclear envelope. Upon Nod factor signaling it migrates into the nucleus, where it will be required for Nod factor induced gene regulation (Kaló et al. 2005).

The signaling pathway downstream the LysM Nod factor receptors up to CCaMK is not only essential for *Rhizobium* induced nodulation, but also for the formation of arbuscular mycorrhizae with fungi belonging to the *Glom-*

eromycetes (Catoira et al. 2000; Kanamori et al. 2006; Saito et al. 2007). In *M. truncatula* it was shown that this part of the signaling pathway is also essential for gene expression associated with mycorrhizae formation. The subset of genes induced during mycorrhization overlaps only partially with *Rhizobium*-induced gene expression suggesting that CCaMK is at a cross point of both signaling pathways (Fig. 4; Manthey et al. 2004; Weidmann et al. 2004). This supports the hypothesis that additional signaling cues are required to maintain specificity (Fig. 4). Some evidence of an additional signaling pathway has indeed been found. For example, mutants that are affected in the components shared by Rhizobium and mycorrhizae still show root hair growth responses upon Nod factor perception that are absent in the Nod factor receptor mutants. This shows that the LysM-receptors can activate a signaling cue independent of the DMI proteins. Further support for a split in the Nod factor signaling pathway is found by different Ca^{2+} responses that are triggered upon nod factor perception. Intracellular oscillation of the Ca^{2+} concentration is dependent on the Nod factor-mycorrhizae common pathway, whereas Nod factor induced extracellular Ca^{2+} influx does not require this signaling (Shaw and Long 2004). Ca^{2+} influx is triggered independently from Ca^{2+} oscillation and is, at least in part, required for Nod factor induced gene expression (Fig. 4; Pingret et al. 1998; Engstrom et al. 2002).

4.3.2
Nod Factor-Controlled Bacterial Infection

It was demonstrated based upon bacterial and plant genetics that bacterial infection requires the most stringent Nod factor signaling. Bacterial mutants that produce Nod factors with structural deficiencies can still trigger symbiotic responses in their potential host plant, though they are regularly hampered in the infection process. This led to the hypothesis that bacterial infection is controlled by a specific receptor, called the Nod factor Entry Receptor (Ardourel et al. 1994). Identification of pea accessions that display genetic variation in stringency of this putative entry receptor enabled a cloning approach. The corresponding locus, named *SYM2*, was identified in the pea accession Afghanistan, where it inhibits infection by *Rhizobium leguminosarum* biovar *viciae* strains lacking the *nodX* nodulation gene (Geurts et al. 1997). NodX O-acetylates pentameric Nod factors at the reducing terminal sugar residue (Firmin et al. 1993) and is only present in strain isolates from the Middle East, where *SYM2* containing peas occur naturally. To clone a Nod Factor entry receptor a synteny-based approach was applied and the *PsSYM2* orthologous region in *M. truncatula* was characterized. This region contains several LysM receptor kinases that were named *MtLYK* (Limpens et al. 2003). This *MtLYK* cluster in *M. truncatula* is syntenic to the *LjNFR1* region in *L. japonicus*, which also contains three genes encoding LysM receptor kinases (Zhu et al. 2006). However, mutations in the *MtLYK3*, the putative ortholog of *LjNFR1*, or knockdown of *MtLYK3* and *MtLYK4* by an RNA interference (RNAi) approach, does not block Nod factor signaling, but specifically effects the infection process (Limpens et al. 2003; Smit et al. 2007). Even so *MtLYK3* and *MtLYK4* are highly homologous to *LjNFR1* their function seems to be specific in the infection process. Therefore these genes can be called a Nod factor Entry receptor. However, since both receptors are specifically involved during bacterial infection, it remains unclear whether a LjNFR1 homolog in *M. truncatula* is also required for early Nod factor induced responses, similar to what was found in *L. japonicus*. In case a heterodimer complex of NFR1 and NFR5-type receptors is formed, such a complex will be present in *M. truncatula* as well, and it has not been found (yet) in genetic screens, probably due to gene redundancy. Based on phylogeny *MtLYK2* could fulfill such a function in *M. truncatula* (Arrighi et al. 2006; Zhu et al. 2006). Alternatively, both proteins do not form a complex, and LjNFR5/MtNFP will have a different – not yet identified – partner.

Besides *MtLYK3* and *MtLYK4*, evidence is emerging that the other NFS genes – as well as the downstream targets *NIN* and *ERN* – are required for proper Rhizobium infection. Incomplete knockdown of the LysM-receptor kinase MtNFP and the LRR-receptor kinase MtDMI2 by RNAi in *M. truncatula* results in inappropriate infection thread formation (Arrighi et al. 2006; Limpens and Geurts, unpublished). Infection structures do not display a thread-like phenotype, though they become more lavished. This is possibly due to the disturbance of tip focused growth, resulting in growth in all directions. The subsequent infection structure gets arrested in the root hair. Further evidence for the occurrence of Nod factor signaling during infection is provided by the *nin* and *ern* knockout phenotypes. Both genes are under transcriptional control of the NFS network and expression of the native genes is required for the infection thread formation in the curled root hair (Schauser et al. 1999; Marsh et al. 2007; Middleton et al. 2007).

The NFS genes remain transcribed until the moment of bacterial release into nodule cells. In the case of *M. truncatula*, which makes indeterminate nodules, this means the NFS genes are expressed in the infection zone of the root nodule (Limpens et al. 2005; Arrighi et al. 2006). In accordance with this is the observation that the Rhizobium *nod* genes also remain active in the infection zone of indeterminate nodules (Sharma and Signer 1990; Schlaman et al. 1991). By using the RNAi approach to mimic allelic series, it was shown that the LRR-receptor kinase is essential for the release of the bacteria into the host cells. When knocking down the expression of this gene in the infection zone of the nodule, the bacteria remain within the infection threads. These threads maintain growth and ultimately will completely fill up the cell (Carpoen et al. 2005; Limpens et al. 2005). The infection phenotype is somewhat reminiscent of the fixation threads found in primitive legumes and Parasponia. Since the expression of the LRR-receptor kinase encoding *MtDMI2* gene in *M. truncatula* is strongly upregulated in the nodule infection zone, it is appealing to speculate that this high expression is essential for bacterial release, and that this evolutionary adaptation has not occurred in those species making fixation threads.

4.4
Signaling during Crack Entry

From the analysis in *M. truncatula* and pea it in becomes apparent that Nod factor signaling is most stringently controlled during rhizobium infection. As described above most legumes are invaded by a similar root hair curling process, whereas some tropical legumes have an alternative way of infection in which the root epidermis is passed via a crack entry mechanism. This crack entry based infection is found often in legume species grow in flooded areas, and can be seen as an adaptation to water logging. To unravel the differences between root hair based infection and crack entry at a molecular level, *Sesbania rostrata* has especially shown to be of excellent scientific value, because it can be infected in either way: crack entry during hydroponic growth conditions and root hair based infection during well-aerated growth conditions. This allows comparison of both programs within a single plant species.

Crack entry in *S. rostrata* is established via infection pockets that are formed at the base of emerging lateral roots. The pockets are formed upon induced cell death of epidermal and outer cortical cells, which requires Nod factor signaling (D'Haeze et al. 2003). The infection pocket contains an external bacterial population and represents the functional equivalent of the bacteria colony entrapped in the root hair curl. Likewise as in a curled root hair, the entrapment of bacteria could be required to restrain the turgor pressure of the cortical cell that will be infected. Also, the formation of a micro-colony could be crucial for the production of higher amounts of Nod factors required for the initiation of the symbiotic program in submerged growth conditions. Due to the large size of a microcolony in an infection pocket (1–3 cortical cells) the signaling is expected to be stronger and perceived by a wider range of cortical cells. One outcome of this is that the nodule primordium is shaped like an open basket around the infection pocket (Goormachtig et al. 2004a).

In contrast to primordium formation, the role of Nod factor signaling in infection thread formation upon crack entry seems to be less important when compared to root hair based infection. For this there are several lines of evidence: (1) Rhizobium mutants that produce altered Nod factors are not blocked in crack entry, whereas they are hampered in root hair based infection (D'Haeze et al. 2000; Goormachtig et al. 2004b). (2) The LRR-receptor kinase of the Nod factor signaling pathway – named *SrSYMRK* in *S. rostrata* – is not essential for infection thread formation during crack entry, whereas it is for root hair based infection (Capoen et al. 2005). (3) Nod factor-deficient rhizobium strains are able to invade via crack entry by co-inoculating a bacterial mutant with defective surface polysaccharides, whereas they are unable to penetrate during root hair based infection (D'Haeze et al. 1998; Den Herder et al. 2007). These data suggest that infection thread formation in the root epidermis forms a serious barrier for rhizobium and that this hurdle can be overcome by producing appropriate Nod factors that are recognized by a Nod factor receptor controlling rhizobium entry. Alternatively, a crack entry mechanism could be used. Once the epidermis is taken, infecting the root cortex and nodule cells seems to be less controlled and independent of specific Nod factor structures.

A key signaling molecule involved in crack entry is ethylene. Ethylene diffusion is 1000 times slower in water compared to air leading to high ethylene concentration in submerged plants. In *M. truncatula* and *L. japonicus* ethylene affects Nod factor signaling at an early stage, because Nod factor induced calcium oscillations are inhibited (Oldroyd et al. 2001). In contrast, an ethylene insensitive mutant (Sickle) is more persistent in root hair infections at high ethylene concentrations, resulting in a super nodulating phenotype (Penmetsa and Cook 1997). *S. rostrata* adapted to flooding conditions by making use of crack entry, which requires high ethylene concentrations to form infection pockets.

5
Perspectives

The capacity to establish a symbiotic relationship with nitrogen fixing rhizobium is an advantageous trait providing legumes unlimited access to a nitrogen source. Consequently, legumes have evolved a life style that is based on high protein contents in leaves and seeds. This is a unique feature among plants, and, in order to keep in pace with the growing world demand for plant derived raw materials, the rhizobium symbiosis must be exploited to its limits. This notion has led to the development of legume model species, and having genome sequences of plant and prokaryotic partners readily available allows the possibility to unravel the interaction in great detail.

Initial research has been focused on those genes involved in Nod factor signaling, since perception of this bacterial signal molecule by the plant forms the main trigger for root nodule development. Strikingly, the number of genes essential for Nod factor signaling that can be identified genetically is low and largely conserved among all legume species studied so far. Furthermore, mutations in these genes mainly affect symbiosis, suggesting that they do not play important roles in other plant processes. It can be expected that in the near future the link between the Nod factor signaling network and common cellular processes will be elucidated, and thereby knowledge will become available as to how plants are able to rewire processes for organ formation.

Homologues of genes affected in non-nodulating mutants are also present in non-legumes. Trans-complementation experiments in legumes using some of these non-legume orthologous genes – *NSP1* of tobacco, and *CCaMK* of lily and rice – have shown that those non-legume genes can function in the Nod factor signaling network (Gleason et al. 2006; Godfroy et al. 2006; Heckmann et al. 2007). This indicates that processes needed for nodule formation could be partly present in non-legume species and suggests that rhizobium has recruited genes involved in general plant development for nodule formation. The finding that five NFS genes are also impaired in arbuscular mycorrhiza interaction supports this hypothesis, suggesting that the interaction with rhizobia has evolved from this more widespread symbiosis. This would also imply that non-legumes may contain the basic signaling pathways used by rhizobium, though they lack certain adaptations of genes that are essential to perceive rhizobia as a symbiotic host. In our opinion such adaptations would be relatively minor, since it has occurred at least twice in evolution. One of the challenges in future research will be to pinpoint these adaptations and to establish a functional network in non-legumes by the introduction of the missing key genes.

References

Ane JM, Kiss GB, Riely BK, Penmetsa RV, Oldroyd GE, Ayax C, Levy J, Debelle F, Baek JM, Kaló P, Rosenberg C, Roe BA, Long SR, Denarie J, Cook DR (2004) *Medicago truncatula* DMI1 required for bacterial and fungal symbioses in legumes. Science 303:1364–1347

Ardourel M, Demont N, Debelle F, Maillet F, de Billy F, Prome JC, Denarie J, Truchet G (1994) *Rhizobium meliloti* lipooligosaccharide nodulation factors: different structural requirements for bacterial entry into target root hair cells and induction of plant symbiotic developmental responses. Plant Cell 6:1357–11374

Arrighi JF, Barre A, Ben Amor B, Bersoult A, Campos Soriano L, Mirabella R, De Carvalho-Niebel F, Journet EP, Ghérardi M, Huguet T, Geurts R, Dénarié J, Rougé P, Gough C (2006) The *Medicago truncatula* lysine motif-receptor-like kinase gene family includes *NFP* and new nodule-expressed genes. Plant Physiol 142:265–279

Barnett MJ, Rushing BG, Fisher RF, Long SR (1996) Transcription start sites for *syrM* and *nodD3* flank an insertion sequence relic in *Rhizobium meliloti*. J Bacteriol 178:1782–1787

Ben Amor B, Shaw SL, Oldroyd GE, Maillet F, Penmetsa RV, Cook D, Long SR, Denarie J, Gough C (2003) The *NFP* locus of *Medicago truncatula* controls an early step of Nod factor signal transduction upstream of a rapid calcium flux and root hair deformation. Plant J 34:495–506

Bender GL, Goydych W, Rolfe BG, Nayudu M (1987) The role of *Rhizobium* conserved and host specific nodulation genes in the infection of the non-legume *Parasponia andersonii*. Mol Gen Genet 210:299–306

Benhizia Y, Benhizia H, Benguedouar A, Muresu R, Giacomini A, Squartini A (2004) Gamma proteobacteria can nodulate legumes of the genus *Hedysarum*. Syst Appl Microbiol 27:462–468

Bright LJ, Liang Y, Mitchell DM, Harris JM (2005) The *LATD* gene of *Medicago truncatula* is required both for nodule and root development. Mol Plant Microbe Interact 18:521–532

Catoira R, Galera C, de Billy F, Penmetsa RV, Journet EP, Maillet F, Rosenberg C, Cook D, Gough C, Denarie J (2000) Four genes of *Medicago truncatula* controlling components of a nod factor transduction pathway. Plant Cell 12:1647–1666

Charron D, Pingret J, Chabaud M, Journet E, Barker DG (2004) Pharmacological evidence that multiple phospholipid signaling pathways link rhizobium nodulation factor perception in *Medicago truncatula* root hairs to intracellular responses, indicating Ca^{2+} spiking and specific ENOD gene expression. Plant Physiol 136:3582–3593

Chen XC, Feng J, Hou BH, Li FQ, Li Q, Hong GF (2005) Modulating DNA bending affects NodD-mediated transcriptional control in *Rhizobium leguminosarum*. Nucleic Acid Res 33:2540–2548

Compaan B, Yang WC, Bisseling T, Franssen H (2001) *ENOD40* expression in the pericycle precedes cortical cell division in Rhizobium–legume interaction and the highly conserved internal region of the gene does not encode a peptide. Plant Soil 230:1–8

D'Haeze W, Gao M, De Rycke R, Van Montagu M, Engler G, Holsters M (1998) Roles for azorhizobial nod factors and surface polysaccharides in intercellular invasion and nodule penetration, respectively. Mol Plant Microbe Interact 11:999–1008

D'Haeze W, Mergaert P, Prome J-C, Holsters M (2000) Nod factor requirements for efficient stem and root nodulation of the tropical legume *Sesbania rostrata*. J Biol Chem 275:15676–15684

D'Haeze W, De Rycke R, Mathis R, Goormachtig S, Pagnotta S, Verplancke C, Capoen W, Holsters M (2003) Reactive oxygen species and ethylene play a positive role in lateral root base nodulation of a semiaquatic legume. Proc Natl Acad Sci USA 100:11789–11794

De Faria SM, Sutherland JM, Sprent JI (1986) A new type of infected cell in root nodules of *Andira* spp. (Leguminosae). Plant Sci 45:143–147

Demont N, Debellé F, Aurelle H, Dénarié J, Promé J-C (1993) Role of *Rhizobium meliloti nodF* and *nodE* genes in the biosynthesis of lipo-oligosaccharidic nodulation factors. J Biol Chem 268:20134–20142

Den Hartog M, Musgrave A, Munnik T (2001) Nod factor-induced phosphatidic acid and diacylglycerol pyrophosphate formation; a role for phospholipase C and D in root hair deformation. Plant J 25:55–65

Den Hartog M, Verhoef N, Munnik T (2003) Nod factor and elicitors activate different phospholipid signaling pathways in suspension-cultured alfalfa cells. Plant Physiol 132:311–317

Den Herder J, Vanhee C, De Rycke R, Corich V, Holsters M, Goormachtig S (2007) Nod factor perception during infection thread growth fine-tunes nodulation. Mol Plant Microbe Interact 20:129–137

Ehrhardt DW, Wais R, Long SR (1996) Calcium spiking in plant root hairs responding to *Rhizobium* nodulation signals. Cell 85:673–681

Endre G, Kereszt A, Kevei Z, Mihacea S, Kaló P, Kiss GB (2002) A receptor kinase gene regulating symbiotic nodule development. Nature 417:962–966

Engstrom EM, Ehrhardt DW, Mitra RM, Long SR (2002) Pharmacological analysis of nod factor-induced calcium spiking in *Medicago truncatula*. Evidence for the requirement of type IIA calcium pumps and phosphoinositide signaling. Plant Physiol 128:1390–401

Escobar MA, Dandekar AM (2003) *Agrobacterium tumefaciens* as an agent of disease. Trends Plant Sci 8:380–386

Esseling JJ, Lhuissier FG, Emons AM (2004) A nonsymbiotic root hair tip growth phenotype in *NORK*-mutated legumes: implications for nodulation factor-induced signaling and formation of a multifaceted root hair pocket for bacteria. Plant Cell 16:933–944

Fellay R, Hanin M, Montorzi G, Frey J, Freiberg C, Golinowski W, Staehelin C, Broughton WJ, Jabbouri S (1998) NodD2 of *Rhizobium* sp. NGR234 is involved in the repression of the *nodABC* operon. Mol Microbiol 27:1039–1050

Feng J, Li Q, Hu HL, Chen XC, Hong GF (2003) Inactivation of the nod box distal half-site allows tetrameric NodD to activate *nodA* transcription in an inducer-independent manner. Nucleic Acids Res 31:3143–3156

Ferraioli S, Tatè R, Rogato A, Chiurazzi M, Patriarca EJ (2004) Development of ectopic roots from abortive nodule primordia. Mol Plant Microbe Interact 17:1043–1050

Firmin JL, Wilson KE, Carlson RW, Davies AE, Downie J (1993) Resistance to nodulation of c.v. Afghanistan peas is overcome by nodX, which mediates an O-acetylation of the *Rhizobium leguminosarum* lipo-oligosaccharide nodulation factor. Mol Microbiol 10:351–360

Geurts R, Heidstra R, Hadri AE, Downie A, Franssen H, Van Kammen A, Bisseling T (1997) Sym2 of *Pisum sativum* is involved in a Nod factor perception mechanism that controls the infection process in the epidermis. Plant Physiol 115:351–359

Gleason C, Chaudhuri S, Yang T, Muñoz A, Poovaiah BW, Oldroyd GED (2006) Nodulation independent of rhizobia induced by a calcium-activated kinase lacking autoinhibition. Nature 441:1149–1152

Godfroy O, Debelle F, Timmers T, Rosenberg C (2006) A rice calcium- and calmodulin-dependent protein kinase restores nodulation to a legume mutant. Mol Plant Microbe Interact 19:495–501

Goormachtig S, Capoen W, Holsters M (2004a) Rhizobium infection: lessons from the versatile nodulation behavior of water-tolerant legumes. Trends Plant Sci 9:518–522

Goormachtig S, Capoen W, James EK, Holsters M (2004b) Switch from intracellular to intercellular invasion during water stress-tolerant legume nodulation. Proc Natl Acad Sci USA 101:6303–6308

Gupta RS (2005) Protein signatures distinctive of alpha proteobacteria and its subgroups and a model for alpha-proteobacterial evolution. Crit Rev Microbiol 31:101–135

Heckmann AB, Lombardo F, Miwa H, Perry JA, Bunnewell S, Parniske M, Wang TL, Downie JA (2007) *Lotus japonicus* nodulation requires two GRAS domain regulators, one of which is functionally conserved in a non-legume. Plant Physiol 142:1739–1750

Heidstra R, Geurts R, Franssen H, Spaink HP, Van Kammen A, Bisseling T (1994) Root hair deformation activity of nodulation factors and their fate on *Vicia sativa*. Plant Physiol 105:787–797

Honma MA, Asomaning M, Ausubel FM (1990) *Rhizobium meliloti nodD* genes mediate host-specific activation of *nodABC*. J Bacteriol 172:901–911

Imaizumi-Anraku H, Takeda N, Charpentier M, Perry J, Miwa H, Umehara Y, Kouchi H, Murakami Y, Mulder L, Vickers K, Pike J, Downie JA, Wang T, Sato S, Asamizu E, Tabata S, Yoshikawa M, Murooka Y, Wu GJ, Kawaguchi M, Kawasaki S, Parniske M, Hayashi M (2005) Plastid proteins crucial for symbiotic fungal and bacterial entry into plant roots. Nature 433:527–531

Kaló P, Gleason C, Edwards A, Marsh J, Mitra RA, Hirsch S, Jakab J, Sims S, Long SR, Rogers J, Kiss GB, Downie JA, Oldroyd GED (2005) Nodulation signaling in legumes requires NSP2, a member of the GRAS family of transcriptional regulators. Science 308:1786–1789

Kanamori N, Madsen LH, Radutoiu S, Frantescu M, Quistgaard EMH, Miwa H, Downie JA, James EK, Felle HH, Haaning LL, Jensen TH, Sato S, Nakamura Y, Tabata S, Sandal N, Stougaard J (2006) A nucleoporin is required for induction of Ca^{2+} spiking in legume nodule development and essential for rhizobial and fungal symbiosis. Proc Natl Acad Sci USA 103:359–364

Kobayashi H, Naciri-Graven Y, Broughton WJ, Perret X (2004) Flavonoids induce temporal shifts in gene-expression of *nod*-box controlled loci in *Rhizobium* sp. NGR234. Mol Microbiol 51:335–347

Kondorosi E, Buire M, Cren M, Iyer N, Hoffman B, Kondorosi A (1991) Involvement of the *syrM* and *nodD3* genes of *Rhizobium meliloti* in *nod* gene activation and in optimal nodulation of the plant host. Mol Microbiol 5:3035–3048

Kowalksa I, Stochmal A, Kapusta I, Janda B, Pizza C, Piacente S, Oleszek W (2007) Flavonoids from barrel medic (*Medicago truncatula*) aerial parts. J Agric Food Chem 55:2645–2652

Krusell L, Madsen LH, Sato S, Aubert G, Genua A, Szczyglowski K, Duc G, Kaneko T, Tabata S, De Bruijn F, Pajuelo E, Sandal N, Stougaard J (2002) Shoot control of root development and nodulation is mediated by a receptor-like kinase. Nature 420:422–426

Lafay B, Bullier E, Burdon JJ (2006) Bradyrhizobia isolated from root nodules of *Parasponia* (*Ulmaceae*) do not constitute a separate coherent lineage. Int J Syst Evol Microbiol 56:1013–1018

Lancelle SA, Torrey JG (1984) Early development of *Rhizobium*-induced root nodules of *Parasponia rigida*, 1. Infection and early nodule initiation. Protoplasma 123:26–37

Lee K-B, Liu C-T, Anzai Y, Kim H, Aono T, Oyaizu H (2005) The hierarchical system of the "Alphaproteobacteria": description of *Hyphomonadaceae* fam. nov., *Xanthobacteraceae* fam. nov. and *Erythrobacteraceae* fam. nov. Int J Syst Evol Microbiol 55:1907–1919

Lerouge P, Roche P, Faucher C, Maillet F, Truchet G, Promé J-C, Dénarié J (1990) Symbiotic host-specificity of *Rhizobium meliloti* is determined by a sulphated and acylated glucosamine oligosaccharide signal. Nature 344:781–784

Levy J, Bres C, Geurts R, Chalhoub B, Kulikova O, Duc G, Journet EP, Ane JM, Lauber E, Bisseling T, Denarie J, Rosenberg C, Debelle F (2004) A putative Ca^{2+} and calmodulin-dependent protein kinase required for bacterial and fungal symbioses. Science 303:1361–1364

Lewis G, Schrire B, MacKinder B, Lock M (eds) (2005) Legumes of the World. The Royal Botanic Gardens, Kew, UK

Liang Y, Mitchella DM, Harris JM (2007) Abscisic acid rescues the root meristem defects of the *Medicago truncatula latd* mutant. Mol Dev 304:297–307

Limpens E, Franken C, Smit P, Willemse J, Bisseling T, Geurts R (2003) LysM domain receptor kinases regulating rhizobial Nod factor-induced infection. Science 302:630–633

Limpens E, Mirabella R, Fedorova E, Franken C, Franssen H, Bisseling T, Geurts R (2005) Formation of organelle-like N_2-fixing symbiosomes in legume root nodules is controlled by DMI2. Proc Natl Acad Sci USA 102:10375–10380

Madsen EB, Madsen LH, Radutoiu S, Olbryt M, Rakwalska M, Szczyglowski K, Sato S, Kaneko T, Tabata S, Sandal N, Stougaard J (2003) A receptor kinase gene of the *LysM* type is involved in legume perception of rhizobial signals. Nature 425:637–640

Manthey K, Krajinski F, Hohnjec N, Firnhaber C, Pühler A, Perlick AM, Küster H (2004) Transcriptome profiling in root nodules and arbuscular mycorrhiza identifies a collection of novel genes induced during *Medicago truncatula* root endosymbioses. Mol Plant Microbe Interact 17:1063–1077

Marie C, Deakin WJ, Viprey V, Kopcińska J, Golinowski W, Krishnan HB, Perret X, Broughton WJ (2003) Characterization of Nops, nodulation outer proteins, secreted via the type III secre tion system of NGR234. Mol Plant Microbe Interact 16:743–751

Marie C, Deakin JW, Ojanen-Reuhs T, Diallo E, Reuhs B, Broughton JW, Perret X (2004) TtsI, a key regulator of *Rhizobium* species NGR234 is required for type III-dependent protein secretion and synthesis of rhamnose-rich polysaccharides. Mol Plant Microbe Interact 17:958–966

Marsh JF, Rakocevic A, Mitra RM, Brocard L, Sun J, Eschstruth A, Long SR, Schultze M, Ratet P, Oldroyd GE (2007) *Medicago truncatula* NIN is essential for rhizobial-independent nodule organogenesis induced by autoactive calcium/calmodulin-dependent protein kinase. Plant Physiol 144:324–335

Maxwell CA, Hartwig UA, Joseph CM, Phillips DA (1989) A chalcone and two related flavonoids released from alfalfa roots induce nod genes of *Rhizobium meliloti*. Plant Physiol 91:842–847

Middleton PH, Jakab J, Penmetsa RV, Starker CG, Doll J, Kaló P, Prabhu R, Marsh JF, Mitra RM, Kereszt A, Dudas B, VandenBosch K, Long SR, Cook DR, Kiss GB, Oldroyd GE (2007) An ERF transcription factor in *Medicago truncatula* that is essential for Nod factor signal transduction. Plant Cell 19:1221–1234

Mitra RM, Gleason CA, Edwards A, Hadfield J, Downie JA, Oldroyd GE, Long SR (2004a) A Ca^{2+}/calmodulin-dependent protein kinase required for symbiotic nodule development Gene identification by transcript-based cloning. Proc Natl Acad Sci USA 101:4701–4705

Mitra RM, Shaw SL, Long SR (2004b) Six nonnodulating plant mutants defective for Nod factor-induced transcriptional changes associated with the legume–rhizobia symbiosis. Proc Natl Acad Sci USA 101:10217–10222

Moulin L, Munive A, Dreyfus B, Boivin-Masson C (2001) Nodulation of legumes by members of the β-subclass of proteobacteria. Nature 411:948–950

Mulligan JT, Long SR (1989) A family of activator genes regulates expression of *Rhizobium meliloti* nodulation genes. Genetics 122:7–18

Murray JD, Karas BJ, Sato S, Tabata S, Amyot L, Szczyglowski K (2007) A cytokinin perception mutant colonized by rhizobium in the absence of nodule organogenesis. Science 15:101–104

Naisbitt T, James EK, Sprent JI (1992) The evolutionary significante of the legume genus *Chamaecrista*, as determined by nodule structure. New Phytol 122:487–492

Nishimura R, Hayashi M, Wu GJ, Kouchi H, Imaizumi-Anraku H, Murakami Y, Kawasaki S, Akao S, Ohmori M, Nagasawa M, Harada K, Masayoshi K (2002) HAR1 mediates systemic regulation of symbiotic organ development. Nature 420:426–429

Oldroyd GED, Engstrom EM, Long SR (2001) Ethylene inhibits the Nod factor signal transduction pathway of *Medicago truncatula*. Plant Cell 13:1835–1849

Oldroyd GED, Long SR (2003) Identification and characterization of nodulation-signaling pathway 2, a gene of *Medicago truncatula* involved in Nod actor signaling. Plant Physiol 131:1027–1032

Ogawa J, Long SR (1995) The *Rhizobium meliloti* groELc locus is required for regulation of early nod genes by the transcription activator NodD. Genes Dev 9:714–729

Pawlowski K, Sirrenberg A (2003) Symbiosis between *Frankia* and actinorhizal plants: root nodules of non-legumes. Ind J Exp Biol 41:1165–1183

Peck MC, Fisher RF, Long SR (2006) Diverse flavonoids stimulate NodD1 binding to *nod* gene promoters in *Sinorhizobium meliloti*. J Bacteriol 188:5417–5427

Penmetsa RV, Cook DR (1997) A legume ethylene-insensitive mutant hyperinfected by its rhizobial symbiont. Science 275:527–530

Peters NK, Frost JW, Long SR (1986) A plant flavone, luteolin, induces expression of *Rhizobium meliloti* nodulation genes. Science 233:917–1008

Peters JW, Szilagyi RK (2006) Exploring new frontiers of nitrogenase structure and mechanism. Curr Opin Chem Biol 10:101–108

Phillips DA, Joseph CM, Maxwell CA (1992) Trigonelline and stachydrine released from alfalfa seeds activate NodD2 protein in *Rhizobium meliloti*. Plant Physiol 99:1526–1531

Pingret JL, Journet EP, Barker DG (1998) Rhizobium Nod factor signaling. Evidence for a G protein-mediated transduction mechanism. Plant Cell 10:659-672

Price NPJ, Talmont F, Wieruszeski JM, Promé D, Promé J-C (1996) Structural determination of symbiotic nodulation factors from the broad host-range Rhizobium species NGR234. Carbohydr Res 289:115-136

Pueppke SG, Broughton WJ (1999) Rhizobium sp. strain NGR234 and R. fredii USDA257 share exceptionally broad, nested host ranges. Mol Plant Microbe Interact 12:293-318

Radutoiu S, Madsen LH, Madsen EB, Felle HH, Umehara Y, Gronlund M, Sato S, Nakamura Y, Tabata S, Sandal N, Stougaard J (2003) Plant recognition of symbiotic bacteria requires two LysM receptor-like kinases. Nature 425:585-592

Riely BK, Lougnon G, Ané JM, Cook DR (2007) The symbiotic ion channel homolog *DMI1* is localized in the nuclear membrane of Medicago truncatula roots. Plant J 49:208-216

Saito K, Yoshikawa M, Yano K, Miwa H, Uchida H, Asamizu E, Sato S, Tabata S, Imaizumi-Anraku H, Umehara Y, Kouchi H, Murooka Y, Szczyglowski K, Downie JA, Parniske M, Hayashi M, Kawaguchi M (2007) NUCLEOPORIN85 is required for calcium spiking, fungal and bacterial symbioses, and seed production in Lotus japonicus. Plant Cell 19:610-624

Sawada H, Kuykendall LD, Young JM (2003) Changing concepts in the systematics of bacterial nitrogen-fixing legume symbionts. J Gen Appl Microbiol 49:155-179

Schauser L, Roussis A, Stiller J, Stougaard J (1999) A plant regulator controlling development of symbiotic root nodules. Nature 402:191-195

Schnabel E, Journet PE, De Carvalho-Niebel F, Duc G, Frugoli J (2005) The *Medicago truncatula SUNN* gene encodes a CLV1-like leucine-rich repeat receptor kinase that regulates nodule number and root length. Plant Mol Biol 58:809-822

Schultze M, Quiclet-Sire B, Kondorosi E, Virelizier H, Glushka JN, Endre G, Géro SD, Kondorosi A (1992) Rhizobium meliloti produces a family of sulfated lipo-oligosaccharides exhibiting different degrees of plant host specificity. Proc Natl Acad Sci USA 89:192-196

Searle IR, Men AE, Laniya TS, Buzas DM, Iturbe-Ormaetxe I, Carroll BJ, Gresshoff PM (2003) Long-distance signaling in nodulation directed by a CLAVATA1-Like Receptor kinase. Science 299:109-112

Sharma SB, Signer ER (1990) Temporal and spatial regulation of the symbiotic genes of Rhizobium meliloti in planta revealed by transposon Tn5-gusA. Genes Dev 4:344-356

Shaw SL, Long SR (2004) Nod factor elicits two separable calcium responses in Medicago truncatula root hair cells. Plant Physiol 131:976-984

Schlaman HR, Horvath B, Vijgenboom E, Okker RJ, Lugtenberg BJ (1991) Suppression of nodulation gene expression in bacteroids of Rhizobium leguminosarum biovar viciae. J Bacteriol 173:4277-4287

Scott KF (1986) Conserved nodulation genes from the non-legume symbiont Bradyrhizobium sp. (Parasponia). Nucleic Acid Res 14:2905-2919

Smit P, Raedts J, Portyanko V, Debellé F, Gough C, Bisseling T, Geurts R (2005) NSP1 of the GRAS protein family is essential for rhizobial Nod factor-induced transcription. Science 308:1789-1791

Smit P, Limpens E, Geurts R, Fedorova E, Dolgikh E, Gough C, Bisseling T (2007) Medicago LYK3, an entry receptor in rhizobial nodulation factor signaling. Plant Physiol 145:183-191

Soltis DE, Soltis PS, Morgan DR, Swensen SM, Mullin BC, Dowd JM, Martin PG (1995) Chloroplast gene sequence data suggest a single origin of the predisposition for symbiotic nitrogen fixation in angiosperms. Proc Natl Acad Sci USA 92:2647-2651

Soltis DE, Soltis PS, Chase MW, Mort ME, Albach DC, Zanis M, Savolainen V, Hahn WH, Hoot SB, Fay MF, Axtell M, Swensen SM, Prince LM, Kress WJ, Nixon KC, Farris JS

(2000) Phylogeny inferred from 18S rDNA, *rbcL*, and *atpB* sequences. Bot J Linnean Soc 133:381–461
Spaink HP, Kondorosi A, Hooykaas PJJ (1998) The Rhizobiaceae. Kluwer Academic Publishers, Norwell, USA
Sprent JI (ed) (2001) Nodulation in legumes. The Royal Botanic Gardens, Kew, UK
Stracke S, Kistner C, Yoshida S, Mulder L, Sato S, Kaneko T, Tabata S, Sandal N, Stougaard J, Szczyglowski K, Parniske M (2002) A plant receptor-like kinase required for both bacterial and fungal symbiosis. Nature 417:959–962
Tirichine L, Imaizumi-Anraku H, Yoshida S, Murakami Y, Madsen LH, Miwa H, Nakagawa T, Sandal N, Albrektsen AS, Kawaguchi M, Downie A, Sato S, Tabata S, Hiroshi H, Parniske M, Kawasaki S, Stougaard J (2006a) Deregulation of a Ca^{2+}/calmodulin-dependent kinase leads to spontaneous nodule development. Nature 441:1153–1156
Timmers ACJ, Auriac MC, Truchet G (1999) Refined analysis of early symbiotic steps of the Rhizobium–Medicago interaction in relationship with microtubular cytoskeleton rearrangements. Development 126:3617–3628
Tirichine L, Sandal N, Madsen LH, Radutoiu S, Albrektsen AS, Sato S, Asamizu E, Tabata S, Stougaard J (2007) A gain-of-function mutation in a cytokinin receptor triggers spontaneous root nodule organogenesis. Science 315:104–107
Trinick MJ (1973) Symbiosis between *Rhizobium* and the non-legume, *Trema aspera*. Nature 244:459–468
Velazquez E, Peix A, Zurdo-Pineiro JL, Palomo JL, Mateos PF, Rivas R, Munoz-Adelantado E, Toro N, Garcia-Benavides P, Martinez-Molina E (2005) The coexistence of symbiosis and pathogenicity-determining genes in *Rhizobium rhizogenes* strains enables them to induce nodules and tumors or hairy roots in *plants*. Mol Plant Microbe Interact 12:1325–1332
Viprey V, Del Greco A, Golinowski W, Broughton WJ, Perret X (1998) Symbiotic implications of type III protein secretion machinery in *Rhizobium*. Mol Microbiol 28:381–389
Wais RJ, Galera C, Oldroyd G, Catoira R, Penmetsa RV, Cook D, Gough C, Denarie J, Long SR (2000) Genetic analysis of calcium spiking responses in nodulation mutants of *Medicago truncatula*. Proc Natl Acad Sci USA 97:13407–13412
Wasson AP, Pellerone FI, Mathesius U (2006) Silencing the flavonoid pathway in *Medicago truncatula* inhibits root nodule formation and prevents auxin transport regulation by rhizobia. Plant Cell 18:1617–1629
Weidmann S, Sanchez L, Descombin J, Chatagnier O, Gianinazzi S, Gianinazzi-Pearson V (2004) Fungal elicitation of signal transduction-related plant genes precedes mycorrhiza establishment and requires the *dmi3* gene in *Medicago truncatula*. Mol Plant Microbe Interact 17:1385–1393
Wopereis J, Pajuelo E, Dazzo FB, Jiang Q, Gresshoff PM, De Bruijn FJ, Stougaard Szczyglowski K (2000) Short root mutant of *Lotus japonicus* with a dramatically altered symbiotic phenotype. Plant J 23:97–114
Yang GP, Debellé F, Savagnac A, Ferro M, Schiltz O, Maillet F, Promé D, Treilhou M, Vialas C, Lindstrom K, Dénarié J, Promé J-C (1999) Structure of the *Mesorhizobium huakuii* and *Rhizobium galegae* Nod factors: a cluster of phylogenetically related legumes are nodulated by rhizobia producing Nod factors with α,β-unsaturated N-acyl substitutions. Mol Microbiol 34:227–237
Zhu H, Riely BK, Burns NJ, An JM (2006) Tracing nonlegume orthologs of legume genes required for nodulation and arbuscular mycorrhizal symbioses. Genetics 172:2491–2499

Functional Genomics of Rhizobia

Anke Becker

Institut für Genomforschung und Systembiologie, Centrum für Biotechnologie, Universität Bielefeld, Postfach 100131, 33501 Bielefeld, Germany
Anke.Becker@Genetik.Uni-Bielefeld.de

1	Introduction	71
2	Rhizobial Genomes	72
3	Transcriptome Studies	77
3.1	Symbiotic Conditions	79
3.2	Nutrient Sources and Stress Conditions	83
4	Proteome Studies	85
5	Metabolome Studies	89
6	Other High Throughput Approaches	90
7	Concluding Remarks and Future Perspectives	91
	References	93

Abstract Complete genome sequences of a number of rhizobia have recently become available and constitute an archive of data which paved the way for postgenomic strategies. This review summarizes data deduced from the genome sequences of rhizobia and related bacteria. It gives a compact survey of the composition and structure of rhizobial multipartite genomes that comprise circular chromosomes and large plasmids. Applications and perspectives of the main *omics* approaches, namely transcriptomics, proteomics and metabolomics, in rhizobial research are discussed. The number of studies applying such profiling methods to investigate the role of regulatory genes as well as responses and adaptation to environmental factors has constantly increased. So far, expression studies at transcriptome and proteome level mainly addressed symbiotic conditions as well as stress conditions and effects of different nutrient sources. Most recently, metabolite profiling was also applied in pilot studies and showed to be a promising approach to learn more about rhizobial metabolism. Apart from profiling methods, several other high throughput strategies for functional analysis of rhizobial genes were established. Examples for various mutagenesis, cloning and gene fusion strategies are given. This review aims to assess the achievements made employing *omics* and other high throughput approaches to broaden our knowledge of the symbiotic rhizobia-plant interaction.

1
Introduction

Rhizobia represent a diverse group of bacteria capable of entering a symbiosis with specific host plants. In the process of this symbiosis the bacteria fix atmospheric nitrogen (reviewed in Brewin 1991, Spaink 2000). Most rhizobia belong to different families of the α-proteobacteria subdivision (Garrity et al. 2002), although recently also symbiotic nitrogen-fixing β-proteobacteria (so-called β-rhizobia) were reported (Chen et al. 2003a; Rasolomampianina et al. 2005; Barrett et al. 2006).

Rhizobia are versatile bacteria that have to face various environmental conditions as free-living bacteria in the soil and during symbiosis with their host plant. The capacity of rhizobia to adapt to changing environmental conditions is extremely important for establishment in the ecosystem and the symbiotic interaction with the host plant which involves specific recognition and progressive differentiation of both bacterial and host cells. In recent years, complete genome sequences of a number of rhizobia have been determined and constitute an archive of new valuable information. Exploitation of this archive requires new strategies and certainly will result in a boost of new knowledge of the genetic basis of the rhizobial lifestyle. Many gene products deduced from the genome sequences have no experimentally proven function or even lack a functional assignment based on similarity measures. Moreover, a considerable number of orphan genes not found in any other sequenced organism were found in rhizobial genomes. Less than 1% of the genes found in rhizobia have been implicated in symbiosis.

In the postgenome era, a broad range of strategies was developed to explore genomic data of all kinds of organisms for functional information. These approaches comprise high throughput methods that allow to monitor gene expression at RNA and protein level in a highly parallel fashion. Such transcriptome and proteome studies have been complemented by metabolomics that aims to analyse the metabolites of an organism. Rhizobial functional genomics has entered the *omics* era just after the first rhizobial genome sequences became available in the year 2000. After the first years of development and expedition, these techniques became standard tools in rhizobial research. The purpose of this review is to assess the achievements made employing *omics* and other high throughput approaches to broaden our knowledge of the symbiotic rhizobia-plant interaction.

2
Rhizobial Genomes

The Rhizobiales, an order of the α-proteobacteria subdivision, include not only nitrogen-fixing plant symbiotic bacteria, but also the plant pathogen *Agrobacterium* and the mammalian pathogens *Bartonella* and *Brucella*. So far, almost 20 complete genome sequences of bacteria from this order have been released. These include the rhizobial genome sequences of *Mesorhizobium loti* MAFF303099 (Phylobacteriaceae) (Kaneko et al. 2000), *Sinorhizobium meliloti* 1021 (Rhizobiaceae) (Barnett et al. 2001; Capela et al.

2001; Finan et al. 2001; Galibert et al. 2001), *Bradyrhizobium japonicum* USDA110 (Bradyrhizobiaceae) (Kaneko et al. 2002), *Rhizobium etli* CFN42 (Rhizobiaceae) (González et al. 2006) and *Rhizobium leguminosarum* bv. viciae strain 3841 (Rhizobiaceae) (Young et al. 2006) (Table 1). In addition sequences of the symbiotic island of *M. loti* strain R7A (Sullivan et al. 2002) and the symbiotic plasmid pNGR234a of *Rhizobium* sp. NGR234 (Freiberg et al. 1997) are available. The databases of *The Joint Genome Institute* (JGI) (http://www.jgi.doe.gov) and *The Institute of Genomic Research* (TIGR) (http://www.tigr.org) report a finished genome project for *Sinorhizobium medica* WSM419 and an ongoing project for *Bradyrhizobium* sp. BTAi1.

Members of the Rhizobiales often possess multipartite genomes comprising circular chromosomes and large plasmids. The genomes of the mammalian pathogens *Bartonella* and *Brucella* are quite small compared to the genomes of the symbiotic or pathogenic plant-associated bacteria. Whereas the *Bartonella* genomes sequenced to date consist of a single circular chromosome of 1.6 to 1.9 Mb (Alsmark et al. 2004), *Brucella* genomes are composed of two circular chromosomes (chromosome I, 2.1 Mb and chromosome II, 1.2 Mb) (Jumas-Bilak et al. 1998; DelVecchio VG et al. 2002; Paulsen et al. 2002). In contrast, the *Agrobacterium tumefaciens* C58 genome comprises a circular and a linear chromosome as well as two plasmids making up a total genome size of 5.6 Mb (Goodner et al. 2001; Wood et al. 2001).

Except for the *B. japonicum* genome that consists of a single circular chromosome, all rhizobial genomes sequenced so far have multipartite architectures (Table 1). The *M. loti* genome comprises a large circular 7.0 Mb chromosome and two plasmids, pMLa (352 kb) and pMLb (208 kb). *S. meliloti* contains a circular 3.65 Mb chromosome and the megaplasmids pSymA (1.35 Mb) and pSymB (1.68 Mb). The genomes of *R. leguminosarum* bv. viciae and *R. etli* differ from the other sequenced rhizobia by the higher number of replicons. Their genomes are composed of a circular chromosome and several large plasmids, varying in number and size among isolates (Laguerre et al. 1992, 1993; Palmer and Young 2000; González et al. 2006).

The majority of the symbiosis-related genes map to plasmids or chromosomal islands. In *M. loti* MAFF303099 a 611 kb chromosomal DNA region was defined as a putative transmittable symbiosis island carrying 580 protein-encoding genes of which 54 genes were assigned to nodulation or nitrogen fixation (Kaneko et al. 2000). Only 31 genes related to nodulation or nitrogen fixation were identified outside this DNA segment and only one of these mapped to plasmid pMLa. The corresponding region from *M. loti* strain R7A comprises 502 kb containing 414 genes and shares 248 kb of DNA with the *M. loti* MAFF303099 symbiosis island (Sullivan et al. 2002). The shared regions contain all genes predicted to be involved in nodulation, nitrogen fixation, and island transfer. In the *B. japonicum* USDA110 genome, two putative chromosomal symbiosis islands, symbiosis island A (680 kb) and symbiosis island B (6 kb), were identified (Kaneko et al. 2002). These two regions might

Table 1 General features of sequenced rhizobial genomes

Organism	Genome size	Genomic architecture	G+C content	cds	tRNA genes	rRNA operons	Refs.
Mesorhizobium loti MAFF303099	7.60 Mb	Chromosomes: 1 Plasmids: 2	62.5%	6752	50	2	Kaneko et al. 2000 http://www.kazusa.or.jp/rhizobase/Mesorhizobium/index.html
Sinorhizobium meliloti 1021	6.69 Mb	Chromosomes: 1 Megaplasmids: 2	62.1%	6207	54	3	Galibert et al. 2001 http://bioinfo.genopole-toulouse.prd.fr/annotation/iANT/bacteria/rhime http://www.cebitec.uni-bielefeld.de/groups/nwt/sinogate
Bradyrhizobium japonicum USDA110	9.11 Mb	Chromosomes: 1 Plasmids: 0	64.1%	8317	50	1	Kaneko et al. 2002 http://www.kazusa.or.jp/rhizobase/Bradyrhizobium/index.html
Rhizobium etli CFN42	6.53 Mb	Chromosomes: 1 Plasmids: 6	60.5%	6034	50	3	González et al. 2006 http://www.ccg.unam.mx/retlidb
Rhizobium leguminosarum bv. viciae strain 3841	7.75 Mb	Chromosomes: 1 Plasmids: 6	60.9%	7236	52	3	Young et al. 2006 http://www.sanger.ac.uk/Projects/R_leguminosarum http://www.genedb.org/genedb/rleguminosarum

be the result of a genomic rearrangement that divided a larger ancient island. Among the 655 protein-encoding genes predicted in both symbiosis islands are clusters of genes related to symbiotic nitrogen fixation (Göttfert et al. 2001; Kaneko et al. 2002).

In *S. meliloti*, *R. etli*, and *R. leguminosarum* bv. viciae the majority of genes directly involved in nodulation and nitrogen fixation are located on plasmids. In case of *S. meliloti* 1021 most of these genes were found on the pSymA megaplasmid (Barnett et al. 2001) whereas many of these genes are located on medium-sized plasmids in *R. leguminosarum* bv. viciae strain 3841 (pRL10 or pSym plasmid, 488 kb) (Young et al. 2006) and *R. etli* CFN42 (p42d or pSym, 371 kb) (González et al. 2006). Nevertheless, additional genes which are known to or might have a role in symbiosis were identified on other replicons. The rhizobial plasmids are not completely dispensable elements (Charles and Finan 1991) and functional relationships of genes situated on different replicons have been reported (González et al. 2006).

The availability of five complete rhizobial genome sequences and sequences of genomes of related non-symbiotic α-proteobacteria opened the venue for comparative considerations of rhizobial genome structures. Extensive synteny was found between the chromosomes of *R. etli*, *R. leguminosarum* and *S. meliloti*, and the circular chromosome of *A. tumefaciens* (González et al. 2006; Young et al. 2006). Synteny is less pronounced between these replicons and the *M. loti* and *B. japonicum* chromosomes, and chromosome I of *Brucella* species. Nevertheless, frequent microsyntenic regions were present in all of these genomes, particularly considering housekeeping genes. Guerrero et al. (2005) analyzed the chromosomes of *A. tumefaciens*, *B. melitensis*, *M. loti* and *S. meliloti*. They predicted about 2000 orthologs in each chromosome of which 80% were syntenic. Syntenic genes encode a high proportion of essential cell functions and show a high level of functional relationships. The higher level of conservation found in syntenic than in non-syntenic gene products suggests resistance of the syntenic genes to sequence variation and rearrangements, and might be explained by their essential character.

Young et al. (2006) revisited the concept of "core" and "accessory" genomes previously raised by Davey and Reanney (1980), Campbell (1981), and Lan and Reeves (2000) who suggested "universal" and "peripheral" genes, "conserved" and "experimental" DNA, "euchromosomal" and "accessory" DNA, or "core" and "auxiliary" genes. Bacteria with large genomes were frequently found among the soil-borne bacteria. A reason for this observation might be the heterogeneous environment in bulk soil, in the rhizosphere and in close association with the plant. To survive in soil such bacteria have to deal with many different substrates and have to adapt to biotic and abiotic stress conditions. Furthermore, symbiotic rhizobia have to interact specifically with their host plant. Young et al. (2006) suggested a core genome representing genes shared among closely related species and an accessory genome as a pool of genes that have a long-term association with one or more bacterial species but are adapted to a "nomadic life". Following this concept, these accessory genes can be expected to be frequently transferred between strains, but are found only in a subset of strains at a time. Such accessory genes may be more frequently found in plasmids and genomic islands.

Currently, the identification of such genes is complicated by the availability of only few rhizobial genomes each representing just a snapshot. The availability of more rhizobial genome sequences for comparative approaches will certainly facilitate this task. Sequencing of reference rhizobial strains cannot give comprehensive insights into rhizobial gene content. Variations in genome size and gene content is common even in closely related bacteria (Stepkowski and Legocki 2001). Guo et al. (2005) applied representational difference analysis (RDA) to identify novel genes in the natural *S. meliloti* strain ATCC 9930. The genome of this strain is about 370 kb larger than that of *S. meliloti* 1021. In this study 85 novel DNA fragments were identified from which 55 fragments showed no significant homologies to database sequences. Distribution of 12 of these fragments in a collection of natural *S. meliloti* strains suggested high rates of gene gain and loss in *S. meliloti* genomes.

Many rhizobia contain additional smaller plasmids, which represent the accessory genome. Analysis of this accessory genome was started in *S. meliloti*. Such plasmids can be widespread in indigenous *S. meliloti* populations (Roumiantseva et al. 2002). The sequence of the *B. melitensis* 144 kb plasmid pSmeSM11a carrying 160 protein-encoding genes was determined (Stiens et al. 2006). This plasmid was isolated from a dominant indigenous *S. meliloti* strain. Interestingly, a 42 kb continuous region that shows a high degree of identity to a region on the pSymA megaplasmid and additional copies of the nodulation genes *nodP* and *nodQ* were identified on the accessory plasmid. The corresponding 42 kb region was missing in pSymA of the *S. meliloti* strain SM11 from which pSmeSM11a was derived, implying transfer of this region from pSymA to pSmeSM11a in this strain (Stiens et al. 2006). This led to the speculation that assembly of symbiotic and accessory plasmids in rhizobial strains may be highly dynamic.

DNA arrays cannot only be applied for expression profiling (see Sect. 3), but also as diagnostic tools for typing of strains and for comparative genomic hybridizations (CGH). Giuntini et al. (2005) used a *S. meliloti* 1021 oligonucleotide microarray to estimate genetic variation at the genomic level in four natural *S. meliloti* strains and *S. meliloti* 1021 and found the largest variation considering pSymA-borne genes. Bontemps et al. (2005) used a phylogenetic microarray containing *nodC* specific probes for classification of rhizobial strains and demonstrated that such an approach allow to type this gene directly from legume nodules.

The key to a better understanding of genome architecture and evolution of rhizobia will be to learn more about the distribution and functions of accessory and orphan genes which make an important contribution to adaptation of a bacterial strain to a specific lifestyle. An important consideration in this respect is the regulation of these accessory genes in rhizobia which have relatively large and complex genomes. From an economic perspective, maintainance of genome size and of the potential to adapt to many different environmental factors requires a strict and specific regulation of gene expression.

3
Transcriptome Studies

Identification of the factors inducing or repressing gene expression is an important step towards a better understanding of the bacterial response to environmental conditions. Therefore, researchers have spent many efforts to study gene expression. Gene fusions to reporter genes like *lacZ* and *gusA* have already been used in the early days of *Rhizobium* research to study gene expression and regulation of a limited number of genes (Glazebrook and Walker 1991). Later, the gene fusion technique was upscaled to a genome-wide approach (Milcamps et al. 1998; Summers et al. 1998; Cowie et al. 2006).

IVET (in vivo expression technology) was applied in rhizobia to identify genes expressed under certain conditions (Oke and Long 1999; Zhang and Cheng 2006). In these studies expression of a positive selection marker required for survival of the cell under the test condition is placed under the control of test DNA fragments. Survival of a clone depends on insertion of a promoter fragment driving expression of the positive selection gene under the test condition.

Gene expression profiling using macroarrays and microarrays is a relatively young approach in this field. The first studies employing these technologies in rhizobia applied targeted arrays carrying a limited numer of probes (Perret et al. 1999; Ampe et al. 2003; Bergès et al. 2003; Barloy-Hubler et al. 2004). Genome-wide arrays allow monitoring of the transcriptional state, the transcriptome, of an organism at a given time. The first microarray, carrying probes for all predicted protein-encoding genes of a rhizobial strain was published for *S. meliloti* 1021 (Rüberg et al. 2003). Meanwhile, different array formats including PCR fragment and oligonucleotide microarrays as well as Affymetrix GeneChips are used for *B. japonicum, M. loti, R. leguminosarum,* and *S. meliloti* (Table 2).

Array hybridization methods still require RNA purified from a high number of bacterial cells, although more recently also amplification of bacterial RNA was applied in transcriptome studies (Moreno-Paz and Parro 2006). This permits expression profiling from a relatively low number of bacterial cells which will be an important issue for analysis of defined symbiotic states in the future. Expression studies by means of array hybridizations compare gene expression under two conditions resulting in relative values, ratios, for the amount of transcripts. Hence, it is important to thoroughly consider the reference condition in such studies and particularly when comparing data from

Table 2 Rhizobial arrays for transcriptome studies

Organism	Type of array	No. of cds represented	No. of IG represented	Refs.
S. meliloti 1021	Spotted PCR fragment macroarray	6063	0	Becker et al. 2004
	Spotted PCR fragment microarray	6207	0	Rüberg et al. 2003 http://www.cebitec.uni-bielefeld.de/groups/nwt/sinogate
	Spotted 70mer oligonucleotide microarray	6208	0	Krol and Becker 2004 http://www.cebitec.uni-bielefeld.de/groups/nwt/sinogate
	Spotted 50mer–70mer oligonucleotide microarray	6208	8080[a]	http://www.cebitec.uni-bielefeld.de/groups/nwt/sinogate
	Affymetrix custom GeneChip, 25mer probes	6270	5788[b]	Barnett et al. 2004
	Affymetrix GeneChip, 25mer probes	8226	0	Affymetrix http://www.affymetrix.com
B. japonicum USDA110	Affymetrix custom GeneChip, 25mer probes	738	yes[c]	Hauser et al. 2006
	Affymetrix custom GeneChip, 25mer probes	8317	yes	Rudolph et al. 2006
M. loti MAFF303099	Spotted PCR fragment macroarray	3832[d]		Uchiumi et al. 2004
Rhizobium sp. NGR234	Spotted PCR fragment macroarray	416[d]		Perret et al. 1999
R. leguminosarum bv. viciae strain 3841	Spotted 70mer oligonucleotide microarray	7413	0	Operon Biotechnologies http://www.operon.com

[a] Intergenic regions > 150 bp (spacing of probes ~ 50–150 nt, both strands)
[b] Intergenic regions > 150 bp (spacing of probes ~ 1–7 nt, both strands)
[c] Intergenic regions > 40 bp (spacing of probes ~ 40 nt, both strands)
[d] PCR fragments containing cds and intergenic regions

cds, predicted coding sequences; IG, predicted intergenic regions different studies. In spite of these technical limitations, arrays have proven to be a promising tool for identification of candidate genes relevant under specific conditions and in characterization of regulatory circuits. New developments in the prediction of promoters (MacLellan et al. 2006) and operon structures (Chandra Janga et al. 2006) may support analysis of transcriptome data for identification of coordinately expressed genes in the future. To date, expression studies in rhizobia mainly addressed symbiotic conditions as well as stress conditions and different nutrient sources.

3.1
Symbiotic Conditions

Transcriptome studies focusing on symbiosis looked at gene expression in symbiotic states compared to free-living bacteria, at induction of gene expression by signal molecules related to symbiosis, and at expression patterns of mutants in genes relevant to symbiosis.

IVET (in vivo expression technology) approaches have been used in S. meliloti to identify genes expressed in the symbiotic state. Expression of the *bacA* gene required for bacteroid differentiation was used as a positive selection marker to identify promoters that are able to drive transcription of this gene during infection and differentiation stages (Oke and Long 1999). Transcription in the free-living state was assessed measuring the activity of a *gusA* reporter gene fused in a tandem structure to the test promoter fragment and *bacA*. 230 fusions were identified that were predominantly expressed in the nodule. Some of these genes, e.g. *nifS* and *tspO*, were also identified as induced in nodule bacteria compared to cultured bacteria in microarray studies whereas other genes appeared to be not differentially expressed or even repressed in these experiments, e.g. *nex18*, *pckA* and *groEL5* (Barnett et al. 2004; Becker et al. 2004). These differences might be explained by different cultivation conditions and the choice of the reference.

In S. meliloti, a similar approach using *exoY* required for successful infection as positive selection marker was applied by Zhang and Cheng (2006). They screened for promoters that were able to drive *exoY* expression during nodule invasion but were inactive in the cultured state in rich medium. Activity of a *gfp* reporter gene downstream of the promoter fragment-*exoY* fusion was determined to assess *ex planta* expression. 183 nodule isolates meeting these criteria were identified. 23 of these carried promoter fusions that were induced by alfalfa root exudates. Among six upstream regions that were inducible by root exudate but not apigenin (a known *nod* gene inducer) were promoter regions of *lsrA*, a *lysR*-type regulatory gene required for symbiosis (Luo et al. 2005) and *dgkA* involved in synthesis of cyclic β-1,2-glucans which are important for nodulation (Breedveld and Miller 1994). Three of the root exudate-inducible genes (SMb21333, SMb21400, and SMc04194 encoding proteins of unknown function) were also identified in microarray studies of gene expression in nodule bacteria compared to cultured bacteria (Barnett et al. 2004). The others were neither identified in such transcriptome studies nor in proteome analysis of bacteroids (Djordjevic et al. 2003). This is not unexpected, since the IVET approach of Zhang and Cheng (2006) focused on the early stages of nodule invasion whereas the proteomic studies looked at purified bacteroids and the microarray-based studies at nodule bacteria from mature nodules that predominantly contain rhizobia in the bacteroid state.

The effect of plant-derived flavonoid inducers on the global gene expressin pattern in *S. meliloti* was addressed in different studies in *Rhizobium* sp. NGR234 (Perret et al. 1999) and *S. meliloti* (Capela et al. 2005; Barnett et al. 2004). Flavonoid-dependent *nod* gene induction in *S. meliloti* is mediated by LysR-type transcription activators of the NodD family. NodD1 and NodD2 are activated by the flavonoid luteolin whereas NodD3 acticity is inducer-independent (Swanson et al. 1993). In a *S. meliloti* strain overexpressing *nodD1* only few genes were induced dependent on luteolin (Barnett et al. 2004; Capela et al. 2005). These mainly comprised genes which have been identified previously as NodD1- and luteolin-dependent induced genes demonstrating that the NodD1 regulon is not much larger than previously determined by reporter gene fusion studies. This is consistent with observations made in *Rhizobium* sp. NGR234 (Perret et al. 1999). Here 147 pNGR234a-borne genes were induced in response to daidzein with the most strongly induced genes clustering in four regions. In both rhizobia, expression of several of these flavonoid-inducible genes is driven by promoter regions containing experimentally proven or putative NodD binding sites (*nod* boxes). In contrast to overexpression of *nodD1*, an enhanced expression level of *nodD3* resulted in more pronounced changes of the global expression pattern in *S. meliloti* (Barnett et al. 2004). More than 100 genes were induced including the regulatory genes *syrM* and *syrA* as well as genes directing biosynthesis of the exopolysaccharide succinoglycan. Approximately half of the almost 100 genes that showed decreased expression levels were genes related to chemotaxis and motility. Barnett et al. (2004) suggested that the largest effect of NodD3 is probably mediated by *syrM* which is positively regulated by NodD3.

Gene expression in later stages of symbiosis were analysed in *S. meliloti* (Barnett et al. 2004; Becker et al. 2004; Capela et al. 2006) and *Rhizobium* sp. NGR234 (Perret et al. 1999) in indeterminate nodules and in *M. loti* (Uchiumi et al. 2004), and *Rhizobium* sp. NGR234 (Perret et al. 1999) in determinate nodules. These studies mainly focused on the bacteroid state. In both nodule types, expression patterns of bacteroids compared to cultured cells in the logarithmic growth phase showed a bias towards repressed genes. This is consistent with an overall slow-down of metabolism in the non-dividing status of mature bacteroids. Most of the induced genes mapped to the symbiotic plasmids or symbiosis islands. In *S. meliloti*, Capela et al. (2006) analysed transcriptome profiles in young nodules five days after inoculation and in nodules induced by a *bacA* mutant. The over-representation of repressed genes observed in mature nodules was not found in young nodules implying that gene expression is more active in stages which do not yet fix nitrogen. Compared to cultured bacteria in the logarithmic growth phase bacteria in *bacA* mutant-induced nodules that are trapped at the infection state showed a general decrease of transcript levels. This is in agreement with rhizobia actively dividing only at the tip of the infection thread (Gage 2004).

Regulons of several regulators involved in symbiosis were investigated by transcriptome studies in *S. meliloti* and *B. japonicum*. These included RpoN (NtrA), NifA-, FixK-, FixJ-, and NtrR-dependent regulation of gene expression in *S. meliloti* (Barnett et al. 2004; Puskas et al. 2004; Bobik et al. 2006). All of these regulators play an important role in symbiosis and microaerobiosis. In *S. meliloti* the σ factor RpoN activates nitrogen fixation genes, ammonium transport and assimilation genes, and C4-dicarboxylate metabolism genes in conjunction with the enhancer-binding proteins NifA, NtrC, and DctD, respectively (Ronson et al. 1987). Comparison of expression profiles of a *S. meliloti rpoN* mutant to the wild type in rich medium showed only few differentially expressed genes (Barnett et al. 2004). Among the induced genes was *glnII* encoding glutamine synthase II and *dctD* coding for a C4-dicarboxylate transport protein confirming previous data.

Expression of genes involved in nitrogen fixation and associated respiration is controlled by the regulators NifA and FixK which underlie regulation by the two-component regulatory system FixLJ under microaerobic conditions. The putative FixJ regulon was characterized in *S. meliloti* cells cultured under microoxic conditions and in nodules (Barnett et al. 2004; Bobik et al. 2006). Further analysis of the FixJ regulon applying microarrays identified a considerable number of novel genes whose transcription was affected by a *fixJ* mutation either in microoxic cultures or in nodules. The list of genes that may be activated directly or indirectly by FixJ currently comprises 122 genes including genes involved in nitrogen metabolism, respiration, transport, and stress responses. About one third of the FixJ target genes are hypothetical. The FixJ regulon was mainly pSymA-specific. Comparison of expression profiles of wild type nodule bacteria to *fixJ* mutant nodule bacteria suggested that most of the bacterial gene induction in nodules depends on the ability to fix nitrogen and thereby on the activity of FixJ. In microaerobiosis, FixJ was also required for induction of almost 75% of the genes with increased expression level in the wild type under this condition (Becker et al. 2004; Bobik et al. 2006). Nevertheless, FixJ-dependent genes activated in symbiosis and microaerobiosis showed only limited overlaps. This raises questions for the regulatory mechanisms in these two conditions and the role of the microoxic environment as a regulatory factor of gene expression during symbiosis.

In symbiosis, *nifA* was required for activation of only 19 genes whereas a *fixK* mutation had a much more pronounced effect on gene activation in microaerobiosis or symbiosis (Bobik et al. 2006). The strict symbiotic specificity of NifA-dependent regulation is probably due to the requirement of an almost anoxic environment present in the nodule. Using a partial-genome Affymetrix GeneChip representing 738 genes including those mapping to the symbiosis island, Hauser et al. (2006) analyzed *nifA*-dependent and *regR*-dependent gene expression in *B. japonicum*. As was the case in the other studies using a transcriptomic array-based approach to analyze regulons, in this study the previously known and well characterized regulons could also be further expanded.

NtrR was originally identified as a regulatory gene affecting transcription of *nod* and *nif* genes in the presence of ammonium (Oláh et al. 2001). A nitrogen regulatory function was therefore assumed for *ntrR* and the upstream gene *ntrP*. A transcriptomic approach showed that the *ntrR* mutation had a drastic effect on the global gene expression pattern (Puskas et al. 2004). Transcript abundance of about 7% of the 6207 predicted protein-coding genes was altered under oxic or microoxic conditions. Recently, Bodogai et al. (2006) reported that NtrP and NtrR form an antitoxin-toxin module which may participate in adaptation to symbiotic and stress conditions and therefore may have a more global role in *S. meliloti* than was assumed first.

Few more regulators controlling processes that are important in symbiosis were studied in *S. meliloti* by a transcriptomic approach. ExoR and ExoS were previously identified as regulators of biosynthesis of succinoglycan which is required for infection of alfalfa nodules (Becker et al. 2000). Global gene expression profiling revealed that these regulators are also involved in regulation of motility (Yao et al. 2004) and suggested that motility and biosynthesis of exopolysaccharides is contrarily regulated in *S. meliloti*. Repression of genes related to motility occurred in concert with upregulation of succinoglycan biosynthesis genes. Requirement of ExoR and ExoS for flagellation was confirmed by mutant studies (Yao et al. 2004).

A similar observation concerning coordinated regulation of motility and exopolysaccharide biosynthesis was made under stress conditions (see below) and in the course of a transcriptome study of the Sin quorum sensing system in *S. meliloti* (Hoang et al. 2004). This system that controls population density-dependent adaptations consists of the N-acyl homoserine lactone synthase gene *sinI* as well as *sinR* and *expR*, both encoding LuxR-type regulators. The Sin system seems to regulate a multidue of genes including genes involved in production of the exopolysaccharide galactoglucan, endoglycolytic degradation of succinoglycan, motility and chemotaxis, and other cellular processes. External supply of Sin system-specific N-acyl homoserine lactones (AHLs) or synthesis of these signal molecules by SinI in the presence of a functional ExpR regulator resulted in induction of genes governing production of the exopolysaccharide galactoglucan (Marketon et al. 2003; Hoang et al. 2004). Regulation of genes involved in motility and chemotaxis was observed in a *sinI* mutant background which blocks synthesis of Sin system-specific AHLs (Hoang et al. 2004). In the *sinI* mutant expression of these genes was positively affected by ExpR. Deficiencies of *sinI* and *expR* mutants in surface swarming motility also pointed to a connection between the Sin quorum sensing system and motility (Gao et al. 2005). Further conclusions drawn from the gene expression profiles concerned the role the LuxR-type regulators (Hoang et al. 2004). Differential expression of most of the Sin system-dependent genes required ExpR suggesting that this protein constitutes the major LuxR-type regulator in this quorum sensing signal transduction.

3.2
Nutrient Sources and Stress Conditions

Adaptations at transcriptome level to various environmental conditions were investigated in numerous studies in *S. meliloti*. These particularly included stress factors playing an important role in soil and during symbiosis.

The gene fusion technique was employed in a genome-wide approach to identify carbon and nitrogen deprivation-induced loci (Milcamps et al. 1998). A mutagenesis of *S. meliloti* with a Tn5 derivative containing the promoterless reporter genes *luxAB* was carried out. Screening of this library identified 21 fusions induced by nitrogen deprivation and 12 fusions induced by carbon-deprivation.

Recently, Cowie et al. (2006) generated a library of 6298 plasmids carrying randomly generated *S. meliloti* DNA fragments fused to *gfp+* and *lacZ* or *gusA* and *rfp* reporter genes, depending on the orientation of the cloned fragment. These plasmids replicate in *Escherichia coli* and can be transferred by conjugation to *S. meliloti*. Since these plasmids cannot replicate in *S. meliloti*, maintenance requires integration into the genome by homologous recombination which allows to determine expression of the gene fusions in *S. meliloti*. Activities of these fusions were assayed in complex medium and minimal medium with either glucose or succinate as sole carbon source. A database of gene expression activities was generated (http://www.sinorhizobium.org/). These examples showed that the use of reporter genes to monitor gene expression is not limited to single gene studies but can also be used for genome-wide transcriptional analysis.

The response of *S. meliloti* to limitation of phosphate as one of the major nutrients has been an early focus of research. High affinity and low affinity phosphate uptake systems were identified (Voegele et al. 1997) and it was found that in *S. meliloti* 1021, the high affinity transporter was required for formation of nitrogen-fixing nodules (Bardin et al. 1996). In many Gram-negative bacteria phosphate-dependent regulation is controlled by a two-component regulatory system consisting of the receptor kinase PhoR and the response regulator PhoB (Wanner 1993).

Summers et al. (1998) generated a *S. meliloti* mutant library using a transposon which generates *lacZ* transcriptional fusions and screened for phosphate stress-inducible genes. This approach led to the identification of six genes induced under phosphate-limiting conditions. The phosphate starvation response was analyzed at the global transcriptional level in the two closely related *S. meliloti* strains, 1021 and 2011, as well as in a *S. meliloti* 1021 *phoB* mutant (Krol and Becker 2004). *phoB*-dependent induction of 98 genes and *phoB*-independent repression of 86 genes in phosphate-stressed cells was observed. 20 new putative PhoB binding sites (PHO boxes) were found in upstream regions of 17 transcriptional units with *phoB*-dependent or partially *phoB*-dependent regulation, indicating a direct regulation of these genes by PhoB. Genome prediction of PhoB regulated promoters based on prediction of PHO box sequences using a position weight matrix algorithm revealed 96 putative Pho regulon members (Yuan et al. 2006a). With few exceptions, these predictions correspond well to data from transcriptional reporter gene fusion experiments (Yuan et al. 2006a) and microarray data (Krol and Becker 2004).

Despite an overall similarity of phosphate stress responses in *S. meliloti* 1021 and 2011, strain 1021 exhibited a moderate constitutive activation of 12 phosphate starvation inducible genes with *phoB*-dependent regulation. Deletion of one nucleotide from the Rm1021 *pstC* gene sequence resulting in truncation of the *pstC* gene product is probably responsible for the deregulated expression of phosphate stress inducible genes in strain 1021 in the absence of phosphate stress (Krol and Becker 2004; Yuan et al. 2006b). This finding shows that strain specific variation in regulation even between very closely related strains may not be a rare case and has to be considered for interpretation of gene expression data.

Responses to limitation of iron have been another focus of transcriptome approaches in rhizobia. Iron is an essential micronutrient and its availability is restricted due to its low solubility under physiological pH and aerobic conditions. Rhizobia have to compete for iron with other soil organisms and have a special high demand for iron in symbiosis, since many enzymes required for nitrogen fixation contain iron compounds as cofactors and the high respiratory demand of bacteroids requires cytochromes and other ferroproteins in abundance (Johnston et al. 2001). Iron uptake is usually strictly regulated because high concentrations of iron can lead to the formation of hydroxyl radicals. Unlike in many Gram-negative and Gram-positive bacteria (Hantke 2001), in *S. meliloti* a central Fur-like repressor system does not operate. The gene product of the *S. meliloti fur* gene activates transcription of the downstream *sit* operon encoding a Mn(II) rather than an Fe(II) uptake system (Chao et al. 2004). Global expression analysis revealed no further induced genes related to uptake of metals. Therefore, Chao et al. (2004) suggested to rename *fur* into *mur*. A similar role of Fur was also found in *R. leguminosarum* (Wexler et al. 2003). RirA was identified as the major regulator of iron uptake genes in *S. meliloti* in a transcriptome study (Chao et al. 2005) and *R. leguminosarum* applying a proteomic approach (Todd et al. 2005).

In *B. japonicum*, Irr was proposed as an important regulator of iron uptake genes since putative Irr binding sites (ICE motifs) were found upstream of a large proportion of iron-regulated genes that were identified by global expression profiling (Rudolph et al. 2006).

Salinity is one of the most serious factors limiting the productivity of agricultural crops. Salt stress may inhibit establishment of symbiosis and also has a depressive effect on nitrogen fixation (Zahran 1999; Nogales et al. 2002). High salt concentrations may have a detrimental effect on rhizobial populations. Adaptation and tolerance to salinity is therefore markedly important for symbiosis. Prolonged exposure of *S. meliloti* to an elevated sodium chloride concentration resulted in inhibition of amino acid biosynthesis, iron uptake, motility and chemotaxis, and activation of genes related to polysaccharide biosynthesis and transport of small molecules (amino acids, amines, peptides, anions and alcohols) (Rüberg et al. 2003).

The response of *S. meliloti* to an osmotic upshift elicited by addition of sodium chloride or sucrose to exponentially growing cultures was studied by monitoring the changes in expression patterns in a time course (Dominguez-Ferreras et al. 2006). Salt stress and hyperosmotic stress had similar effects on gene expression patterns in *S. meliloti*, provoking the induction of a large number of genes of unknown function and the repression of many chromosomal genes coding for proteins of known function. A strong replicon bias in the pattern of the osmotic stress response was observed. About 65% of the induced genes mapped to the megaplasmids whereas 85% of the repressed genes were on the chromosome. About 80% of the osmoresponsive pSymB-borne genes were upregulated, suggesting a special role of this plasmid in osmoadaptation. The time course of adaptation to the initial shock followed a general pattern of two phases: an initial reaction to the sudden exposure and subsequent cellular adaptation to prolonged growth under the stress condition. Compared to the number of genes showing a differential expression pattern in phase 2, the number of genes displaying altered expression was much higher in phase 1 with a bias to repressed genes.

Adaptation to different abiotic stress factors, e.g. phosphate limitation, iron limitation, and osmotic shock, share common features. Genes governing flagellum production and chemotaxis, were frequently repressed and genes involved in biosynthesis of exopolysaccharides were upregulated. Genes related to iron uptake were repressed under phosphate and osmotic stress conditions.

4
Proteome Studies

Although, gene expression profiling at transcription level has impressively proven to be a valuable strategy to learn more about regulatory networks and about actvities of genes in the biological context, a more comprehensive understanding how genes govern the properties of a bacterial cell in a specific environment can only be achieved by studying all levels, the transcriptome, proteome and metabolome. Gene function is clearly associated with the biochemical activities of the gene products and their posttranslational modifications. The combination of high resolution 2-D polyacrylamide gel electrophoresis (2D-PAGE) that has a high performance in protein separation by isoelectric point and molecular weight with efficient methods for protein identification allow large-scale analysis of the protein state of bacteria. Methods that have become standard in protein identification are N-terminal sequencing, peptide mass fingerprinting (PMF) as well as determination of peptide amino acid sequences by post source decay (PSD) or progressing fragmentation (MS/MS) (Gevaert and Vandekerckhove 2000).

Early studies suffered from the lack of complete rhizobial genome sequences which complicated identification of proteins. A major landmark towards functional proteomics in rhizobia was a large scale examination of the proteome of *S. meliloti* 1021 under a range of conditions including complex and nutrient-depleted media (Djordjevic et al. 2003). 1180 gene products derived from 810 genes (13.1% of the predicted protein-encoding genes) were identified in this study. These proteins could be assigned to 53 metabolic pathways. The identification of 27 putative nodule-specific, 35 putative nutrient stress-specific and seven novel proteins underlined the power of such protein profiling strategies that can also provide further information on posttranslational processing and modifications.

Protein profiling was applied to study conditions such as different growth phases and media (Dainese-Hatt et al. 1999; Guerreiro et al. 1999; Münchbach et al. 1999; Encarnacion et al. 2003; Kajiwara et al. 2003), nutrient deprivation (Djordjevic et al. 2003; Todd et al. 2005), symbiotic state (Natera et al. 2000; Djordjevic et al. 2003; Djordjevic 2004; Sarma and Emerich 2005, 2006), flavonoid induction (Guerreiro et al. 1997; Chen et al. 2000), plasmid-encoded functions (Guerreiro et al. 1998; Chen et al. 2000), regulatory networks (Chen et al. 2000, 2003b, 2005; Guerreiro et al. 2000; Gao et al. 2005; Todd et al. 2005; Cantero et al. 2006) and salt tolerance (Shamseldin et al. 2005).

A number of conditions studied at transcriptome level were also investigated by proteomic approaches. A comparison of data generated by both strategies is only possible to a limited extent, since not in all cases the genome sequence was available at the time of the study and experimental design, such as growth conditions and strains used, differed in most cases.

The proteome of bacteroids was investigated in *S. meliloti* (Natera et al. 2000; Djordjevic et al. 2003; Djordjevic 2004) and *B. japonicum* (Sarma and Emerich 2005, 2006). Generally, downregulated proteins exceeded the number of upregulated proteins in bacteroids compared to cultured cells. This is in agreement with observations made in transcriptome studies of nodule bacteria. Nevertheless, only very limited overlaps were found between transcripome and proteome data of nodule bacteria. Housekeeping proteins derived from genes found to be repressed according to transcriptome data were detected in bacteroids. A high rate of new transcription of these genes comparable to transcription in growing cultured cells may not be required due to reduced degradation of proteins and induction of chaperonins in the non-dividing bacteroids. In free-living cells *S. meliloti* and *B. japonicum* predominantly expressed proteins involved in housekeeping functions (Djordjevic et al. 2003; Sarma and Emerich 2006). Candidates for nodule-specific proteins were identified in all studies. Concerning central carbon and nitrogen metabolism pathways, a broad consistency was found between proteomic and biochemical data in *S. meliloti* (Djordjevic 2004). The range of ABC-type transporters that was identified in *S. meliloti* bacteroids included a high percentage of putative amino acid and inorganic ion transporters suggesting that nutrient exchange in the environment of the nodule is highly specialized. Bacterial proteins prominent in the nodule were involved in adaptation to stress factors.

A comparison of the proteomes of cultured *B. japonicum* cells to bacteroids also emphasized the distinctive and subtle adaptation of bacteroid metabolism to the nodule environment, but also showed similarities between the two states (Sarma and Emerich 2006). Proteins involved in fatty acid, nucleic acid and cell surface synthesis were more abundant in cultured cells whereas proteins related to nitrogen metabolism were more prominent in bacteroids. The authors concluded four patterns of metabolic changes in both cell types. I, Pathways that are very flexible such as carbon metabolism that changed in terms of utilization of different carbon sources, catabolic enzymes and transporters. II, Pathways with limited flexibility including amino acid metabolism. III, Pathways that showed significant variations, such as nucleic acid and fatty acid metabolism, nitrogen metabolism, protein degradation and stabilization. IV, Pathways that remain unchanged, such as those related to metabolism of small molecules.

Other conditions studied by transcriptomic and proteomic approaches were responses to flavonoids, quorum-sensing signals, iron depletion and salt stress. Flavonoid-induced proteins were assayed in *R. leguminosarum* bv. trifolii (Guerreiro et al. 1997) and in *S. meliloti* (Chen et al. 2000). In *R. leguminosarum* bv. trifolii the global pattern of proteins remained unaltered in response to flavonoid treatment. Only four inducible proteins were found and two of these, NodB and NodE, were identified by N-terminal sequencing. Transcriptome expriments in *Rhizobium* sp. NGR234 and *S. meliloti* (Perret et al. 1999; Capela et al. 2005; Barnett et al. 2004) also indicated that the effect of flavonoids on global gene expression patterns is a limited and specialized response.

Chen et al. (2000) examined the contribution of pSymA-encoded functions in the intracellular regulation in *S. meliloti*. In the course of this study that compared protein profiles of pSymA-cured cells to those of the wild type, cells were grown in the presence or absence of luteolin. Although pSymA encodes about 1400 proteins only 60 differences were found between the wild type and its pSymA-cured derivative. The data suggested that pSymA contributes to regulation of genes mapping to the other two replicons of *S. meliloti*. Proteome analysis was also used to identify plasmid-encoded functions in *R. leguminosarum* bv. trifolii (Guerreiro et al. 1998). As in *S. meliloti* plasmid-cured derivatives of *R. leguminosarum* bv. trifolii also showed only few changes in the global protein patterns. Consistent with conclusions drawn from the genomic analysis of the seven *R. etli* CFN42 replicons (González et al. 2006) and from the proteome examination of the plasmid-cured *S. meliloti* strain (Chen et al. 2000) evidences for regulatory interactions between replicons were also found in *R. leguminosarum* bv. trifolii. Here, synthesis of 39 proteins was consistently affected in either plasmid a-, plasmid c-, or plasmid e-cured derivatives (Guerreiro et al. 1998).

Responses to quorum sensing signals were studied in *S. meliloti* by global expression profiling at RNA (Hoang et al. 2004) and at the protein level (Chen et al. 2003b; Gao et al. 2005). 55 proteins showed altered expression depending on the AHL synthase SinI and the LuxR-type regulator ExpR (Gao et al. 2005). This corresponds to 5% of the approximately 1500 proteins resolved by 2D-PAGE in this work. 75% of these proteins also changed expression in response to two of the known *S. meliloti* specific AHLs (C14-HSL and C16:1-HSL) suggesting that most of the changes observed in the *sinI* and *expR* mutants were direct effects of the Sin quorum sensing regulatory network. Only two of these proteins, ExpE6 and ExpA4, showed patterns consistent with trancriptome (Hoang et al. 2003) and earlier reporter gene fusion data (Marketon et al. 2003). In the transcriptome study about 150 genes coresponding to approximately 2.4% of all predicted protein-encoding genes of *S. meliloti* showed a *sinI*- or *expR*-dependent expression pattern (Hoang et al. 2003). This limited overlap might be explained by posttranscriptional quorum-sensing regulation which has not yet been identified and studied in *S. meliloti*, but also by several other factors, such as differences in the experimental design including culture conditions and growth phases analyzed.

Limited overlap was also found between transcriptome and proteome studies of salt tolerance in *S. meliloti* (Rüberg et al. 2003; Shamseldin et al. 2005; Dominguez-Ferreras et al. 2006). In these studies *S. meliloti* was grown in the presence of either 380 mM (Rüberg et al. 2003), 300 mM and 400 mM (Dominguez-Ferreras et al. 2006), or 428 mM sodium chloride (Shamseldin et al. 2005). In the latter study, four overexpressed and six downregulated proteins were identified. The four overexpressed proteins were identified as a hypothetical protein of the family of extracytoplasmic solute receptors (SMb20724), a putative carboxynospermidine decarboxylase (SMb21631, NspC), a protein probably involved in amino acid transport and metabolism (SMb21630) and an aminoacyl-tRNA synthase (SMc00908, IleS), which have previously been connected to salt tolerance in other rhizobia (Shamseldin et al.

2005). Only one gene, SMb20724, induced in a microarray study after prolonged exposure to 400 mM sodium chloride (Dominguez-Ferreras et al. 2006) corresponds to one of the induced proteins. Similarly only one gene, SMb21133 encoding a conserved hypothetical exported protein, identified as repressed in the microarray study, corresponds to one of the repressed proteins.

Protein profiling also demonstrated to be useful in functional characterization of transcriptional regulators. Todd et al. (2005) focused on the role of the RirA regulator in control of iron uptake proteins in *R. leguminosarum* bv. viciae. Compared to the wild type about 100 proteins were present at higher levels in a *rirA* mutant, but only 10 of these were identified by PMF. Among these were some putative periplasmic binding proteins that might be involved in iron binding and proteins that probably are involved in synthesis of iron sulfur clusters. Although, the identity of only few altered proteins was determined, protein profiling demonstrated that RirA is a major iron-responsive regulator in *R. leguminosarum* bv. viciae. This is in congruency with the results from transcriptome studies of a *rirA* mutant in *S. meliloti* (Chao et al. 2005) and demonstrates the central role of RirA in iron-responsive regulation in different rhizobia.

5
Metabolome Studies

More recently, the analysis of complex metabolite mixtures applying combinations of gas chromatography and mass spectrometry (GC-MS) (Fiehn et al. 2000), liquid chromatography and mass spectrometry (LC-MS) (Huhmann and Sumner 2002), or capillary electrophoresis and mass spectrometry (CE-MS) (Soga et al. 2003) was introduced to functional genomics in rhizobia and their host plants. Barsch et al. (2004) established metabolite analyses for cultured *S. meliloti* cells in a pilot study. Following mechanical cell lysis in methanol, hydrophilic compounds were analyzed by GC-MS. Metabolite profiles of cells grown with glucose, mannit, or succinate as sole carbon source differed significantly. The authors succeeded in identification of 65 compounds.

It was demonstrated that metabolite profiling can serve as a valuable method in characterization of bacterial mutants (Barsch et al. 2004). Compared to the wild type a leucine auxotrophic mutant accumulated 2-isopropylmalate which is the substrate of isopropylmalate isomerase. But the auxotrophic strain carried a mutation in *leuB* encoding a 3-isopropylmalate dehydrogenase, which is the enzyme catalyzing the step downstream of the reaction catalyzed by the isopropyl malate isomerase. This suggests that the reversible reaction catalyzed by the isopropylmalate isomerase favours the production of 2-isopropylmalate if 3-isopropylmalate cannot be further converted.

As a first step towards a better understanding of nodule metabolism characteristic metabolites of alfalfa root nodules were profiled in wild type nodules and nodules induced by *S. meliloti* symbiotic mutants (Barsch et al. 2006). Metabolite profiles of bacteroid-free pseudonodules induced by a *S. meliloti* succinoglycan-deficient mutant indicated that early nodule developmental processes were accompanied by photosynthate translocation but no massive organic acid formation. Nitrogenfixation-deficient nodules induced by a *S. meliloti nifH* mutant were characterized by reduced levels of characteristic amino acids involved in ammonium fixation. Elevated levels of starch and sugars in these nodules implied that plant sanctions preventing a transformation from a symbiotic to a potentially parasitic interaction are not strictly realized via photosynthate supply. Instead, metabolite profiles suggested that alfalfa plants reacted to nitrogen fixation-deficient bacteroids with decreased synthesis of organic acid and an early induction of senescence.

6
Other High Throughput Approaches

Several other strategies for functional analysis of *S. meliloti* genes on a genomic scale have been established. These include several high throughput mutagenesis approaches (Luo et al. 2005; Cowie et al. 2006; Pobigaylo et al. 2006) but also cloning of all predicted protein-encoding open reading frames, the ORFeome, of *S. meliloti* 1021 (Schroeder et al. 2005) (http://www.bioinformatics.wsu.edu/kahn). This clone library is based on a versatile set of plasmids. Using integrase recombination coding regions can be shuttled between different plasmids in vitro or in vivo. This efficient strategy can be applied to generate operon fusions, mutants, and protein expression plasmids for functional analyses.

Mutants were also generated by the plasmid integration approach which primarily aimed to generate transcriptional reporter gene fusions (Cowie et al. 2006). Genomic integration of plasmids carrying intragenic fragments of coding regions resulted in gene disruption. The inability to recover recombinants for a specific plasmid construct points to an essential function of the interrupted gene under the given growth condition. This concept was applied to identify genes essential for growth on complex medium. 101 clones that failed to generate recombinants were identified. Many of the targets were genes predicted to encode essential proteins, such as amino-acyl tRNA synthases, DNA and RNA polymerase subunits, ribosomal proteins, genes related to replication, transcription and translation, cell division and protein translocation. Moreover, this strategy identified 16 genes of unknown function as being probably essential.

A plasmid integration strategy was also employed by Luo et al. (2005) to generate mutants in 90 predicted novel LysR-type regulatory genes. Screening of these mutants for symbiotic phenotypes revealed two new LysR-type regulatory genes, *lsrA* and *lsrB*, that are involved in symbiosis. The *lsrA* mutant elicited only white nodules that exhibited no nitrogenase activity whereas the *lsrB1* mutant induced a mixture of pink and white nodules. Considering the neighboring genes, the authors suggested that *lsrA* and *lsrB* might be involved in controlling redox potential inside bacteroids, which is essential for effective nitrogen fixation.

Pobigaylo et al. (2006) established a signature-tagged mutagenesis (STM) strategy to analyze competitiveness and symbiotic ability of *S. meliloti* mutants. Signature-tagged mutants can be used to screen for specific phenotypes in mutant pools on a large scale. A set of 412 different mini-Tn5 transposons each carrying two individual sequence tags was generated. These tags are short DNA segments that contain a 23 bp variable central region flanked by invariant "arms" including priming sites for amplification and labeling of the central portions by PCR. When the tagged transposons are used for mutagenesis, each individual mutant of a set can be distinguished from every other mutant based on the different tags carried by the transposon in its genome. Identification and quantification of the mutants is performed by amplification and labeling of the tag sequences of mutants from a set and comparative hybridization of a microarray carrying the variable tag sequences. The broad host range for transposition of the mTn5 transposon, the high number of signature tags, and the tag-specific microarray makes the mTn5-STM transposon set a powerful and easy-to-use tool that can be applied to a broad range of rhizobia. Due to a promoterless *gusA* gene carried by the transposon, expression studies can be also carried out in strains carrying the *gusA* gene of the transposon in sense orientation of the targeted gene.

12 000 *S. meliloti* mutants were constructed and insertion sites for more than 5000 mutants were determined (http://www.cebitec.uni-bielefeld.de/groups/nwt/sinogate). 44% of all predicted protein-coding genes carry at least one of these ∼5000 mapped transposons. This library therefore represents one of the most comprehensive collections of defined mutants in *S. meliloti*. Pilot experiments performed to verify the novel signature-tagged transposon set in combination with a microarray hybridization approach proved the reliability of this system for identification of attenuated mutants (Pobigaylo et al. 2006).

7
Concluding Remarks and Future Perspectives

Driven by the availability of complete genome sequences, during the recent years researchers have applied *omics* techniques to a constantly growing extent to study rhizobia. The number of studies using profiling methods allow to reflect on the contribution such approaches have made so far and on the share these strategies may have in rhizobial research in the future. Nowadays, most studies of responses and adaptation to environmental factors and of regulatory genes make use of profiling methods either at the RNA or at the protein level. In this respect, these techniques have become standard.

Although, widely regarded as "global" profiling methods, currently none of the *omics* techniques is really comprehensive. The most comprehensive snapshot of bacterial cells under a given condition at a specific point of time seems to be achieved at the transcriptome level using array hybridizations. Nevertheless, in most transcriptome studies not all previously known regulated genes were re-identified. Low abundant or highly unstable transcripts might be missed. But the failure to re-identify a gene previously characterized as differentially expressed by gene fusion techniques may also be explained by differences in the methods themselves. Gene fusion techniques measure a combination of transcription of the reporter gene fusion, stability of the hybrid transcript, its translation as well as the stability and activity of the reporter gene product. Arrays measure the current state of transcript abundance in the cells, usually including an additional step that converts RNA into cDNA and they rely on hybridization. Therefore, factors such as efficiency of cDNA synthesis and labelling as well as cross-hybridization affect array data. Moreover, most arrays contain only probes designed for protein-encoding genes. Non-coding RNAs with enzymatic or regulatory functions are usually not considered, but may be represented on arrays that include also probes for intergenic regions. These RNA species probably fulfill important functions in rhizobia, but are still widely unknown and ignored by the standard postgenomic approaches.

Regarding proteome analysis methods, only a limited fraction of proteins corresponding to the whole number of protein-encoding genes can be resolved by 2D-PAGE. Since regulation may occurr at different levels and genes or their products can be missed in transcriptome and proteome profiling due to constraints in both technologies, they should be regarded as complementary rather than alternative strategies.

Most recently, also metabolite profiling has entered rhizobial research. Analysis of metabolites is complicated by the large variation of their properties, the extremely high turnover rate of some metabolites, and difficulties in extraction in terms of time scale and separation from the growth medium. Nevertheless, in the future postgenomic research will strongly profit from integration of transcriptomic, proteomic and metabolic data.

All profiling strategies, independent of the level that is monitored, have to be regarded as descriptive approaches that initialize further studies by the identification of candidate genes and generation of hypotheses. Data resulting from this kind of high throughput studies fill data bases with a wealth of information. Furthermore, an increasing number of genomic resources, such as mutant, gene fusion, and plasmid clone libraries become available. To what extent these resources will drive or facilitate research in the field of rhizobial research remains to be seen in the future years.

References

Alsmark CM, Frank AC, Karlberg EO, Legault B-A, Ardell DH, Canback B, Eriksson A-S, Naslund AK, Handley SA, Huvet M, La Scola B, Holmberg M, Andersson SG (2004) The louse-borne human pathogen *Bartonella quitina* is a genomic derivative of the zoonotic agent *Bartonella henselae*. Proc Natl Acad Sci USA 101:9716–9721

Ampe F, Kiss E, Sabourdy F, Batut J (2003) Transcriptome analysis of *Sinorhizobium meliloti* during symbiosis. Genome Biol 4:R15

Bardin S, Dan S, Osteras M, Finan TM (1996) A phosphate transport system is required for symbiotic nitrogen fixation by *Rhizobium meliloti*. J Bacteriol 178:4540–4547

Barloy-Hubler F, Cheron A, Hellegouarch A, Galibert F (2004) Smc01944, a secreted peroxidase induced by oxidative stresses in *Sinorhizobium meliloti* 1021. Microbiology 150:657–664

Barnett MJ, Fisher RF, Jones T, Komp C, Abola AP, Barloy-Hubler F, Bowser L, Capela D, Galibert F, Gouzy J, Gurjal M, Hong A, Huizar L, Hyman RW, Kahn D, Kahn ML, Kalman S, Keating DH, Palm C, Peck MC, Surzycki R, Wells DH, Yeh KC, Davis RW, Federspiel NA, Long SR (2001) Nucleotide sequence and predicted functions of the entire *Sinorhizobium meliloti* pSymA megaplasmid. Proc Natl Acad Sci USA 98:9883–9888

Barnett MJ, Toman CJ, Fisher RF, Long SR (2004) A dual-genome symbiosis chip for coordinate study of signal exchange and development in a prokaryote-host interaction. Proc Natl Acad Sci USA 47:16636–16641

Barrett CF, Parker MA (2006) Coexistence of *Burkholderia*, *Cupriavidus*, and *Rhizobium* sp. nodule bacteria on two *Mimosa* spp. in Costa Rica. Appl Environ Microbiol 72:1198–1206

Barsch A, Patschkowski T, Niehaus K (2004) Comprehensive metabolite profiling of *Sinorhizobium meliloti* using gas chromatography-mass spectrometry. Funct Integr Genomics 4:219–230

Barsch A, Tellstrom V, Patschkowski T, Küster H, Niehaus K (2006) Metabolite profiles of nodulated alfalfa plants indicate that distinct stages of nodule organogenesis are accompanied by global physiological adaptations. Mol Plant-Microbe Int 19:998–1013

Becker A, Niehaus K, Pühler A (2000) The role of rhizobial extracellular polysaccharides (EPS) in the *Sinorhizobium meliloti* – alfalfa symbiosis. In: Triplett E (ed) Nitrogen Fixation in Prokaryotes: Molecular and Cellular Biology. Horizon Scientific Press, Norfolk, UK, pp 433–447

Becker A, Bergès H, Krol E, Bruand C, Rüberg S, Capela D, Lauber E, Meilhoc E, Ampe A, de Bruijn FJ, Fourment J, Francez-Charlot A, Kahn D, Küster H, Liebe C, Pühler A, Weidner S, Batut J (2004) Global changes in gene expression in *Sinorhizobium meliloti* 1021 under microoxic and symbiotic conditions. Mol Plant-Microbe Int 17:292–303

Bergès H, Lauber E, Liebe C, Batut J, Kahn D, de Bruijn FJ, Ampe F (2003) Development of *Sinorhizobium meliloti* pilot macroarrays for transcriptome analysis. Appl Environ Microbiol 69:1214–1219

Bobik C, Meilhoc E, Batut J (2006) Fix J: a major regulator of the oxygen limitation response and late symbiotic functions of *Sinorhizobium meliloti*. J Bacteriol 188:4890–4902

Bodogai M, Ferenczi D, Bashtovyy D, Miclea P, Papp P, Dusha I (2006) The *ntrPR* operon of *Sinorhizobium meliloti* is organized and functions as a toxin-antitoxin module. Mol Plant-Microbe Int 19:811–822

Bontemps C, Golfier G, Gris-Liebe C, Carrere S, Talini L, Boivin-Masson C (2005) Microarray-based detection and typing of the Rhizobium nodulation gene *nodC*: potential of DNA arrays to diagnose biological function of interest. Appl Environ Microbiol 71:8042–8048

Breedveld MW, Miller KJ (1994) Cyclic beta-glucans of members of the family Rhizobiaceae. Microbiol Rev 58:145–161

Brewin NJ (1991) Development of the legume root nodule. Annu Rev Cell Biol 7:191–226

Campbell A (1981) Evolutionary significance of acessory DNA elements in bacteria. Annu Rev Microbiol 35:55–83

Cantero L, Palacios JM, Ruiz-Argueso T, Imperial J (2006) Proteomic analysis of quorum sensing in *Rhizobium leguminosarum* biovar viciae UPM791. Proteomics 6 Suppl 1:S97–S106

Capela D, Barloy-Hubler F, Gouzy J, Bothe G, Ampe F, Batut J, Boistard P, Becker A, Boutry M, Cadieu E, Dréano S, Gloux S, Godire T, Goffeau A, Kahn D, Lelaure V, Masuy D, Pohl T, Portelle D, Pühler A, Purnelle B, Ramsperger U, Thébault P, Vandenbol M, Weidner S, Galibert F (2001) Analysis of the chromosome sequence of the legume symbiont *Sinorhizobium meliloti*. Proc Natl Acad Sci USA 98:9877–9882

Capela D, Carrere S, Batut J (2005) Transcriptome-based identifiaction of the *Sinorhizobium meliloti* NodD1 regulpn. Appl Environ Microbiol 71:4010–4013

Capela D, Filipe C, Bobik C, Batut J, Bruand C (2006) *Sinorhizobium meliloti* differentiation during symbiosis with alfalfa: a transcriptomic dissection. Mol Plant-Microbe Int 19:363–372

Chao T-C, Becker A, Buhrmester J, Pühler A, Weidner S (2004) The *Sinorhizobium meliloti fur* gene regulates in dependence of Mn(II) the transcription of the *sitABCD* operon encoding a metal type transporter. J Bacteriol 186:3609–3620

Chao T-C, Buhrmester J, Hansmeier N, Pühler A, Weidner S (2005) Role of the regulatory gene *rirA* in the transcriptional response of *Sinorhizobium meliloti* to iron limitation. Appl Environ Microbiol 71:5969–5982

Chandra Janga S, Lamboy WF, Huerta A, Moreno-Hagelsieb G (2006) The distinctive signatures of promoter regions and operon junctions across prokaryotes. Nucl Acids Res 34:3980–3987

Charles TC, Finan TM (1991) Analysis of a 1600-kilobase *Rhizobium meliloti* megaplasmid using defined deletions generated in vivo. Genetics 127:5–20

Chen H, Higgins J, Orensik IJ, Hynes MF, Natera S, Djordjevic MA, Weinman JJ, Rolfe BG (2000) Proteome analysis demonstrates complex replicon and luteolin interactions in pSyma-cured derivatives of *Sinorhizobium meliloti* strain 2011. Electrophoresis 21:3833–3842

Chen WM, James EK, Prescott AR, Kierans M, Sprent JI (2003a) Nodulation of Mimosa spp. by the beta-proteobacterium *Ralstonia taiwanensis*. Mol Plant-Microbe Int 16:1051–1061

Chen H, Teplitski M, Gao M, Robinson JB, Bauer WD (2003b) Proteomic analysis of wild type *Sinorhizobium meliloti* responses to N-acyl homoserine lactone quorum-sensing signals and the transition to stationary phase. J Bacteriol 185:5029–5036

Chen H, Gao K, Kondorosi E, Kondorosi A, Rolfe BG (2005) Functional genomic analysis of global regulator NolR in *Sinorhizobium meliloti*. Mol Plant-Microbe Int 18:1340–1352

Cowie A, Cheng J, Sibley CD, Fong Y, Zaheer R, Patten CL, Morton RM, Golding GB, Finan TM (2006) An integrated approach to functional genomics: construction of a novel reporter gene fusion library for *Sinorhizobium meliloti*. Appl Environ Microbiol 72:7156–7167

Dainese-Hatt P, Fischer HM, Hennecke H, James P (1999) Classifying symbiotic proteins from *Bradyrhizobium japonicum* into functional groups by proteome analysis of altered gene expression levels. Electrophoresis 20:3514–3520

Davey RB, Reanney DC (1980) Extrachromosomal genetic elements and the adaptive evolution of bacteria. Evol Biol 13:113–147

DelVecchio VG, Kapatral V, Redkar RJ, Patra G, Mujer C, Los T, Ivanova N, Anderson I, Bhattacharyya A, Lykidis A, Reznik G, Jablonski L, Larsen N, D'Souza M, Bernal A, Mazur M, Goltsman E, Selkov E, Elzer PH, Hagius S, O'Callaghan D, Letesson JJ, Haselkorn R, Kyrpides N, Overbeek R (2002) The genome sequence of the facultative intracellular pathogen *Brucella melitensis*. Proc Natl Acad Sci USA 99:443–448

Djordjevic MA (2004) *Sinorhizobium meliloti* metabolism in the root nodule: a proteomic perspective. Proteomics 4:1859–1872

Djordjevic MA, Chen HC, Natera S, Van Noorden G, Menzel C, Taylor S, Renard C, Geiger O, Weiller GF (2003) A global analysis of protein expression profiles in *Sinorhizobium meliloti*: discovery of new genes for nodule occupancy and stress adaptation. Mol Plant-Microbe Int 16:508–524

Dominguez-Ferreras A, Perez-Arnedo R, Becker A, Olivares J, Soto MJ, Sanjuan J (2006) Transcriptome profiling reveals the importance of plasmid pSymB for osmoadaptation of *Sinorhizobium meliloti*. J Bacteriol 188:7617–7625

Encarnacion S, Guzman Y, Dunn MF, Hernandez M, del Carmen Vargas M, Mora J (2003) Proteome analysis of aerobic and fermentative metabolism in *Rhizobium etli* CE3. Proteomics 3:1077–1085

Fiehn O, Kopka J, Dormann P, Altmann T, Trethewey RN, Willmitzer L (2000) Metabolite profiling for plant functional genomics. Nat Biotechnol 18:1157–1161

Finan TM, Weidner S, Chain P, Becker A, Wong K, Cowie A, Buhrmester J, Vorhölter F-J, Golding B, Pühler A (2001) The complete sequence of the *Sinorhizobium meliloti* pSymB megaplasmid. Proc Natl Acad Sci USA 98:9889–9894

Freiberg C, Fellay R, Bairoch A, Broughton WJ, Rosenthal A, Perret X (1997) Molecular basis of symbiosis between Rhizobium and legumes. Nature 387:394–401

Gage DJ (2004) Infection and invasion of roots by symbiotic, nitrogen-fixing rhizobia during nodulation of temperate legumes. Microbiol Mol Biol Rev 68:280–300

Galibert F, Finan TM, Long SR, Pühler A, Abola P, Ampe F, Barloy-Hubler F, Barnett MJ, Becker A, Boistard P, Bothe G, Boutry M, Bowser L, Buhrmester J, Cadieu E, Capela D, Chain P, Cowie A, Davis RW, Dréano S, Federspiel NA, Fisher RF, Gloux S, Godrie T, Goffeau A, Golding B, Gouzy J, Gurjal M, Hernandez-Lucas I, Hong A, Huizar L, Hyman RW, Jones T, Kahn D, Kahn ML, Kalman S, Keating D, Kiss E, Komp C, Lelaure V, Masuy D, Palm C, Peck MC, Pohl T, Portetelle D, Purnelle B, Ramsperger U, Surzycki R, Thébault P, Vandenbol M, Vorhölter F-J, Weidner S, Wells DH, Wong K, Yeh K-C, Batut J (2001) The composite genome of the legume symbiont *Sinorhizobium meliloti*. Science 293:668–672

Gao M, Chen H, Eberhard A, Gronquist MR, Robinson JB, Rolfe BG, Bauer WD (2005) *sinI*- and *expR*-Dependent Quorum Sensing in *Sinorhizobium meliloti*. J Bacteriol 187:7931–7944

Garrity GM, Johnson KL, Bell JA, Searles DB (2002) Taxonomic Outline of the Procaryotes. In: Bergey's Manual of Systematic Bacteriology, Second Edition, Release 5.0. Springer-Verlag, New York, http://dx.doi.org/10.1007.bergeysoutline200310, pp 34–63

Gevaert K, Vandekerckhove J (2000) Protein identification in proteomics. Electrophoresis 21:1145–1154

Giuntini E, Mengoni A, De Filippo C, Cavalieri D, Aubin-Horth N, Landry CR, Becker A, Bazzicalupo M (2005) Large-scale genetic variation of the symbiosis-required megaplasmid pSymA revealed by comparative genomic analysis of *Sinorhizobium meliloti* natural strains. BMC Genomics 6:158

Glazebrook J, Walker GC (1991) Genetic techniques in *Rhizoium meliloti*. Meth Enzymol 204:398–419

González V, Santamaría PB, Hernández-González I, Medrano-Soto A, Moreno-Hagelsieb G, Chandra Janga S, Ramírez MA, Jiménez-Jacinto V, Collado-Vides J, Dávila G (2006) The partitioned *Rhizobium etli* genome: genetic and metabolic redundancy in seven interacting replicons. Proc Natl Acad Sci USA 103:3834–3839

Goodner B, Hinkle B, Gattung S, Miller N, Blanchard M, Qurollo B, Goldman BS, Cao YW, Askenazi M, Halling C, Mullin L, Houmiel K, Gordon J, Vaudin M, Iartchouk O, Epp A, Liu F, Wollam C, Allinger M, Doughty D, Scott C, Lappas C, Markelz B, Flanagan C, Crowell C, Gurson J, Lomo C, Sear C, Strub G, Cielo C, Slater S (2001) Genome sequence of the plant pathogen and biotechnology agent *Agrobacterium tumefaciens* C58. Science 294:2323–2328

Göttfert M, Röthlisberger S, Kündig C, Beck C, Marty R, Hennecke H (2001) Potential symbiosis-specific genes uncovered by sequencing a 410-kilobase DNA region of the *Bradyrhizobium japonicum* chromosome. J Bacteriol 183:1405–1412

Guerreiro N, Redmond JW, Rolfe BG, Djordjevic MA (1997) New *Rhizobium leguminosarum* flavonoid-induced proteins revealed by proteome analysis of differentially displayed proteins. Mol Plant-Microbe Int 10:506–516

Guerreiro N, Stepkowski T, Rolfe BG, Djordjevic MA (1998) Determination of plasmid-encoded functions in *Rhizobium leguminosarum* biovar *trifolii* using proteome analysis of plasmid-cured derivatives. Electrophoresis 19:1972–1979

Guerreiro N, Djordjevic MA, Rolfe BG (1999) Proteome analysis of the model microsymbiont *Sinorhizobium meliloti*: isolation and characterisation of novel proteins. Electrophoresis 20:818–825

Guerreiro N, Ksenzenko VN, Djordjevic MA, Ivashina TV, Rolfe BG (2000) Elevated levels of synthesis of over 20 proteins results after mutation of the *Rhizobium leguminosarum* exopolysaccharide synthesis gene *pssA*. J Bacteriol 182:4521–4532

Guerrero G, Peralta H, Aguilar A, Díaz R, Villalobos MA, Medrano-Soto A, Mora J (2005) Evolutionary, structural and functional relationships revealed by comparative analysis of syntenic genes in Rhizobiales. BMC Evol Biol 5:55

Guo H, Sun S, Finan TM, Xu J (2005) Novel DNA sequences from natural strains of the nitrogen-fixing symbiotic bacterium *Sinorhizobium meliloti*. Appl Environ Microbiol 71:7130–7138

Hantke K (2001) Iron and metal regulation in bacteria. Curr Opin Microbiol 4:172–177

Hauser F, Lindemann A, Vuilleumier S, Patrignani A, Schlapbach R, Fischer HM, Hennecke H (2006) Design and validation of a partial-genome microarray for transcriptional profiling of the *Bradyrhizobium japonicum* symbiotic gene region. Mol Genet Genomics 275:55–67

Hoang HH, Becker A, González JE (2004) The LuxR homolog ExpR, in combination with the Sin quorum sensing system, plays a central role in *Sinorhizobium meliloti* gene expression. J Bacteriol 186:5460–5472

Huhmann DV, Sumner LW (2002) Metabolicprofiling of saponins in *Medicago sativa* and *Medicago truncatula* using HPLC coupled to an electrospry ion-trap mass spectrometer. Phytochemistry 59:347–360

Johnston AW, Yeoman KH, Wexler M (2001) Metals and the rhizobial-legume symbiosis-uptake, utilization and signalling. Adv Microb Physiol 45:113–156

Jumas-Bilak E, Michaux-Charachon S, Bourg G, O'Callaghan D, Ramuz M (1998) Differences in chromosome number and genome rearrangements in the genus *Brucella*. Mol Microbiol 27:99–106

Kajiwara H, Kaneko T, Ishizaka M, Tajima S, Kouchi H (2003) Protein profile of symbiotic bacteria *Mesorhizobium loti* MAFF303099 in mid-growth phase. Biosci Biotechnol Biochem 67:2668–2673

Kaneko T, Nakamura Y, Sato S, Asamizu E, Kato T, Sasamoto S, Watanabe A, Idesawa K, Kawashima K, Kimura T, Kishida Y, Kiyokawa C, Kohara M, Matsumoto M, Matsuno A, Mochizuki Y, Nakayama S, Nakazaki N, Shimpo S, Sugimoto M, Takeuchi C, Yamada M, Tabata S (2000) Complete genome structure of the nitrogen-fixing symbiotic bacterium *Mesorhizobium loti*. DNA Res 7:331–338

Kaneko T, Nakamura Y, Sato S, Minamisawa K, Uchiumi T, Sasamoto S, Watanabe A, Idesawa K, Iriguchi M, Kawashima K, Kohara M, Matsumoto M, Shimpo S, Tsuruoka H, Wada T, Yamada M, Tabata S (2002) Complete genomic sequence of nitrogen-fixing symbiotic bacterium *Bradyrhizobium japonicum* USDA110. DNA Res 9:189–197

Krol E, Becker A (2004) Global transcriptional analysis of phosphate stress responses in *Sinorhizobium meliloti* strains 1021 and 2011. Mol Gen Genomics 272:1–17

Laguerre G, Mazurier SI, Amarger N (1992) Plasmid profiles and restriction fragment polymorphism of *Rhizobium leguminosarum* bv. viciae in field populations. FEMS Microbiol Ecol 101:17–26

Laguerre G, Geniaux E, Mazurier SI, Casartelli RR, Amarger N (1993) Conformity and diversity among field isolates of *Rhizobium leguminosarum* bv. *viciae*, and bv. *phaseoli* revealed by DNA hybridization using chromosome and plasmid probes. Can J Microbiol 39:412–419

Lan RT, Reeves PR (2000) Intraspecies variation in bacterial genomes: the need for a species genome concept. Trends Microbiol 8:396–401

Luo L, Yao S-Y, Becker A, Rüberg S, Yu Q-Q, Zhu J-B, H Cheng HP (2005) Identification of two new *Sinorhizobium meliloti* LysR-like transcriptional regulators required in nodulation. J Bacteriol 187:4562–4572

MacLellan SR, MacLean AM, Finan TM (2006) Promoter prediction in rhizobia. Microbiology 152:1751–1763

Marketon MM, Llamas I, Glenn SA, Eberhard A, González JE (2003) Quorum sensing controls exopolysacchride production in *Sinorhizobium meliloti*. J Bacteriol 185:325–331

Milcamps A, Ragatz DM, Lim P, Berger KA, de Bruijn FJ (1998) Isolation of carbon- and nitrogen-deprivation-induced loci of *Sinorhizobium meliloti* 1021 by Tn5-*luxAB* mutagenesis. Microbiology 144:3205–3218

Moreno-Paz M, Parro V (2006) Amplification of low quantity bacterial RNA for microarray studies: time-course analysis of *Leptospirillum ferrooxidans* under nitrogen-fixing conditions. Environ Microbiol 8:1064–1073

Münchbach M, Dainese P, Staudenmann W, Narberhais F, James P (1999) Proteome analysis of heat shock protein expression in *Bradyrhizobium japonicum*. Eur J Biochem 264:39–48

Natera SHA, Guerreiro N, Djordjevic MA (2000) Proteome analysis of differentially displayed proteins as a tool for the investigation of symbiosis. Mol Plant-Microbe Int 13:995–1009

Nogales J, Campos R, BenAbdelkhalek H, Olivares J, Lluch C, Sanjuan J (2002) *Rhizobium tropici* genes involved in free-living salt tolerance are required for the establishment of efficient nitrogen-fixing symbiosis with *Phaseolus vulgaris*. Mol Plant-Microbe Int 15:225–232

Oke V, Long SR (1999) Bacterial genes induced within the nodule during the Rhizobium-legume symbiosis. Mol Microbiol 32:837–849

Oláh B, Kiss E, Györgypál Z, Borzi J, Cinege G, Csanádi G, Batut J, Kondorosi Á, Dusha I (2001) Mutation in the ntrR gene, a member of the vap gene family, increases the symbiotic efficiency of Sinorhizobium meliloti. Mol Plant-Microbe Int 14:887–894

Palmer KM, Young JPW (2000) Higher diversity of Rhizobium leguminosarum biovar viciae populations in arable soils than in grass soils. Appl Environ Microbiol 66:2445–2450

Paulsen IT, Seshadri R, Nelson KE, Eisen JA, Heidelberg JF, Read TD, Dodson RJ, Umayam L, Brinkac LM, Beanan MJ, Daugherty SC, Deboy RT, Durkin AS, Kolonay JF, Madupu R, Nelson WC, Ayodeji B, Kraul M, Shetty J, Malek J, Van Aken SE, Riedmuller S, Tettelin H, Gill SR, White O, Salzberg SL, Hoover DL, Lindler LE, Halling SM, Boyle SM, Fraser CM (2002) The Brucella suis genome reveals fundamental similarities between animal and planz pathogens and symbionts. Proc Natl Acad Sci USA 99:13148–13153

Perret X, Freiberg C, Rosenthal A, Broughton WJ, Fellay R (1999) High-resolution transcriptional analysis of the symbiotic plasmid of Rhizobium sp. NGR234. Mol Microbiol 32:415–425

Pobigaylo N, Wetter D, Szymczak S, Schiller U, Kurtz S, Meyer F, Nattkemper TW, Becker A (2006) Construction of a large sequence signature-tagged miniTn5 transposon library and its application to mutagenesis of Sinorhizobium meliloti. Appl Environ Microbiol 72:4329–4337

Puskás L, Nagy Z, Kelemen JZ, Rüberg S, Bodogai M, Becker A, Dusha I (2004) Wide-range transcriptional modulating effect of ntrR under microaerobiosis in Sinorhizobium meliloti. Mol Gen Genomics 272:275–289

Rasolomampianina R, Bailly X, Fetiarison R, Rabevohitra R, Bena G, Ramaroson L, Raherimandimby M, Moulin L, De Lajudie P, Dreyfus B, Avarre JC (2005) Nitrogen-fixing nodules from rose wood legume trees (Dalbergia spp.) endemic to Madagascar host seven different genera belonging to alpha- and beta-Proteobacteria. Mol Ecol 14:4135–4146

Roumiantseva ML, Andronov EE, Sharypova LA, Dammann-Kalinowski T, Keller M, Young JPW, Simarov BV (2002) Diversity of Sinorhizobium meliloti from the Central Asian alfalfa gene center. Appl Environ Microbiol 68:4694–4697

Rudolph G, Semini G, Hauser F, Lindemann A, Freiberg M, Hennecke H, Fischer H-M (2006) The iron control element, acting in positive and negative control of iron-regulated Bradyrhizobium japonicum genes, is a target for the Irr protein. J Bacteriol 188:733–744

Rüberg S, Tian ZX, Krol E, Linke B, Meyer F, Wang Y, Pühler A, Weidner S, Becker A (2003) Construction and validation of a DNA microarray for genome-wide expression profiling in Sinorhizobium meliloti. J Biotechnol 106:255–268

Ronson CW, Nixon BT, Albright LM, Ausuble FM (1987) Rhizobium meliloti ntrA (rpoN) gene is required for diverse metabolic functions. J Bacteriol 169:2424–2431

Sarma AD, Emerich DW (2005) Global protein expression pattern of Bradyrhizobium japonicum bacteroids: a prelude to functional proteomics. Proteomics 5:4170–4184

Sarma AD, Emerich DW (2006) A comparative proteomic evaluation of culture grown vs nodule isolated Bradyrhizobium japonicum. Proteomics 6:3008–3028

Schroeder BK, House BL, Mortimer MW, Yurgel SN, Maloney SC, Ward KL, Kahn ML (2005) Development of a functional genomics platform for Sinorhizobium meliloti: construction of an ORFeome. Appl Environ Microbiol 71:5858–5864

Shamseldin A, Nyalwidhe J, Werner D (2005) A proteomic approach towards the analysis of salt tolerance in *Rhizobium etli* and *Sinorhizobium meliloti* strains. Curr Microbiol 52:333–339

Soga T, Ohashi Y, Ueno Y, Naraoka H, Tomita M, Nishioka T (2003) Quantitative metabolome analysis using capillary electrophoresis mass spectrometry. J Proteome Res 2:488–494

Spaink H (2000) Root nodulation and infection factors produced by rhizobial bacteria. Annu Rev Microbiol 54:257–288

Stepkowski T, Legocki AB (2001) Reduction of bacterial genome size and expansion resulting from obligate intracellular life style and adaptation to soil habitat. Acta Biochim Po 48:367–381

Stiens M, Schneiker S, Keller M, Kuhn S, Pühler A, Schlüter A (2006) Sequence analysis of the 144-kilobase plasmid pSmeSM11a, isolated from a dominant *Sinorhizobium meliloti* strain identified during long-term field release experiment. Appl Environ Microbiol 72:3662–3672

Sullivan JT, Trzebiatowski JR, Cruickshank RW, Gouzy J, Brown SD, Elliot RM, Fleetwood DJ, McCallum NG, Rossbach U, Stuart GS, Weaver JE, Webby RJ, de Bruijn FJ, Ronson CW (2002) Comparative sequence analysis of the symbiosis island of *Mesorhizobium loti* strain R7A. J Bacteriol 184:3086–3095

Summers ML, Elkins JG, Elliott BA, McDermott TR (1998) Expression and regulation of phosphate stress inducible genes in *Sinorhizobium meliloti*. Mol Plant-Microbe Int 11:1094–1101

Swanson JA, Mulligan JT, Long SR (1993) Regulation of *syrM* and *nodD3* in *Rhizobium meliloti*. Genetics 134:435–444

Todd JD, Sawers G, Johnston AWB (2005) Proteomic analysis reveals the wide-ranging effects of the novel, iron-responsive regulator RirA in *Rhizobium leguminosarum* bv. viciae. Mol Gen Genet 273:197–206

Uchiumi T, Ohwada T, Itakura M, Mitsui H, Nukui N, Dawadi P, Kaneko T, Tabata S, Yokoyama T, Tejima K, Saeki K, Omori H, Hayashi M, Maekawa T, Sriprang R, Murooka Y, Tajima S, Simomura K, Nomura M, Suzuki A, Shimoda Y, Sioya K, Abe M, Minamisawa K (2004) Expression islands clustered on the symbiosis island of the *Mesorhizobium loti* genome. J Bacteriol 186:2439–2448

Voegele RT, Bardin S, Finan TM (1997) Characterization of the *Rhizobium* (*Sinorhizobium*) *meliloti* high- and low-affinity phosphate uptake systems. J Bacteriol 179:7226–7232

Wanner BL (1993) Gene regulation by phosphate in enteric bacteria. J Cell Biochem 51:47–54

Wexler M, Todd JD, Kolade O, Bellini D, Hemmings AM, Sawers G, Johnston AW (2003) Fur is not the global regulator of iron uptake genes in *Rhizobium leguminosarum*. Microbiology 149:1357–1365

Wood DW, Setubal JC, Kaul R, Monks DE, Kitajima JP, Okura VK, Zhou Y, Chen L, Wood GE, Almeida NF Jr, Woo L, Chen Y, Paulsen IT, Eisen JA, Karp PD, Bovee D Sr, Chapman P, Clendenning J, Deatherage G, Gillet W, Grant C, Kutyavin T, Levy R, Li MJ, McClelland E, Palmieri A, Raymond C, Rouse G, Saenphimmachak C, Wu Z, Romero P, Gordon D, Zhang S, Yoo H, Tao Y, Biddle P, Jung M, Krespan W, Perry M, Gordon-Kamm B, Liao L, Kim S, Hendrick C, Zhao ZY, Dolan M, Chumley F, Tingey SV, Tomb JF, Gordon MP, Olson MV, Nester EW (2001) The genome of the natural genetic engineer *Agrobacterium tumefaciens* C58. Science 294:2317–2323

Yao S-Y, Luo L, Har KJ, Becker A, Rüberg S, Yu G-Q, Zhu JB, Cheng H-P (2004) The *Sinorhizobium meliloti* ExoR and ExoS proteins regulate both succinoglycan and flagella production. J Bacteriol 186:6042–6049

Young JPW, Crossman LC, Johnston AWB, Thomson NR, Ghazoui ZF, Hull KH, Wexler M, Curson ARJ, Todd JD, Poole PS, Mauchline TH, East AK, Quail MA, Churcher C, Arrowsmith C, Cherevach I, Chillingworth T, Clarke K, Cronin A, Davis P, Fraser A, Hance Z, Hauser H, Jagels K, Moule S, Mungall K, Norbertczak H, Rabbinowitsch E, Sanders M, Simmonds M, Whitehead S, Parkhill J (2006) The genome of *Rhizobium leguminosarum* has recognizable core and accessory components. Genome Biol 7:R34

Yuan Z-C, Zaheer R, Morton R, Finan TM (2006a) Genome prediction of PhoB regulated promoters in Sinorhizobium meliloti and twelve proteobacteria. Nucl Acids Res 34:2686–2697

Yuan Z-C, Zaheer R, Finan TM (2006b) Regulation and properties of PstSCAB, a high-affinity, high-velocity phosphate transport system of Sinorhizobium meliloti. J Bacteriol 188:1089–1102

Zahran HH (1999) *Rhizobium*-legume symbiosis and nitrogen fixation under severe conditions and in an arid climate. Microbiol Mol Biol Rev 63:968–989

Zhang X-S, Cheng H-P (2006) Identification of *Sinorhizobium meliloti* early symbiotic genes by use of a positive functional screen. Appl Environ Microbiol 72:2738–2748

Part II
Actinorhizal Symbioses

Evolution and Diversity of *Frankia*

Philippe Normand (✉) · Maria P. Fernandez

UMR CNRS 5557 Ecologie Microbienne, IFR41 Bio Environnement et Santé,
Université Lyon 1, 69622 CEDEX Villeurbanne, France
Philippe.Normand@univ-lyon1.fr

1	Introduction	103
2	Taxonomy	104
3	Phylogeny	111
4	Evolution and Relation with Plants	113
5	Diversity (Link with Soil Factors, Geography of Host Plant)	117
6	Conclusion and Perspectives	119
	References	120

Abstract *Frankia* is an actinobacterium that induces symbiotic nodules on the root of 25 genera of woody dicotyledonous plants (Benson and Silvester 1993). It took a century of unsuccessful efforts before it could be isolated in pure culture and its taxonomic status thus remained disputed for a long time. Pure cultures can be grouped according to host infectivity into (1) strains infective on *Alnus*, (2) strains infective on *Casuarina*, (3) strains infective on Elaeagnaceae, (4) unisolated strains and (5) uninfective isolates. These strains have been grouped into genospecies and there are between 3–7 *Alnus*-infective, 1 *Casuarina*-infective and 8–12 *Elaeagnus*-infective species. The phylogeny of these species has been reconstructed using *rrs* sequences and *Frankia* is now ascribed to the *Frankiaceae*, one of the six families of the suborder Frankinae. Genus *Frankia* is subdivided into phylogenetic clusters: #1-*Frankia alni* and related (contains *Casuarina*-infective and most *Alnus*-infective strains), #2-Unisolated strains present on Rosaceous, *Datisca*, *Coriaria* and *Ceanothus*, #3-Elaeagnaceae-infective strains and #4-a large group of "atypical" strains.

1
Introduction

Frankia is a nitrogen-fixing actinobacteria that induces symbiotic nodules on the root of 25 genera of woody dicotyledonous plants (Benson and Silvester 1993). It took a century of unsuccessful efforts before it could be isolated in pure culture and its taxonomic status thus remained disputed for a long time, not to mention its phylogeny and biodiversity. The evolution of bacteria could only be speculated about 20 years ago because of a lack of solid data, of a lack of consensus on their significance and on the interpretations that could be drawn from them. Several widely divergent taxonomic outlines, published in the successive editions of the Bergey's Manual were the result of that situation. *Frankia* for instance was one of the 30 genera in the 16th section "Actinomycetes" based on a few common morphological features: branched hyphae and spores (Holt 1994). This started to change with the emergence and diffusion of the concept of semantides (Zuckerkandl and Pauling 1965), that macromolecules carried a message about their evolutionary origin. Woese extended this idea into a working approach and in a series of papers (Fox et al. 1980; Woese 1987) proposed a method based on the sequencing of the *rrs* gene also called 16S rRNA and a conceptual framework for the evolution of bacteria in general and the actinobacteria in particular (Stackebrandt and Woese 1981).

These tools were also applied to the study of the symbionts present in the root nodules of alder and other woody plants, many of which could not be cultivated in vitro, together with more classical taxonomical approaches. These provided an emerging picture that permitted researchers to group the isolates according to a coherent model that now integrates host plant evolution and phylogenetic neighbors. This review aims at providing a global picture of the evolution of the symbiotic actinobacterium *Frankia*.

2
Taxonomy

Many people had observed nodules on the roots of several plants, among others the Italian anatomist Marcello Malpighi (1675) who in the 17th century had concluded that they were galls caused by insects. The Russian scientist Michael S. Woronin (1866) also studied the swellings ("Anschwellungen") on both the Legume lupine and the Actinorhizal alder roots and concluded following microscopic observations that they were filled with microbial cells. However, the identity of the microbe present in the actinorhizal root nodules remained elusive, consisting of hyphae, spores and vesicles, this last structure being without equivalent in the microbial world. Most important, the identity was difficult to determine because of a lack of isolate able to fulfill Koch's postulates.

Attempts to isolate *Frankia* in pure culture had been undertaken for several decades before Pommer (1956) succeeded in isolating a microbe that had unique morphological features, reminiscent of those found in nodules (Fig. 1), yet slightly distinct (hyphae, multilocular sporangia much larger than in nodules, and vesicles smaller than in nodules) to confirm the isolation in pure culture of this slow-growing actinobacterium. Indeed, it took 2–3 weeks to obtain 0.6 mm colonies using the glucose-asparagin agar previously advertised by Waksman as appropriate for actinomycetes (1950).

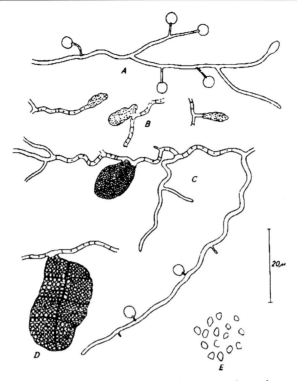

Abb. 1. Der Wurzelknöllchen-Endophyt von *Alnus glutinosa* auf Glukose-Asparagin-Agar. Erläuterungen im Text.

Fig. 1 Drawings by Pommer (1959) illustrating the different morphological features of *Frankia*: **A** unseptate hyphae 0.5–1 μm thick; **B** thinner side-hyphae 0.3–0.6 μm thick and 2 μm long that becomes bubble-shaped ("blasenförmig"); **C** septations divide the "hyphae" into "pellets" ("Körnchen"); **D** these "pellets" get bigger and more separated from one another until they reach the so-called bacteroid stage enclosed in a hose ("Schläuche"); **E** the "bacteroids" are released. Pommer did not use words such as spores or sporangia, yet he talked about hyphae and septations, typical actinomycetal features and should have expected spores, instead he expected *Rhizobium* features (bacteroids). All these features are typical of cluster 1 *Frankia*

However, this strain was lost before it could be distributed to other laboratories for independent confirmation. More than 20 years elapsed before Callaham (1978) obtained an isolate able to fulfill Koch's postulates that was sent to several laboratories for independent studies. The history of failed attempts (see review by Baker and Torrey 1979) highlights the fact that there were conflicting interpretations about the nature of the causative microbe, a bacterium, a fungi, a slime mold, etc. The advent of electron microscopy (Becking et al. 1964) helped create a consensus that actinobacteria were involved, which in turn resulted in targeting slow-growing actinobacteria in subsequent isolation attempts.

Taxonomy is a discipline with many pitfalls, with a history of errors and corrections, of conflicting views. The paper by Becking (1970) that described family Frankiaceae (Brunchorst 1886) with ten species (Table 1) of unisolated symbionts in genus *Frankia* is often considered as an example of an error, based as it was on in planta morphology, on cross-inoculation groups and on the stated belief that *Frankia* was an obligate symbiont. Cross-inoculation with pure cultures later showed that isolated strains had broader host ranges than those of endophyte suspensions and that the in planta morphology was under control of the host plant (Lalonde 1979).

Nevertheless, the paper by Becking (1970) was quite thorough and remained for many years the foundation that several subsequent authors challenged.

The list of authorities for the species name *Frankia alni* given by Becking (1970) is in itself a recapitulation of the research that had taken place on the causative microorganism: *Frankia alni* (Woronin) von Tubeuf 1895, 118; (*Schinzia alni* Woronin 1866, 6; *Entorrhiza alni* Weber 1884, 378; *Plasmodiophora alni* (Woronin) Möller 1885, 105; *Frankia subtilis* Brunchorst 1886, 174; *Frankiella alni* (Woronin) Maire and Tison 1909, 242; *Aktinomyces alni* Peklo 1910, 505; *Actinomyces alni* von Plotho 1941, 4; *Nocardia alni* Waksman 1941, 35; *Proactinomyces alni* Krassil'nikov 1949, 74 and 1959, 131; *Streptomyces alni* Fiuczek 1959, 285). Woronin's name is given each time to highlight the fact he was the first to describe the then unisolated microbe to which he gave the name *Schinzia alni*, a name that was also given to *Schinzia*

Table 1 List of species of genus *Frankia* proposed by Becking (1970) together with vesicle morphology. The shape of vesicles has been described quite differently by other subsequent authors, in particular those in *Casuarina* nodules have been said to be "hyphae" rather than club-shaped, which is linked to the fact that the symbiont does not need to protect nitrogenase from oxygen diffusion because the plant does this through the synthesis of hemoglobin contrary to other host plants such as *Alnus*

Plant order	Plant family	Host plant genus	*Frankia* species	Vesicles (size, shape)
Fagales	Betulaceae	*Alnus*	alni (type)	3–8 µm, spherical
Rhamnales	Elaeagnaceae	*Elaeagnus*	elaeagni	2–4 µm, spherical
	Rhamnaceae	*Ceanothus*	ceanothi	1.5–3 µm, spherical
	Discariaceae	*Discaria*	discariae	4 µm, spherical
Myricales	Myricaceae	*Myrica*	brunchorstii	7.5–12.5 × 1.6–2.4 µm^2, club
Casuarinales	Casuarinaceae	*Casuarina*	casuarinae	3–4 × 0.6–1.5 µm^2, club
Coriariales	Coriariaceae	*Coriaria*	coriariae	9–12 × 1.2 µm^2, club
Rosales	Rosaceae	*Dryas*	dryadis	1.5–5 × 1.5–2 µm^2, club
		Purshia	purshiae	ND
		Cercocarpus	cercocarpi	ND

leguminosarum, the symbiont of Legumes. This name that is still in use today as a synonym of a fungus, *Entorrhiza*, belonging to the Ustilaginales (http://www.binran.spb.ru/biodiv/fungi/cryptobasidiales.htm), reflecting the common sense view that similar symptoms, root nodules, had to be caused by related microbes. The other authorities are listed because they gave names to the microbe present. In the case of von Tubeuf (1895), he gave the name *Frankia alni* to honor the very famous plant physiologist A.B. Frank who had coined a few years earlier the term "symbiosis" (Frank 1885). However, he gave it to what he thought was a parasitic fungi causing a root disease ("Pflanzenkrankheiten durch kryptogame Parasiten verursacht"). The other synonyms given above were given because *Frankia* was considered successively a fungus (*Entorrhiza*), a plasmodium (*Plasmodiophora*), and finally a ray-fungus or actinomycete (*Aktinomyces, Actinomyces, Nocardia, Streptomyces*).

The first reproducible isolation in 1978 of a *Frankia* strain from *Comptonia peregrina* (Callaham et al. 1978) was rapidly followed by many others. Isolation attempts were made with all known actinorhizal plants, as those plants in symbiosis with *Frankia* came to be known. Soon, hundreds of isolates were obtained and more classical taxonomical approaches could be used. Lechevalier (Lechevalier, Lechevalier 1984; Lechevalier, Ruan 1984) proposed to integrate several criteria (ecology, infectivity, morphology, cell chemistry, physiology, serology, DNA homology and 16S rRNA catalogs) and ended up proposing two types, "A" and "B" that correspond to Elaeagnaceae-infective (i.e. cluster 3, see below) and *Alnus*-infective (cluster 1) strains, respectively. This classification scheme was very coarse, yet it contained a distinction useful to this day.

Baker (1987) proposed a functional classification scheme (Fig. 2), albeit not a "taxonomical" one, based on the single most important phenotypic feature, host infectivity. He discerned four groups, those strains infective on (1) *Alnus* and *Myrica*, (2) on *Casuarina* and *Myrica*, (3) Elaeagnaceae and *Myrica* and (4) those infective only on *Elaeagnus*. This classification was based on the finding that *Myrica* spp. was a promiscuous host, which turned out to

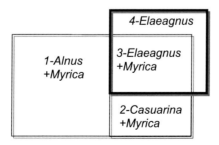

Fig. 2 Schematic of strain grouping based on infection of host plants (Baker 1987)

be true for some species, such as *M. pensylvanica* and *M. cerifera* placed in a new genus *Morella* (Baird 1969; Huguet et al. 2005), but not for others such as *Myrica gale*, this species being nodulated only by *Alnus*-infective strains (Huguet et al. 2005).

Lalonde et al. (1988) proposed another taxonomical scheme based on that of Lechevalier (type A vs. type B) but using the names *alni* and *elaeagni* to refer to host plants, and further subdivided *F. alni* into *pommerii* and *vandickyi*, based on the phenotypic distinction between *Frankia* symbionts producing in planta no spores (Sp^-) and others producing in planta numerous spores (Sp^+), both coexisting in the same locale (van Dijk 1978; van Dijk et al. 1988). This distinction was applied to isolates (Normand and Lalonde 1982) but was later dropped due to uncertainties as to what was a real "spore-positive" strain, to the existence of intermediate spore formers, and has not been reexamined with molecular tools since.

The golden yardstick of species delineation is DNA homology (Wayne et al. 1987), mentioned by Lechevalier (1984) but not used then because among other reasons it is time-consuming to obtain sufficient amounts of DNA from slow-growing *Frankia* and impossible from uncultivated symbionts. Several scientists later set to delineate species among the hundreds of isolates obtained, however it is impossible to this day to propose a unified vision because not all isolates were used by different teams. An (An et al. 1983, 1985, 1987) was the first to use the technique on *Frankia*: he analyzed 19 isolates and proposed one species with nine Alnus-infective isolates (cluster 1), one with a single *Elaeagnus*-infective (cluster 3) isolate and several unclustered isolates. Fernandez et al. (1989) proposed nine species among the 43 isolates analyzed, three in the group of *Alnus*-infective (cluster 1), five in the group of *Elaeagnus*-infective (cluster 3) and one in the group of *Casuarina*-infective (cluster 1) strains. Akimov and coworkers (Akimov and Dobritsa 1992; Akimov et al. 1991) studied a group of strains similar in part to the one studied by Fernandez et al. (1989) and found five genospecies in the strains infective on *Alnus* and four genospecies in the strains infective on *Elaeagnus*, however there were few strains tested in both studies therefore it is impossible to say if these numbers add to those previously described or not except in the case of *Frankia alni* strains CpI1 and ArI3. Lumini et al. (1996) expanded on this approach with other *Elaeagnus*-infective strains, in particular those able to cross the infectivity barrier between *Elaeagnus* and *Alnus*, and they described three further species (Table 2).

These species being based only on DNA hybridization are called genomospecies or genospecies. The only character systematically studied has been host infectivity, in many instances only considering the host species of origin. Thus, it can be said that there are from three to more than seven genomospecies infective on *Alnus* and *Myrica* (counting four possible supplementary by Akimov et al. 1992), and from eight to more than 12 genomospecies infective on Elaeagnaceae and *Gymnostoma* (counting four possible supplemen-

Table 2 Synopsis of strains tested for DNA homology (Fernandez et al. 1989; Lumini et al. 1996), and 16S sequence is available for phylogeny approaches

Species by Fernandez et al. (1989)	Reference strain	Other strains tested	16S available	Species by Lumini et al. (1996)	Infectivity
Genomic species 1	AcoN24d	CpI1, ArI3, ACN1^{AG}	ACN14a		*Alnus, Morella*
Genomic species 2	AV22C	AVN17o	AVN17s		*Alnus, Morella*
Genomic species 3	ARgP5^{Ag}			ARgP5^{Ag}	*Alnus, Morella*
Genomic species 4	Ea1–12	HR27$_{14}$, Ea1$_2$, Ea2$_6$, Ea3$_3$	Ea1$_2$	Ea33	Elaeagnaceae, *Morella*
Genomic species 5	TX31e^{HR}	EAN1pec, HRX401a	EAN1pec	K1, HR773, Hr611	Elaeagnaceae, *Morella*
Genomic species 6	EUN1f				Elaeagnaceae, *Morella* (+ *Alnus*)
Genomic species 7	HRN18a				Elaeagnaceae, *Morella* (+ *Alnus*)
Genomic species 8	Ea50-1				Elaeagnaceae, *Morella*
Genomic species 9	CeD (ORS020606)	ORS020607, Cj1-82, AllI1	CeD, CcI3		*Casuarina*
Genomic species 10				E1, E13, E15	Elaeagnaceae, *Morella* (+ *Alnus*)
Genomic species 11				HrI1, 2.1.7, 2.1.2	Elaeagnaceae, *Morella* (+ *Alnus*)
Genomic species 12			SCN10a	SCN10a	Elaeagnaceae, *Morella*

tary by Akimov et al. 1992). There is thus a minimum of 12 species in genus *Frankia*, probably at least 20 in the available isolates. All in all, there are probably close to a hundred species of *Frankia*, distributed in the four clusters (below).

The strains infective on *Casuarina* are very close to one another, having DNA homology values above 70% (Fernandez et al. 1989) and conserved PCR-RFLP patterns in the *rrs* and *nif* operons (Rouvier et al. 1992). However, those strains were all isolated directly or indirectly from a single host species, *C. equisetifolia* and from areas all around the world where the family is not native. When native (Australian) populations of four *Casuarina* and *Allocasuarina* species were studied by direct molecular characterization of nodules, five distinct groups were described that were consistently associated with the plant species (Rouvier et al. 1992; Simonet et al. 1999). The question remains if the isolated strains from introduced areas (genomic species 9) and the four new genotypes described from Australian nodules are close enough to be grouped in a unique species. In the absence of isolates from Australia, this question may have to wait for other approaches such as AFLP or MLSTadapted to direct nodule characterization to find an answer. This genomospecies probably should be given a name because it has a very strong phenotypic feature, that of being able to form nodules on *Casuarina*, such a name would in all likelihood be *F. casuarinae*.

Those strains present on the root nodules of Rosaceous + *Datisca* + *Coriaria* plants have sometimes been construed as an iceberg tip of diversity and particularly hard to isolate despite numerous attempts with techniques that had worked with the other host plants. Becking (1970), for instance, had proposed there were at least three *Frankia* species in rosaceous plants, one for each host plant genus he considered (*Dryas, Purshia, Cercocarpus*) and one each for *Coriaria* and *Datisca* (Table 1). However, Mirza et al. (1994) first showed that the symbionts of *Datisca* could nodulate *Coriaria*, had similar 16S rRNA genes and were thus probably the same bacterium. Bosco et al. (1994) added to that *Dryas*, the 16S rRNA sequence of which being close to those of *Datisca* and *Coriaria* symbionts. The symbiont present in Western US populations of Rosaceae, *Ceanothus* (Rhamnaceae) and *Datisca glomerata* can be similarly clustered (Vanden Heuvel et al. 2004) although it is hard to expand this finding to other locales. These can also be nodulated by cluster 3 under laboratory conditions. This is evocative of cluster 3 strains being cosmopolitan or promiscuous, universal soil microbes that will infect whatever actinorhizal plant grows near them, while cluster 2 strains would be selective, but more competitive on *Ceanothus* and on the other plants known to be nodulated by cluster 2 plants (Rosaceous-*Datisca*-*Coriaria*).

The reason only one species name has been given is that there are no phenotypic data to correlate to the genomic species delineation. *Frankia alni*, the first species named by von Tubeuf (1895) and published by Becking (1970) remains valid because it is the type species of the genus and because the phenotypic features then given remain valid except for the term "obligate symbiont": hyphae, 0.3–0.5 µm filamentous, branched and septate, branching not always associated with cell wall formation (...), spherical vesicles, normally 3.0–5.0 µm in diameter (...), polyhedral-shaped cells resulting from complete cell division (...); host plants : (...) all *Alnus* species (...) form root nodules of coralloid structure and up to 5–8 cm in diameter (...) nitrogen fixing capacity. Other genomospecies may correspond to this description and will be named when sufficient phenotypic data becomes available.

The question of naming the other species regularly crops up but it would require either a serious effort at studying phenotypic features, or else an acceptance of genotypic data as solid enough to base species name on it. In any case, this will only be possible when sufficient sequence data has accumulated and statistical approaches are used, which will happen in the coming years.

All *Frankia* strains are today grouped into genus *Frankia*, a lone genus of the Frankiaceae as defined by Normand et al. (1996), member together with five other families (Acidothermaceae, Geodermatophilaceae, Sporichthyaceae, Kineosporiaceae and Microsphaeraceae) of suborder Frankinae, in order Actinomycetales (Normand and Benson, in press), according to the paradigm that taxonomy must be congruent with phylogeny (discussed below).

3
Phylogeny

Hahn et al. (1989) were the first to use *rrs* (16S rRNA) sequences to address the question of the relations between *Frankia* and its neighbors. They came to the conclusion that *Frankia* was indeed close to *Geodermatophilus*, a dry soil dweller and to *Blastococcus*, a Baltic sea microbe but away from *Dermatophilus*, a skin pathogen that has similar multilocus sporangia with which it was lumped in the previous taxonomical treatments (Lechevalier and Lechevalier 1984, 1989). Hahn et al. (1989) sequenced only one *Frankia* strain, Ag45 and obtained a 16S rRNA "catalog" of another strain but could not get an overall view of the structure of the genus. They nevertheless validated the concept that *Frankia* belonged to a group of related soil microbes, the Frankiaceae with three genera, *Frankia*, *Geodermatophilus* and "*Blastococcus*", the brackets being due to the fact that the genus had not been properly described but this has been done since (Urzi et al. 2004).

The availability of nearly complete *rrs* genes allowed one to address the question of the structure of the genus (Fig. 3). Normand et al. (1996) used their own sequences as well as sequences obtained by others and deposited in Genbank. The two main conclusions from that work were that *Frankia* could be subdivided into four coherent clusters (1: *Alnus* and *Casuarina* infective strains; 2: *Rosaceaous*-, *Datiscaceae*- and *Coriariaceae*-unisolated symbionts; 3: *Elaeagnus*-infective strains; 4: a divergent group of ineffective or non-infective strains) and that the closest phyletic neighbor was not *Geodermatophilus* as described previously but a recently isolated thermal spring microbe, *Acidothermus cellulolyticus*. This last surprising finding was confirmed by a different genetic marker, *recA* (Marechal et al. 2000), a marker that had been one of the first used to confirm 16S rRNA results (Lloyd and Sharp 1993).

The scheme was reexamined (Wolters et al. 1997) with the aim of positioning supplementary strains, in particular ineffective ones from *A. glutinosa* in The Netherlands. They obtained the same general topology with the following differences: 1 – the *Casuarina*-infective strain was not close to cluster 1

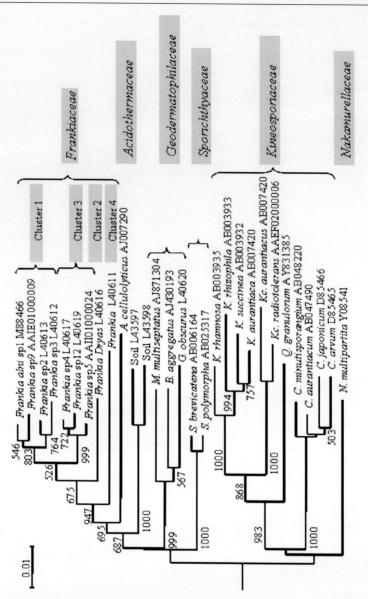

Fig. 3 Phylogenetic tree of the Frankinae (Normand and Benson, in press) by the Neighbor-Joining method (Saitou and Nei 1987) including all validly described genera and species of the suborder Frankinae

sequences; 2 – cluster 1a (*Frankia alni, Casuarina*-infective) was distinct from cluster 1b (*Alnus*-infective genomic species 2 and 3); 3 – presence of a supplementary cluster 5 containing ineffective strains. The first finding is disputable given that there are only 1% difference between *Frankia alni* and *Casuarina*-

infective CeD in the *rrs* sequences, much less than between *Frankia alni* and genomic species 2 and 3 (cluster 1b) and derives in all likelihood from the use of partial *rrs* sequences known to cause artefactual distortions. The second conclusion, that clusters 1a and 1b would be distinct depends on the tree reconstruction method, the two subclusters had low bootstrap support anyway. If the two last conclusions are accepted, it follows that there would be three clusters out of the five proposed that would contain *Alnus*-infective strains.

Jeong et al. (1999) agreed with the general outline with a minor divergence in the numeration system used where "clades" are numbered in the order that they are postulated to have appeared (clade#1 Rosaceaous unisolated = cluster 2, #2 Alnus = cluster 1, #3 Elaeagnus = cluster 3). They added an analysis of the host plants phylogeny based on RbcL and concluded that there had been several emergences of symbiotic ability and concluded that there had been coevolution although they thought the bacteria had appeared before the plants. The scheme was also examined by Clawson (2004) who also used *glnII* sequences as confirmation. Their conclusions were similar with the same basal position of cluster 2 (clade 1). The *glnII* sequences yielded the same topology.

A difference between these studies is the question of cluster 4. It is composed of AgB1-9 an isolate from *Alnus* in India that does reinfect its original host plant but does not fix nitrogen, of PtI1 an isolate from *Purshia tridentata* that does not reinfect its host plant, does not hybridize to a *nif* probe and which sequence probably contains several errors, and of Cn7 and Dc2 both obtained from Pakistan, both having phenotypic characteristics comparable to PtI1. This broad cluster was found not to have a significant bootstrap value, and the sequences in it all contain probable errors, for instance positions where all other Frankinae are different and conserved. Such errors complicate analyses and the question of whether cluster 4 is coherent cannot be resolved beyond doubt. This cluster 4 may be less a cluster in a phylogenetic sense than a catch-all category for those strains that cannot be grouped in the three other well-defined clusters.

The structure of family Frankiaceae has also evolved from that proposed by Hahn et al. (1989) with two sister genera, *Geodermatophilus* and *Blastococcus*, to one with only one genus, *Frankia* (Normand et al. 1996) with another family, Geodermatophilaceae (Normand 2006) created to accommodate *Geodermatophilus* and *Blastococcus* as well as the recently described *Modestobacter*.

4
Evolution and Relation with Plants

The study of symbiosis focuses mostly on understanding the mechanisms involved and on the evolutionary pathways that have led to the present situation. This approach has been given new impetus by the genome sequencing programs underway. The following section aims at describing insights gained on evolution of *Frankia* by the study of individual genes.

The rate of evolution of the *rrs* gene has been evaluated by correlation with the fossil record for a number of evolutionary events involving bacteria and set at 1 to 5×10^{-8} substitutions per site per year which translates to 50 MY for each 1% of distance (Ochman and Wilson 1987) or 2% according to Moran et al. (1993) when working with faster-evolving endosymbiotic bacteria. The *Frankia rrs* distances have been measured by Normand et al. (1996), by Jeong et al. (1999) and by Clawson et al. (2004) and different conclusions were reached. If one takes the Ochman's metric, which has been cited hundreds of times and is now recognized as the best estimate, it follows that at about 350 MY ago, at about the time primitive land plants appeared, there was an explosive radiation that saw the emergence from a group of soil actinomycetes, on the one hand of *Acidothermus*, and on the other hand of the distant ancestor of all *Frankia*, all these bacteria having 6.6% difference in their *rrs* (Table 3). This distant nitrogen-fixing ancestor underwent a second radiation about 100 MY ago when the first dicotyledonous plant families started to appear in the pollen record, among others the actinorhizal Myricaceae and the Betulaceae. Clawson et al. (2004) dispute this view and based on a GlnII marker calibrated on the *E. coli-S. typhimurium* divergence by Turner and Young (2000) postulate an emergence of *Frankia* clusters a long time before that of the host plants (263–285 MY). This refutation is not robust being based simply on one supplementary marker that appears to have been laterally transferred (Pesole et al. 1995). Resolving that dispute in a convincing manner would require a much better calibration of the molecular clocks with several supplementary points in the fossil record. Moran's more rapid estimate is argued as stemming from the small population and short generation

Table 3 Average distances between *rrs* from groups of strains

Pair	Distance in %	Age of event (MY)
F. alni vs. *F. "casuarinae"*	1.1	55
Cluster 1 range	1.1–1.8	55–90
Cluster 2 range	0.9–2.1	45–105
Cluster 3 range	0.1–1.4	5–70
Cluster 1 vs. Cluster 2	2.7	135
Cluster 1 vs. Cluster 3	1.6	80
Cluster 3 vs. Cluster 2	2.2	110
Cluster 4 vs. other Clusters	4.9	245
Acidothermus vs. *Frankia*	6.6	330
Soil sequences vs. *Frankia*	6.8	340

time of obligate symbiotic bacteria (1993), which may also be the case of cluster 2 *Frankia* strains. The timeline proposed by Normand et al. (1996) appears the most coherent that sets the emergence of *Frankia* clusters at 100–200 MY, which corresponds to the emergence of the oldest actinorhizal plant genera *Myrica* and *Alnus*.

The host plants in symbiosis with *Frankia* belong to 25 genera of dicotyledons, collectively called "actinorhizal". These plants belong to eight families and the question of the phylogeny of these host plants has been debated with the aim of determining whether there had been one emergence of the ability to enter a symbiosis with *Frankia* followed by several losses or several independent emergences. Soltis et al. (1995) concluded that the Rosid clade contained all families with symbiotic representatives, and that symbiotic ability had developed several times independently, four times with rhizobia and four times with the genus *Frankia*. Whatever the case, there is little evidence of coevolution (Fig. 4) with the several instances of crossing the infectivity barriers first mentioned by Becking (1970). For instance, cluster 3 *Frankia*

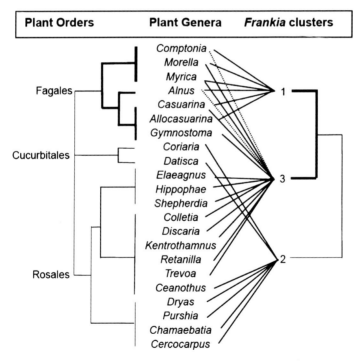

Fig. 4 Correspondence between the plant genera *rbcL* (*left*) and *Frankia* 16S rRNA (*right*) phylogenies. *Heavy lines* – infection via root hairs, *lighter lines* – intercellular penetration (Clawson et al. (2004), web.uconn.edu/mcbstaff/benson/Frankia/PhylogenyFrankia.htm). The *oblique lines* join the plant and bacterial taxa when they can establish symbiotic structures, *dashed lines* indicate laboratory only symbioses not found in nature

can infect *Morella* (Myricaceae), *Gymnostoma* (Casuarinaceae), *Alnus* as well as the Elaeagnaceae and the Rhamnaceae. Cluster 2 *Frankia* can infect *Coriaria* (Coriariaceae), *Datisca* (Datiscaceae), *Ceanothus* (Rhamnaceae). There is evidence of cospecialization only in the *Casuarina* and *Allocasuarina* lineages where Simonet et al. (1999) found slightly different restriction patterns in field-collected nodules from five species tested.

The gene *nifH* that codes for dinitrogenase reductase was the second gene to be sequenced after the 16S rRNA and there are at the moment 31 sequences in Genbank. This gene is very conserved and yet absent from all other actinomycetes except *Frankia*, it is thus linked to symbiosis and it was deemed important to understand its origin and explore the possibility of a horizontal gene transfer (HGT) from another nitrogen-fixer and thus account for the lone status of *Frankia* as a lone N_2-fixing actinobacterium. Their availability also provided the opportunity to use them to follow *Frankia* strains in the complex soil environments. The results of those studies were that *nif* genes had been most probably inherited vertically from the universal actinobacterial ancestor; this implies that it has been lost in all other lineages. The recent finding of *Micromonospora* in *Casuarina equisetifolia* nodules seemed to challenge that vision but the *nif* genes recovered by PCR in that bacterium were very close to those of *Frankia*, meaning they were acquired relatively recently by HGT (Valdes et al. 2005). The direction of that transfer remains unclear, however the prevalence of *nifH* in *Frankia* and apparent rarity in *Micromonospora* would argue in favor of a transfer toward *Micromonospora*.

The Rhamnaceae species *Ceanothus* is a group of widespread bush in Western USA that shares the habitat of several other actinorhizal plants belonging to the Rosaceae: (*Purshia, Chamaebatia, Cercocarpus*) and *Datisca*. Several studies (Ritchie and Myrold 1999; Oakley et al. 2004; Clawson et al. 2004, Vandenheuvel et al. 2004) have concluded that all these plants were nodulated by *Frankia* belonging to cluster 2 that were recalcitrant to growing in culture medium. Only one study (Murry et al. 1997) concluded that the symbiont belonged to cluster 3, which was in line with the characterization of Rhamnaceae (*Discaria*) symbionts and isolates (Clawson et al. 1998) from South America as belonging to cluster 3. These studies point to the major role played by geography in explaining the distribution of *Frankia*, with cluster 3 underrepresented in arid hot Western USA. That explanation should be confirmed by trapping on soils from both North and South America with an array of actinorhizal Rhamnaceae.

A parallel event was postulated by Navarro et al. (1997) to explain the presence of cluster 3 strains on the Casuarinaceae *Gymnostoma*, contrary to *Casuarina* and *Allocasuarina* that are nodulated by cluster 1 strains. The split was also linked to the breakup of the Gondwana supercontinent which brought *Casuarina* to diversify in the drier Australian continent while *Gymnostoma* diversified in the wetter Melanesia during the Tertiary era beginning

65 MY ago (http://earth.leeds.ac.uk/~eargah/Gond.html) (Raven and Axelrod 1974).

The emerging vision we have of *Frankia* being the only causative agent of actinorhizal plants nodules may have to be revisited in a similar way to what has occurred in Legumes with first *Bradyrhizobium*, then *Methanobacter* (Sy et al. 2001), *Ralstonia* and *Burkholderia* (Moulin et al. 2001), opening new avenues of research on LGT. The first such deviation from the *Frankia*-only dogma was the demonstration of nodule induction on *Alnus glutinosa* by *Penicillium nodositatum*, a eukaryote (Sequerra et al. 1997). Gauthier et al. (1981) had shown that *Casuarina* harbored a cluster 3 "atypical" *Frankia* hitch-hiker, before an isolate able to fulfill Koch's postulate on *Casuarina* was isolated (Diem et al. 1983). This was expanded to *Micromonospora*, present in *Casuarina* nodules (Valdes et al. 2005). Finally, it was shown that *Agrobacterium* could induce on *Elaeagnus umbellata* structures (Berg et al. 1992) indistinguishable from actinorhizae. The characterization of complete genomes of *Frankia* and other bacteria will help pinpoint which genes present in these other taxa are involved in interactions with actinorhizal plants.

The hypothesis of several emergences of symbiosis is very hard to confirm or infirm with sequence data alone, which is why Pawlowski et al. (2003) analyzed gene expression patterns and concluded that there were stronger similarities between legumes and intracellularly infected actinorhizal plants (*Alnus*) than between actinorhizal plants of two different phylogenetic subgroups (*Alnus/Datisca*). Given that we still do not know which molecular effectors are involved in *Frankia*, it is not yet possible to determine if all *Frankia* lineages have the same determinants but it is certainly a perspective of study in the coming years.

5
Diversity (Link with Soil Factors, Geography of Host Plant)

The widespread presence of *Frankia* strains has been reported in many soils having various physical and chemical characteristics, even in the absence of a compatible host plant (Smolander and Sundman 1987; Huss-Danell and Frej 1986). Accordingly, actinorhizal plants are often nodulated outside their original geographic range and *Frankia* strains have been detected in early successional soils before the host plant becomes established (McCray Batzli et al. 2004).

Nevertheless some studies reported that actinorhizal plants were not nodulated and concluded on the apparent absence of compatible *Frankia* strains in the soils tested. Among the factors thought to influence the abundance and the diversity of *Frankia* in soils, pH, aeration and moisture are the more commonly cited (Jamann et al. 1992; Smolander 1990; Dawson et al. 1989; Huguet et al. 2004; McCray Batzli et al. 2004).

For example, Jamann et al. (1992; 1993) used the highly variable *nifD-nifK* intergenic region to differentiate strains isolated by trapping on *Elaeagnus* from soils differing in pH. Trapping on *Elaeagnus* permitted one to show that the diversity of strains, as measured by the Shannon-Weaver diversity index was lowest at acidic pH. Acidity is known to influence availability of nutrients, it also reduces the number of nodules obtained on *Alnus* (Quispel 1958) and physiology of the plant in general. *Alnus* (Lumini et al. 1994), *Gymnostoma* (Jaffré et al. 1994) and other actinorhizal plants are known to be relatively resistant to adverse soil conditions, which is why they are used to revegetate degraded soils such as mine spoils.

Frankia strains trapped on *Alnus incana* were used to assess the impact of soil pollutants on microbial diversity (Ridgway et al. 2004). More polluted soils were found to result in reduced plant growth, and had a distinct *Frankia* community including the cosmopolitan cluster 3 strains. Soil type, which comprises differential concentration of metals, was also found to play a major role in the distribution of cluster 3 *Frankia* strains infective on *Gymnostoma* in New Caledonia (Navarro et al. 1999).

The full genetic diversity of *Frankia* may not yet have been revealed due to limits brought about by the approaches used. Indeed, the time-consuming isolation step not only strongly limits sampling, it also induces biases, under-representing non- or poorly nodulating strains and selecting those isolates that are easily culturable. A previous facilitating but bias-inducing step often seen is the use of an intermediate host inoculated in the laboratory by crushed nodules from the field. Indeed, it was reported that the genetic diversity observed in natural populations can be totally different from that observed following greenhouse inoculations with soil or field nodules (Huguet et al. 2005).

Most of the recent attempts to describe *Frankia* diversity used direct detection in nodules or soil of PCR-amplified DNA fragments, but these methods are often based on one or two conserved genes (generally *rrs* and *nif* genes) and cannot but fail to describe the full extent of biological diversity.

For instance, the tools described above have been used to study the rhizosphere *Frankia* strains without the trapping step bias. A primer was designed to target specifically *Frankia*'s *rrs* to the exclusion of *Acidothermus*, *Geodermatophilus* and other Actinobacteria. This study was designed and undertaken before it was realized that two specific primers with at least two discriminating nucleotides at the 3' end were necessary to ensure that a specific amplification would be obtained from complex environments. The *Frankia* specific primer was used together with a universal one on DNA from an *Alnus viridis* rhizophere. The unexpected result was that not a single *Frankia* sequence was recovered, either in the immediate rhizosphere nor in the soil further away, rather a number of distant sequences illustrating the need for two specific primers, but also two unknown very close phylogenetic neighbors (Fig. 3; Normand and Chapelon 1997). A parallel finding using *nifH* primers was described for the rhizosphere of *Dryas drummondii* (Deslippe and Egger 2006). This illustrates that *Frankia* is a slow growing bacterium, present at low abundance even in soils where it is expected to have a selective advantage and were it not for its ability to induce nodules on plants whence it can be isolated, it would not be known. There are thus probably hundreds of such actinobacterial species that thrive slowly in diverse soils and rhizospheres. Such was probably the phenotype of the ancestor of all *Frankia* known today.

Symbiosis determinants are not known at the moment, yet a few genes are known to play a role in symbiosis such as those coding for Fur, Cat, Shc, Hup, Lectins. The study of their phylogeny, which has been undertaken following acquisition of the newly sequenced genomes may shed light on the evolutionary history of the clade.

6
Conclusion and Perspectives

In the coming years, a serious study of the link between phenotype and genomospecies will probably be undertaken, if only to permit naming species but also to explore more seriously the question of coevolution. Naming species is not only done to please taxonomists, it also means the link between phenotype and genotype has been seriously considered by a array of complementary methods. Given that it is difficult to do DNA hybridization with slow-growing strains or unisolated symbionts, and considering the poor state of strain collections, the most promising approach is probably whole-genome PCR-based approaches such as AFLP (Stackebrandt et al. 2002).

Another approach being developed is MLST, which aims at determining if different loci have the same evolutionary history, and whether there have been instances of HGT or sexuality (Barker et al. 2005; Nemoy et al. 2005; Stampone et al. 2005). This when done on *Sinorhizobium* permitted one to show, using Tajima's (1989) D parameter that the *nod* locus had been subjected to more homogenizing selective pressures than neutral housekeeping loci, as expected, but that the *exo* locus involved in the synthesis of an exopolysaccharide that has a Nod-phenotype, had undergone diversifying selection, as though the two locus played opposite roles, one linked to triggering a universal receptor, the other evading defense-related metabolites under control of fast-evolving plant determinants (Bailly et al. 2006). MLST is usually performed on PCR-targeted and sequenced genes but DNA arrays can help speed things up and get full genome coverage. This approach should thus allow us to discern which loci are involved in plant interactions and evaluate the importance of sex in the evolution of *Frankia*. Measuring K_a/K_s (rate of silent mutations/rate of non-silent mutations) can also help discern patterns of selection and to determine in silico which loci are involved in plant interactions (Hurst 2002).

The number of genomes in the databank as well as that in projects is increasing exponentially and we can see the day when any newly isolated strain will be sequenced as a first intention before doing any other characterization. Metagenomes are also coming and they will change the picture of *Frankia* phylogeny, evolution and ecology.

Of course, the most decisive step to further our understanding of evolution of *Frankia* will be to develop a genetic approach. This will allow us to understand where the nodulation genes originated from and which evolutionary pressures these have been subjected to.

References

Akimov V, Dobritsa S (1992) Grouping of *Frankia* strains on the basis of DNA relatedness. Syst Appl Microbiol 15:372–379

Akimov V, Dobritsa S, Stupar O (1991) Grouping of *Frankia* strains by DNA–DNA homology: how many genospecies are in the genus *Frankia*? In: Polsinelli M, Vincenzini M (eds) Nitrogen Fixation. Kluwer Academic Publishers, Dordrecht, The Netherlands, pp 635–636

An C, Riggsby W, Mullin B (1985) Relationships of *Frankia* isolates based on deoxyribonucleic acid homology studies. Int J Syst Bacteriol 35:140–146

An C, Riggsby W, Mullin B (1987) DNA relatedness of *Frankia* isolates ArI4 and EuI1 to other actinomycetes of cell wall type III. Actinomycetes 20:50–59

An C, Wills J, Riggsby W, Mullin B (1983) Deoxyribonucleic acid base composition of 12 *Frankia* isolates. Can J Bot 61:2859–2862

Bailly X, Olivieri I, De Mita S, Cleyet-Marel JC, Bena G (2006) Recombination and selection shape the molecular diversity pattern of nitrogen-fixing *Sinorhizobium* sp. associated to *Medicago*. Mol Ecol 15:2719–2734

Baird JR (1969) A taxonomic revision of the plant family Myricaceae of North America, North of Mexico. The University of North Carolina, Chapel Hill

Baker D (1987) Relationships among pure-cultured strains of Frankia based on host specificity. Physiol Plant 70:245–248

Baker D, Torrey J (1979) The isolation and cultivation of actinomycetous root nodule endophytes. In: Gordon JC, Wheeler CT, Perry DA (eds) Symbiotic Nitrogen Fixation in the Management of Temperate Forests. Forest Research Laboratory, Oregon State University, Corvallis, pp 38–56

Barker M, Thakker B, Priest FG (2005) Multilocus sequence typing reveals that *Bacillus cereus* strains isolated from clinical infections have distinct phylogenetic origins. FEMS Microbiol Lett 245:79–184

Becking J, De Boer W, Houwink A (1964) Electron microscopy of the endophyte of *Alnus glutinosa*. Antonie van Leeuwenhoek J Microbiol Serol 30:343–376

Becking JH (1970) Frankiaceae *fam. nov.* (Actinomycetales) with one new combination and six new species of the genus *Frankia* Brunchorst 1886, 174. Int J Syst Bacteriol 20:201–220

Benson DR, Silvester WB (1993) Biology of *Frankia* strains, actinomycete symbionts of actinorhizal plants. Microbiol Rev 57:293–319

Berg RH, Liu L, Dawson JO, Savka M, Farrand S (1992) Induction of pseudoactinorhizae by the plant pathogen *Agrobacterium rhizogenes*. Plant Physiol 98:777–779

Bosco MS, Jamann S, Chapelon C, Simonet P, Normand P (1994) *Frankia* microsymbiont in *Dryas drummondii* nodules is closely related to the microsymbiont of *Coriaria* and genetically distinct from other characterized *Frankia* strains. In: Hegazi NA, Fayez M, Monib M (eds) Nitrogen Fixation with Non-legumes. The American University in Cairo Press, Cairo, pp 173–183

Callaham D, Del Tredici P, Torrey J (1978) Isolation and cultivation in vitro of the actinomycete causing root nodulation in *Comptonia*. Science 199:899–902

Clawson ML, Bourret A, Benson DR (2004) Assessing the phylogeny of *Frankia*-actinorhizal plant nitrogen-fixing root nodule symbioses with *Frankia* 16S rRNA and glutamine synthetase gene sequences. Mol Phylogenet Evol 31:131–138

Clawson ML, Carú M, Benson DR (1998) Diversity of *Frankia* strains in root nodules of plants from the families Elaeagnaceae and Rhamnaceae. Appl Environ Microbiol 64:3539–3543

Dawson JO, Kowalski DG, Dart PJ (1989) Variation with soil depth, topographic position and host species in the capacity of soils from an Australian locale to nodulate *Casuarina* and *Allocasuarina* seedlings. Plant Soil 118:1–11

Deslippe J, Egger K (2006) Molecular diversity of *nifH* genes from bacteria associated with high arctic dwarf shrubs. Microbiol Ecol 51:516–525

Diem HG, Dommergues Y (1983) The isolation of *Frankia* from nodules of *Casuarina*. Can J Bot 61:2822–2825

Fernandez M, Meugnier H, Grimont P, Bardin R (1989) Deoxyribonucleic acid relatedness among members of the genus *Frankia*. Int J Syst Bacteriol 39:424–429

Fox GE, Stackebrandt E, Hespell RB, Gibson J, Maniloff J, Dyer TA, Wolfe RS, Balch WE, Tanner RS, Magrum LJ, Zablen LB, Blakemore R, Gupta R, Bonen L, Lewis BJ, Stahl DA, Luehrsen KR, Chen KN, Woese CR (1980) The phylogeny of prokaryotes. Science 209:457–463

Frank AB (1885) Über die auf Wurzelsymbiose beruhende Ernährung gewisser Bäume durch unterirdische Pilze. Ber Dtsch Bot Ges 3:128–145

Gauthier D, Diem HG, Dommergues Y (1981) Infectivité et effectivité des souches de *Frankia* isolées de nodules de *Casuarina equisetifolia* et *Hippophaë rhamnoides*. CR Acad Sci Ser III 293:489–491

Hahn D, Lechevalier M, Fischer A, Stackebrandt E (1989) Evidence for a close phylogenetic relationship between members of the genera *Frankia*, *Geodermatophilus*, and "*Blastococcus*" and emendation of the family Frankiaceae. Syst Appl Microbiol 11:236–242

Holt JW (1994) Bergey's Manual of Determinative Bacteriol, 9th edn. Williams & Wilkins, Baltimore

Huguet V, Mergeay M, Cervantes E, Fernandez MP (2004) Diversity of *Frankia* strains associated to *Myrica gale* in Western Europe: impact of host plant (*Myrica* vs. *Alnus*) and of edaphic factors. Environ Microbiol 6:1032–1041

Huguet V, Gouy M, Normand P, Zimpfer JF, Fernandez MP (2005) Molecular phylogeny of Myricaceae: a reexamination of host-symbiont specificity. Mol Phylogenet Evol 34:557–568

Huguet V, Land EO, Casanova JG, Zimpfer JF, Fernandez MP (2005) Genetic diversity of *Frankia* microsymbionts from the relict species *Myrica faya* (Ait.) and *Myrica rivasmartinezii* (S.) in Canary Islands and Hawaii. Microbiol Ecol 49:617–625

Hurst LD (2002) The K_a/K_s ratio: diagnosing the form of sequence evolution. Trends Genet 18:486

Huss-Danell K Frej A-K (1986) Distribution of *Frankia* in soil from forest and afforestation sites in northern Sweden. Plant Soil 90:407–418

Jaffré T, Rigault F, Sarrailh J-M (1994) La végétalisation des anciens sites miniers. Bois For Trop 242:45–57

Jamann S, Fernandez M, Moiroud A (1992) Genetic diversity of Elaeagnaceae-infective *Frankia* strains isolated from various soils. Acta Oecolog 13:395–405

Jamann S, Fernandez M, Normand P (1993) Typing method for N_2-fixing bacteria based on PCR-RFLP application to the characterization of *Frankia* strains. Mol Ecol 2:17–26

Jeong S, Ritchie N, Myrold D (1999) Molecular phylogenies of plants and Frankia support multiple origins of actinorhizal symbioses. Mol Phylogenet Evol 13:493–503

Lalonde M (1979) Immunological and ultrastructural demonstration of nodulation of the European *Alnus glutinosa* (L.) Gaertn. host plant by an actinomycetal isolate from the North American *Comptonia peregrina* (L.) Coult. root nodule. Bot Gaz 140(S):S35–S43

Lalonde M, Simon L, Bousquet J, Séguin A (1988) Advances in the taxonomy of *Frankia*: recognition of species *alni* and *elaeagni* and novel subspecies *pommerii* and *vandijkii*. In: Bothe H, Newton WE (eds) Nitrogen Fixation: Hundred Years After. Gustav Fischer, Stuttgart, pp 671–680

Lechevalier M (1984) The taxonomy of the genus *Frankia*. Plant Soil 78:1–6

Lechevalier M, Lechevalier H (1984) Taxonomy of *Frankia*. In: Ortiz-Ortiz LB, Bojalil LF, Yakoleff V (eds) Biological, Biochemical and Biomedical Aspects of Actinomycetes. Academic Press, New York, pp 575–582

Lechevalier M, Lechevalier H (1989) Genus *Frankia* Brunchorst, 1886, 174. In: Williams ST, Sharpe ME, Holt JE (eds) Bergey's Manual of Systematic Bacteriology. Williams and Wilkins, Baltimore, Maryland, pp 2410–2417

Lechevalier M, Ruan J (1984) Physiology and chemical diversity of *Frankia* spp. isolated from nodules of *Comptonia peregrina* (L.) Coult. and *Ceanothus americanus* L. Plant Soil 78:5–22

Lloyd AT, Sharp PM (1993) Evolution of the *recA* gene and the molecular phylogeny of bacteria. J Mol Evol 37:399–407

Lumini E, Fernandez MP, Bosco M (1996) PCR-RFLP and total DNA homology revealed three related species among broad host-range *Frankia* strains. FEMS Microbiol Ecol 21:303–311

Lumini E, Bosco M, Puppi G, Isopi R, Frattegiani M, Buresti E, Favilli F (1994) Field performance of *Alnus cordata* Loisel (italian alder) inoculated with *Frankia* and VA-mycorrhizal strains in mine spoil afforestation plots. Soil Biol Biochem 26:659–661

Malpighi M (1675) *Anatome plantarum* Regiae Societati (J. Martyn), London, UK

Marechal J, Clement B, Nalin R et al. (2000) A *recA* gene phylogenetic analysis confirms the close proximity of *Frankia* to *Acidothermus*. Int J Syst Evol Microbiol 50:781–785

McCray Batzli J, Zimpfer JF, Huguet V, Smyth CA, Fernandez M, Dawson JO (2004) Distribution and abundance of infective, soilborne *Frankia* and host symbionts *Shepherdia*, *Alnus*, and *Myrica* in a sand dune ecosystem. Can J Bot 82:700–709

Mirza M, Akkermans W, Akkermans A (1994) PCR-amplified 16S rRNA sequence analysis to confirm nodulation of *Datisca cannabina* L. by the endophyte of *Coriaria nepalensis* Wall. Plant Soil 160:147–152

Moran N, Munson M, Baumann P, Ishikawa H (1993) A molecular clock in endosymbiotic bacteria is calibrated using the insect hosts. Proc R Soc London Ser B Biol Sci 253:167–171

Moulin L, Munive A, Dreyfus B, Boivin-Masson C (2001) Nodulation of legumes by members of the beta-subclass of Proteobacteria. Nature 411:948–950

Murry MA, Konopka AS, Pratt SD, Vandergon TL (1997) The use of PCR-based typing methods to assess the diversity of *Frankia* nodule endophytes of the actinorhizal shrub *Ceanothus*. Physiol Plant 99:714–721

Navarro E, Nalin R, Gauthier D, Normand P (1997) The nodular microsymbionts of *Gymnostoma* spp. are *Elaeagnus*-Infective *Frankia* strains. Appl Environ Microbiol 63:1610–1616

Navarro E, Jaffre T, Gauthier D, Gourbiere F, Rinaudo G, Simonet P, Normand P (1999) Distribution of *Gymnostoma* spp. microsymbiotic *Frankia* strains in New Caledonia is related to soil type and to host-plant species. Mol Ecol 8:1781–1788

Nemoy LL, Kotetishvili M, Tigno J, Keefer-Norris A, Harris AD, Perencevich EN, Johnson JA, Torpey D, Sulakvelidze A, Morris JG Jr, Stine OC (2005) Multilocus sequence typing versus pulsed-field gel electrophoresis for characterization of extended-spectrum beta-lactamase-producing *Escherichia coli* isolates. J Clin Microbiol 43:1776–1781

Normand P (2006) *Geodermatophilaceae* fam. nov., a formal description. Int J Syst Evol Microbiol 56:2277–2278

Normand P, Benson D. The Frankiaceae^{VP}. In: Goodfellow M, Garrity G, Kampfer P (eds) The Bergey's Manual of Systematic Bacteriology. Bergey's Manual Trust, Springer (in press)

Normand P, Chapelon C (1997) Direct characterization of *Frankia* and of close phyletic neighbors from an *Alnus viridis* rhizosphere. Physiol Plant 99:722–731

Normand P, Lalonde M (1982) Evaluation of *Frankia* strains isolated from provenances of two *Alnus* species. Can J Microbiol 28:1133–1142

Normand P, Orso S, Cournoyer B et al. (1996) Molecular phylogeny of the genus *Frankia* and related genera and emendation of the family Frankiaceae. Int J Syst Bacteriol 46:1–9

Oakley B, North M, Franklin JF, Hedlund BP, Staley JT (2004) Diversity and distribution of Frankia strains symbiotic with *Ceanothus* in California. Appl Environ Microbiol 70:6444–6452

Ochman H, Wilson AC (1987) Evolution in bacteria: evidence for a universal substitution rate in cellular genomes. J Mol Evol 26:74–86

Pawlowski K, Swensen S, Guan C et al. (2003) Distinct patterns of symbiosis-related gene expression in actinorhizal nodules from different plant families. Mol Plant Microbe Interact 16:796–807

Pesole G, Gissi C, Lanave C, Saccone C (1995) Glutamine synthetase gene evolution in bacteria. Mol Biol Evol 12:189–197

Pommer EH (1956) Beiträge zur Anatomie und Biologie der Wurzelknöllchen von *Alnus glutinosa* Gaertn. Flora 14:603–634

Pommer EH (1959) Uber die Isolierung des Endophyten aus den Wurzelknöllchen *Alnus glutinosa* Gaertn. und uber erfolgreiche re-Infektions Versuche. Ber Dtsch Bot Ges 72:138–150

Quispel A (1958) Symbiotic nitrogen fixation in non-leguminous plants. IV. The influence of some environmental conditions on different phases of the nodulation process in *Alnus glutinosa*. Acta Bot Neerl 7:191–204

Raven PH, Axelrod DI (1974) Angiosperm biogeography and past continental movements. Ann Miss Bot Gard 61:539–673

Ridgway KP, Marland LA, Harrison AF et al. (2004) Molecular diversity of *Frankia* in root nodules of *Alnus incana* grown with inoculum from polluted urban soils. FEMS Microbiol Ecol 50:255–263

Ritchie NJ, Myrold DD (1999) Geographic distribution and genetic diversity of *Ceanothus*-infective *Frankia* strains. Appl Environ Microbiol 65:1378–1383

Rouvier C, Nazaret S, Fernandez MP, Picard B, Simonet P, Normand PP (1992) *rrn* and *nif* intergenic spacers and isozyme patterns as tools to characterize *Casuarina*-infective *Frankia* strains. Acta Oecol 13:487–495

Saitou N, Nei M (1987) The neighbor-joining method: a new method for reconstructing phylogenetic trees. Mol Biol Evol 4:406–425

Sequerra J, Marmeisse R, Valla G, Normand P, Capellano A, Moiroud A (1997) Taxonomic position and intraspecific variability of the nodule forming *Penicillium nodositatum* inferred from RFLP analysis of the ribosomal intergenic spacer and random amplified polymorphic DNA. Mycol Res 101:465-472

Simonet P, Navarro E, Rouvier C, Reddell P, Zimpfer J, Dommergues Y, Bardin R, Combarro P, Hamelin J, Domenach AM, Gourbiere F, Prin Y, Dawson JO, Normand P (1999) Co-evolution between *Frankia* populations and host plants in the family Casuarinaceae and consequent patterns of global dispersal. Environ Microbiol 1:525-533

Smolander A (1990) *Frankia* populations in soils under different tree species with special emphasis on soils under *Betula pendula*. Plant Soil 121:1-10

Smolander A, Sundman V (1987) *Frankia* in acid soils of forests devoid of actinorhizal plants. Physiol Plant 70:297-303

Soltis DE, Soltis PS, Morgan DR, Swensen SM, Mullin BC, Down JM, Martin PG (1995) Chloroplast gene sequence data suggest a single origin of the predisposition for symbiotic nitrogen fixation in angiosperms. Proc Natl Acad Sci USA 92:2647-2651

Stackebrandt E, Woese C (1981) Towards a phylogeny of the actinomycetes and related organisms. Curr Microbiol 5:197-202

Stackebrandt E, Frederiksen W, Garrity GM, Grimont PAD, Kämpfer P, Maiden MCJ, Nesme X, Rossello-Mora R, Swings J, Trüper HG, Vauterin L, Ward AC, Whitman WB (2002) Report of the ad hoc committee for the re-evaluation of the species definition in bacteriology. Int J Syst Evol Microbiol 52:1043-1047

Stampone L, Del Grosso M, Boccia D, Pantosti A (2005) Clonal spread of a vancomycin-resistant *Enterococcus faecium* strain among bloodstream-infecting isolates in Italy. J Clin Microbiol 43:1575-1580

Sy A, Giraud E, Jourand P, Garcia N, Willems A, De Lajudie P, Prin Y, Neyra M, Gillis M, Boivin-Masson C, Dreyfus B (2001) Methylotrophic *Methylobacterium* bacteria nodulate and fix nitrogen in symbiosis with legumes. J Bacteriol 183:214-220

Turner SL, Young JP (2000) The glutamine synthetases of rhizobia: phylogenetics and evolutionary implications. Mol Biol Evol 17:309-319

Urzi C, Salamone P, Schumann P, Rohde M, Stackebrandt E (2004) *Blastococcus saxobsidens* sp. nov., and emended descriptions of the genus *Blastococcus* Ahrens and Moll 1970 and *Blastococcus aggregatus* Ahrens and Moll 1970. Int J Syst Evol Microbiol 54:253-259

Valdes M, Perez NO, Estrada-de Los Santos P et al. (2005) Non-*Frankia* actinomycetes isolated from surface-sterilized roots of *Casuarina equisetifolia* fix nitrogen. Appl Environ Microbiol 71:460-466

Van Dijk C (1978) Spore formation and endophyte diversity in root nodules of *Alnus glutinosa* (L.) Vill. New Phytol 81:601-615

Van Dijk C, Sluimer A, Weber A (1988) Host range differentiation of spore-positive and spore-negative strain types of *Frankia* in stands of *Alnus glutinosa* and *Alnus incana* in Finland. Physiol Plant 72:349-358

Vanden Heuvel BD, Benson DR, Bortiri E, Potter D (2004) Low genetic diversity among *Frankia* spp. strains nodulating sympatric populations of actinorhizal species of Rosaceae, *Ceanothus* (Rhamnaceae) and *Datisca glomerata* (Datiscaceae) west of the Sierra Nevada (California). Can J Microbiol 50:989-1000

Von Tubeuf K (1895) Pflanzenkrankheiten durch kryptogame Parasiten verursacht. Verlag Julius Springer, Berlin

Waksman SA (1950) The Actinomycetes, their nature, occurrence, activities, and importance. Percaline Rouge, Waltham, Mass., USA

Wayne LG, Brenner DJ, Colwell RR, Grimont PAD, Kandler O, Krichevsky MI, Moore LH, Moore WEC, Murray RGE, Stackebrandt E, Starr MP, Trüper HG (1987) Report of the ad hoc committee on reconciliation of approaches to bacterial systematics. Int J Syst Bacteriol 37:463-464

Woese CR (1987) Bacterial evolution. Microbiol Rev 51:221-271

Wolters DJ, Van Dijk C, Zoetendal EG, Akkermans ADL (1997) Phylogenetic characterization of ineffective *Frankia* in *Alnus glutinosa* (L.) Gaertn. nodules from wetland soil inoculants. Mol Ecol 6:971-981

Woronin MS (1866) Über die bei der Schwarzerle (*Alnus glutinosa*) und bei der gewöhnlichen Garten-Lupine (*Lupinus mutabilis*) auftretenden Wurzelanschwellungen. Mémoires de l'Academie Impériale des Sciences de St. Petersbourg, VII Ser. 10:1-13

Zuckerkandl E, Pauling L (1965) Evolutionary divergence and convergence in proteins. In: Bryson V, Vogel HJ (eds) Evolving Genes and Proteins. Academic Press, New York, pp 97-166

Induction of Actinorhizal Nodules by *Frankia*

Katharina Pawlowski

Department of Botany, Stockholm University, 10691 Stockholm, Sweden
pawlowski@botan.su.se

1	Introduction	127
2	How *Frankia* Enters the Plant	130
2.1	Intracellular Infection	130
2.2	Intercellular Infection	132
2.3	Infected Cells	134
3	Nodule Structure	136
3.1	How to Deal with the Oxygen Dilemma of Nitrogen Fixation	137
4	Symbiotic Signaling	139
5	Autoregulation of Nodule Number	142
6	Concluding Remarks and Future Perspectives	145
	References	146

Abstract Actinomycetous soil bacteria of the genus *Frankia* can induce the formation of nitrogen-fixing root nodules on a diverse group of host plants from eight angiosperm families, collectively called actinorhizal plants. Nodule induction involves the colonization of the root surface, followed by the elicitation of changes in the plant that lead to nodule primordium formation and to the entry of bacteria into the root. Like in legume–rhizobia symbioses, bacteria can enter the plant root either intracellularly through a curled root hair, or intercellularly without root hair involvement, and the entry mechanism is determined by the host plant species. Mature actinorhizal nodules are coralloid structures consisting of multiple nodule lobes each of which represents a modified lateral root without root cap, a superficial periderm, and infected cells in the expanded cortex. In this review, an overview of infection mechanisms and nodule structure is given; comparisons with the corresponding mechanisms in legume symbioses are presented. Recent results on the perception of bacterial signal factors are described.

1
Introduction

Nitrogen-fixing root nodule symbioses evolved 50–100 million years ago (Kistner and Parniske 2002). Phylogenetic analysis has shown that all plants able to enter a root nodule symbiosis belong to a single clade (Eurosid I; see chapter by Normand and Fernandez, in this volume), i.e., they go back to a common ancestor (Soltis et al. 1995). Within the Eurosid I clade, rhizobial symbioses are supposed to have evolved four times independently, namely three times within the legume family and once for *Parasponia* (Doyle 1998; Fig. 1). Similarly, three or four independent origins have been suggested for actinorhizal symbioses (Swensen 1996; Benson et al. 2004; Fig. 1). Altogether, the phylogenetic data suggest that the common ancestor of the Eurosid I clade had acquired a unique property based on which a root nodule symbiosis could develop, and that such a development occurred seven times, four times for rhizobia and three times for *Frankia* symbioses. The symbiosis of the Betulaceae, Casuarinaceae, and Myricaceae (Fagales; genera *Alnus, Casuarina, Allocasuarina, Gymnostoma, Myrica*, and *Comptonia*) is supposed to have a common origin; similarly, the symbiois of the Datiscaceae and Coriaraceae (Cucurbitales; genera *Datisca* and *Coriaria*) and the symbiosis of the Rosaceae, Rhamnaceae, and Elaeagnaceae (Rosales; genera *Cercocarpus, Chamaebatia, Cowania, Dryas, Purshia, Ceanothus, Colletia, Discaria, Kentrothamnus, Retanilla, Talguenea, Trevoa, Elaeagnus, Hippophae, Shepherdia*).

Phylogenetically, nitrogen-fixing *Frankia* strains fall into three major clusters (Fig. 1; Benson et al. 2004; see chapter by Normand and Fernandez, in this volume). Cluster 1 strains have been isolated from nodules of *Alnus* (Betulaceae), *Casuarina, Allocasuarina* (Casuarinaceae) as well as *Myrica, Comptonia*, and *Morella* (Myricaceae; Huguet et al. 2004, 2005) species. Cluster 2 strains have been isolated from nodules of *Hippophae, Elaeagnus*, and *Shepherdia* (Elaeagnaceae; Clawson et al. 1998), *Discaria* and *Trevoa* (Rhamnaceae; Clawson et al. 1998), as well as *Gymnostoma* (Casuarinaceae; Navarro et al. 1997) and *Myrica* (Myricaceae; Huguet et al. 2004, 2005). Very rarely, cluster 2 strains have been shown to be able to induce nodules on *Alnus* (Betulaceae; Baker 1987; Bosco et al. 1992; Lumini et al. 1996) and *Casuarina* (Casuarinaceae; Gauthier et al. 1999) and clade 1 strains to induce nodules of *Gymnostoma* (Casuarinaceae; Zhang et al. 1984; Zhang and Torrey 1985). Cluster 3 strains have been found in nodules of actinorhizal Rosaceae, *Datisca, Coriaria*, and *Ceanothus* (Vanden Heuvel et al. 2004). Strains of clusters 1 and 2 can be isolated and used for reinfection experiments and thus have been studied primarily. Cluster 3 strains capable of reinfecting the host plants cannot be isolated (see, e.g., Mirza et al. 1992).

Figure 1 shows that some so-called promiscuous genera, *Myrica* and *Gymnostoma*, can be infected by a broader range of strains than the other genera in the Myricaceae and Casuarinaceae, respectively. Phylogenetic studies and fossil analysis indicate that the promiscuous genera represent more ancient symbiosis than those of more recently derived genera, such as *Casuarina, Allocasuarina* or *Alnus*, which are infected by a narrower range of strains. Consequently, it has been proposed that actinorhizal symbioses proceed evolutionarily toward more specialization, leading to a greater probability of losing the symbiosis in more recently derived lineages (Maggia and Bousquet

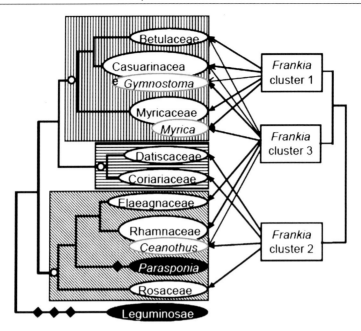

Fig. 1 Simplified scheme of the phylogenetic relationship between and actinorhizal plants based on Benson and Clawson (2000). Groups of plants infected by rhizobia are labeled by inverse print. *Boxes* indicate the three main groups of actinorhizal plants: Fagales (*vertical stripes*), Cucurbitales (*horizontal stripes*), and Rosales (*diagonal stripes*). The phylogenetic relationship between the three clades of symbiotic *Frankia* is included. Some actinorhizal genera (*Gymnostoma*, *Myrica*, *Ceanothus*) that differ in microsymbiont specificity from the rest of the family are indicated. *Thick arrows* connect *Frankia* clades with the plant group members that the clades are commonly associated with. *Thin arrows* indicate that members of that clade have been isolated from, or detected in, an effective or ineffective nodule of a member of the plant group at least once. Host specificity exists within the *Frankia* clades, i.e., not all members of a *Frankia* clade can nodulate all plants associated with that clade. *Circles* indicate the three putative origins of the ability to enter an actinorhizal symbiosis (Swensen 1996; Benson et al. 2004); *black rhombs* indicate the four putative origins of the ability to enter a symbiosis with rhizobia (Doyle 1998)

1994). However, it should be pointed out that this promiscuity is mostly restricted to greenhouse experiments and not commonly observed in the field (Huang et al. 1985; Clawson and Benson 1999; Navarro et al. 1999).

Bacterial nitrogen fixation is catalyzed by the nitrogenase enzyme complex which is irreversibly denatured by oxygen. However, nitrogen fixation has a high energy demand requiring oxygen for respiratory processes in the nodules. This leads to the so-called oxygen dilemma of nitrogen fixation, necessitating a tight control of oxygen distribution in nodules. In legumes, this control is achieved by the host plant. However, in contrast with rhizobia, *Frankia* can fix N_2 ex planta under aerobic conditions, providing oxygen protection for nitrogenase by forming specialized cell types, vesicle, surrounded by multilayered envelopes consisting almost entirely of hopanoids, bacterial steroid lipids (Benson and Silvester 1993; Berry et al. 1993). The number of lipid monolayers in the vesicle envelope is correlated with the oxygen tension (Parsons et al. 1987). Hence, *Frankia* can contribute to the protection of nitrogenase from oxygen. In different actinorhizal plant families, diverse strategies of plant and microsymbiont are combined to achieve this goal, involving nodule morphology, vesiclestructure, plant and bacterial metabolism, and the formation of oxygen-binding proteins in infected cells.

With one exception, *Datisca glomerata*, all actinorhizal plants are perennial trees or woody shrubs. Actinorhizal nodules are perennial organs; i.e., the activity of the meristems of nodule lobes ceases in fall and recommences during bud burst. Actinorhizal root nodules from different plant families can be very dissimilar; in particular, the morphology of root nodules of the Cucurbitales is distinct from all others. All actinorhizal nodules are coralloid organs consisting of multiple lobes, each lobe representing a modified lateral root. The formation of an actinorhizal nodule primordium is always induced in the root pericycle, as is the case with lateral roots. Nodule induction can occur via either intracellular or intercellular infection by a compatible *Frankia* strain.

2
How *Frankia* Enters the Plant

The same *Frankia* isolate can infect different actinorhizal hosts by either root hair infection or intercellular penetration (Miller and Baker 1986; Racette and Torrey 1989). Hence, the infection mechanism is determined solely by the host plant.

2.1
Intracellular Infection

Actinorhizal host plant species from the Fagales are infected by *Frankia* through intracellular penetration of deformed root hairs (Berry and Sunell 1990). Nodules of these host plants also share some anatomical characteristics (Swensen and Mullin 1997), appear among the oldest fossil records for actinorhizal plant species (Benson and Clawson 2000), and show strong similarities with legume nodules with regard to the expression patterns of homologous genes (Pawlowski et al. 2003). Like in legume–rhizobia symbioses, during intracellular infection, the first visible reaction of the plant to the presence of compatible *Frankia* bacteria is root hair deformation.

Infected root hairs exhibit extensive root hair deformation, comparable to root hairs in legume–rhizobial symbioses (Callaham et al. 1979; Fig. 2A). When a *Frankia* hypha is trapped in a curled root hair, it is internalized by

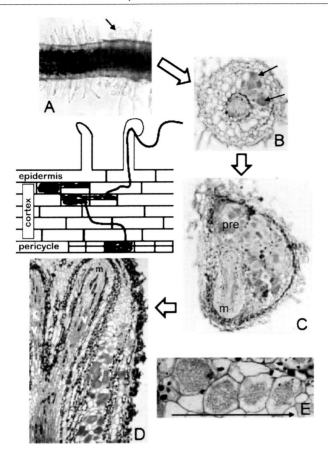

Fig. 2 The intracellular infection pathway in symbioses of actinorhizal Fagales (scheme). Products excreted by *Frankia* lead to root hair deformation. A *Frankia* hyphae enters the plant in an infection thread-like structure formed in a curled root hair. Cell divisions are induced in the root cortex, leading to the formation of the prenodule. Prenodule cells are infected by branching infection thread-like structures. Concomitantly, the nodule primordium is induced in the root pericycle. Infection thread-like structures grow transcellularly from the prenodule to the primordium and infect primordium cells.
A *Alnus rubra* seedling root after 24 h of incubation with culture filtrate from *Frankia* ArI3; the *arrow* points at a deformed root hair (Ribeiro, unpublished). **B** *A. glutinosa* root cross section with prenodule; *thin arrows* point at infected prenodule cells filled with *Frankia* hyphae. **C** *A. glutinosa* root cross section with prenodule and young nodule lobe; *m*—meristem, *pre*—prenodule. **D** Longitudinal section of a branching *Casuarina glauca* nodule lobe; *v*—vascular system, *m*—nodule meristem. **E** Detail of a *C. glauca* nodule section: basipetal growth of infection thread-like structures leads to the formation of files of infected cells; the *arrow* points toward the apical meristem. Sections depicted in **B** and **C** are stained with Ruthenium Red and Toluidine Blue, and the sections in **D** and **E** with Toluidine Blue

invagination of the host plasma membrane and deposition of new primary wall material around the invading hypha (Berry et al. 1986). The newly synthesized host cell wall that encapsulates *Frankia* forms a structure within the root hair that is analogous to the infection thread found in legume–rhizobial symbioses (Berg 1999). In contrast with infection threads in legumes, actinorhizal infection thread-like structure lack an infection thread matrix.

During formation of the infection thread-like structure, cell divisions are triggered in the root cortex subjacent to the infected root hair. Some of the newly divided cells expand and subsequently become infected by *Frankia*, forming the so-called the prenodule(Fig. 2B). During infection of plant cells, infection thread-like structures encapsulating *Frankia* hyphae branch in them until they form a dense network. Infected and uninfected prenodule cells share the same differentiation as the corresponding cells of mature nodule lobes, and *Frankia* can differentiate for nitrogen fixation in infected prenodule cells (Laplaze et al. 2000). So far, the intracellular infection pathway of actinorhizal plants resembles that of legumes. However, while prenodule cells become infected, a nodule lobe primordium is initiated in the root pericycle near the prenodule (Fig. 2B). The number of nodule primordia initiated per prenodule can vary from one to several (Callaham and Torrey 1977). As the nodule lobe primordium expands, infection thread-like structures grow across cortical cells from the prenodule into the base of the primordium and infect primordium cells by extensive branching within these cells. Finally, a nodule lobe develops from the nodule primordium. Within the nodule lobe, infection thread-like structures grow basipetally and infect new cortical cells, leading to the formation of files of infected cells (Fig. 2D).

In legume–rhizobia symbioses, infection thread growth has been examined in detail. Like root hairs and pollen tubes, infection threads show tip growth which requires the polarization of the cytoplasm of a cell that is to be passed by an infection thread. This polarization leads to the formation of so-called pre-infection thread structures (PITs; van Brussel et al. 1992) representing cytoplasmic bridges through which the infection thread will grow. Such cytoplasmic bridges have also been observed in intracellularly infected actinorhizal nodules, indicating that here, the growth of infection thread-like structures follows the same mechanism (Berg 1999).

2.2
Intercellular Infection

In actinorhizal Rosales, infection by *Frankia* takes place via intercellular colonization of the host root. This has been explicitly shown for members of the Elaeagnaceae(*Elaeagnus*; Miller and Baker 1986; *Shepherdia*, Racette and Torrey 1989) and Rhamnaceae families (*Ceanothus*; Liu and Berry 1991a, 1991b; *Discaria*; Valverde and Wall 1999a). The infection pathway was never studied in Rosaceae, also not in the Cucurbitales (Datiscaceae and Coriariaceae), since the microsymbionts of these families cannot be cultured (see chapter by Normand and Fernandez, in this volume). It is assumed that actinorhizal species of the Rosaceae, Datiscaceae, and Coriariaceae are infected intercellularly since their mature nodules lack two features typical of intracellularly infected species, namely, prenodules and the formation of files of infected cells in the cortex of mature nodule lobes.

During intercellular infection, *Frankia* filaments invade the root cortex by penetrating between adjacent epidermal cells and later cortical cells through the middle lamella (Fig. 3; Miller and Baker 1985; Liu and Berry 1991a; Valverde and Wall 1999a). No root hair deformation takes place. The host cells secrete pectin-rich wall material into the intercellular spaces that also stains intensely with Coomassie Blue (Fig. 3; Liu and Berry 1991a; Valverde and Wall 1999a). During the intercellular colonization of the outer root cortex, the formation of a nodule primordium is induced in the root pericycle. Intracellular invasion of primordium cells by the symbiont commences some

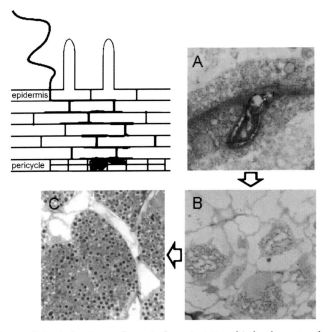

Fig. 3 The intercellular infection pathway (scheme). A *Frankia* hypha enters between two epidermal cells. During colonization of the intercellular spaces of the root cortex, the formation of a nodule primordium is induced in the root pericycle opposite a protoxylem pole. Nodule primordium cells are infected from the apoplast (**A** *Discaria trinervis*). Infected cortical cells are filled with branching hyphae in infection thread-like structures from the center outward (**B** *D. trinervis*). Once infected cells have been filled with branching hyphae, vesicles develop (**C** vesicles are recognizable as *dark-blue dots*; *D. trinervis*). Photographs were kindly provided by L.G. Wall

days later; infection thread-like structures are formed directly in the cells taking up *Frankia* hyphae from the apoplast. Infected cells become hypertrophic. Also in mature nodule lobes, infection of cortical cells occurs by internalization of *Frankia* hyphae coming from the apoplast (Fig. 3A). The occurrence of *Frankia* hyphae in the intercellular spaces between cortical cells of the postmeristematic regions was observed in *Discaria trinervis* (Valverde and Wall 1999a), while infection thread-like structures growing from cell to cell were not observed (Valverde and Wall, personal communication). No cytoplasmic bridges preceding the formation of infection thread-like structures in infected cells have been observed in intercellularly infected plants.

As stated above, actinorhizal members of the Cucurbitales (Datiscaceae and Coriariaceae) are assumed to be infected intercellularly based on the lack of prenodules. Whether or not files of infected cortical cells are formed, as is typical for nodules of intracellularly infected plants (Fig. 2E), cannot be determined in Cucurbitales where the infected cells form a continuous patch in the cortex, not interspersed with uninfected cells (see below). Meanwhile it could be shown that in the cortex of mature nodule lobes of *Datisca glomerata*, infection threadsproceed from cell to cell, and no infection of cortical cells takes place by internalization of *Frankia* hyphae from the apoplast (Berg 1999; Silvester et al. 1999). Hence, nodules of Cucurbitales seem to share features of both intracellularly and intercellularly infected actinorhizal plants. It is still unclear how *Frankia* hyphae enter *D. glomerata*roots; however, the fact that the sister species *D. cannabina* hardly forms any root hairs (Akkermans, personal communication) makes it unlikely that an intracellular infectionmechanism without prenodule formation has evolved in the *Datisca* genus. Altogether, it is likely that the infection mechanism in Cucurbitales is different from the mechanisms in both Fagales (intracellular) and Rosales (intercellular). Interestingly, in Cucurbitales nodules cortical cells become multinucleate prior to and during the infection process (Newcomb and Pankhurst 1982; Hafeez et al. 1984), a phenomenon not observed in any other actinorhizal species or in legumes.

2.3
Infected Cells

Within the host cells, *Frankia* hyphae are separated from the host cytoplasm by the invaginated plasma membrane and the infection thread wall during all stages of development (Berry and Sunell 1990; Berg 1999). In contrast with legumes, no infection thread matrix exists. The infection thread wall is formed by the host cell. It is rich in pectins, with a higher proportion of methyl-esterified pectins in the zone of *Frankia* colonization, which might result in a low level of calcium cross-linking (Liu and Berry 1991b). Besides calcium, boron is another element known to be involved in plant cell wall stabilization (Fleischer et al. 1998), and also to be necessary for normal infection and nodule development in legume–rhizobial symbioses (Bolaños et al. 1996; Redondo-Nieto et al. 2001). A study of boron deficiency in the establishment of the *Discaria* symbiosis suggested that boron is necessary for both partners, permitting normal *Frankia* development, host cell wall development, root infection, and nitrogen fixation (Bolaños et al. 2002). Not much is known about the protein components of the infection thread wall. A host-derived arabinogalactan protein (AGP) epitope was observed at the symbiotic interface during *Frankia* proliferation in *Alnus glutinosa* nodules (Berry et al. 2002). AGPs have been implicated in cell–cell adhesion and signaling. In legume nodules, AGPs have been found in the host–microsymbiont interface (ENOD5; Scheres et al. 1990; Frühling et al. 2000; MtENOD16 and MtENOD20; Greene et al. 1998). Genes encoding putative cell wall proteins, namely, serine proteases of the subtilisin family (Ribeiro et al. 1995; Laplaze et al. 2000b) and glycine- and histidine-rich proteins (Pawlowski et al. 1997), were found to be expressed specifically in infected cells, suggesting that the corresponding proteins localized to the infection thread wall.

In infected cells, infection thread-like structures encapsulating *Frankia* hyphae branch and proliferate until in cross section the whole cell seems filled with *Frankia* material (Fig. 2E). Then, *Frankia* differentiates for symbiotic nitrogen fixation. In most cases, this involves the formation of vesicles; however, in nodules of *Casuarina*, no vesicle formation takes place (Fig. 4A). Here, the induction of nitrogen fixation as determined by *Frankia* nitrogenase structural gene expression occurs in the hyphae (Berg and McDowell 1987; Gherbi et al. 1997). While in free-living conditions, *Frankia* vesicles are always spherical and septate, in planta vesicle morphology as well as their position within the infected cell depend on the host plant species. For example, in *Alnus* nodules, large spherical septate vesicles are formed in the outer parts of the infected cells (Fig. 4B), while in *Elaeagnus* nodules, large spherical septate

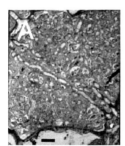 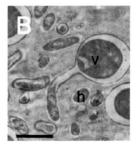 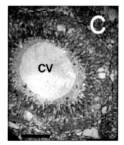

Fig. 4 Transmission electron micrographs of mature infected cells of **A** *Casuarina glauca*, **B** *Alnus glutinosa*, and **C** *Datisca glomerata*. In *C. glauca*, no vesicle differentiation takes place, while in *A. glutinosa*, spherical septate vesicles (*v*) are formed by hyphae (*h*) in the outer part of the cytoplasm. Lanceolate vesicles of *D. glomerata* are localized in radial orientation around the central vacuole (*cv*). *Size bars* denote 2.5 µm (**A,B**) or 25 µm. TEM photographs were kindly provided by N. Koteeva and K. Demchenko

vesicles are formed at an even distribution within infected cells (Sasakawa et al. 1988). In nodules of some Rosaceae, namely *Cercocarpus* and *Dryas*, nonseptate elliptical vesicles are formed. In *Myrica* and *Comptonia*, vesicles represent club-shaped hyphal endings, while in *Datisca* and *Coriaria*, long finger-shaped vesicles in radial orientation form a hollow sphere around the central vacuole of each infected cell (Fig. 4C; Baker and Mullin 1992). Altogether, *Frankia* vesicles in nodules can be considered a case of symbiosis-specific bacterial differentiation comparable to the formation of bacteroids in legume nodules.

3
Nodule Structure

All actinorhizal nodules represent coralloid structures consisting of multiple nodule lobes. Each lobe represents a modified lateral root (Fig. 5A,B) without root cap, with a superficial periderm and infected cells in the expanded cortex. Due to the activity of the apical meristem, the infected cells in the cortex of a nodule lobe are arranged in a developmental gradient (Fig. 5B). The meristematic zone consists of small dividing cells that do not contain bacteria. The zone of infection contains enlarging cortical cells. Some of them

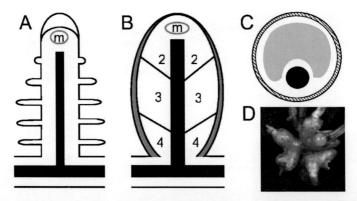

Fig. 5 Schematic structure of mature nodule lobes. **A** Longitudinal section of a lateral root for comparison. The vascular system is given in *black*, and the apical meristem (*m*) is indicated. **B** Longitudinal section of a nodule lobe; the periderm is depicted in *gray*, the zonation of the cortex is indicated. *m*—meristem; *2*—zone of infection; *3*—zone of nitrogen fixation; *4*—zone of senescence. **C** Cross section through a nodule lobe from *Datisca* sp. or *Coriaria* sp.; the suberized periderm is hatched and the patch of infected cells is depicted in *green*. **D** Photograph of a multilobed *Casuarina glauca* nodule with nodule roots growing out of the tips of most lobes. When the apical meristem has re-differentiated to form a nodule root, new primordia are induced on both sides, letting it appear as if nodule roots project from between two nodule lobes

become infected and in turn enlarge more than uninfected cells, while being gradually filled with *Frankia* hyphae from the center outward (Fig. 2B; Lalonde 1979; Schwintzer et al. 1982; Berry and Sunell 1990) or, in the case of *Datisca* and *Coriaria*nodules, from the periphery inward (Hafeez et al. 1984; Newcomb and Pankhurst 1982). When they are completely filled with hyphae, provesicles (terminal swellings on hyphae or on short side branches) are formed (Fontaine et al. 1984). In the zone of nitrogen fixation, vesicledifferentiation has finished and bacterial *nif* (nitrogen fixation) gene expression is induced, leading to the formation of the nitrogenase enzyme complex (Huss-Danell and Bergman 1990). The zone of senescence is characterized by the occurrence of cortical cell senescence and degradation of the host cytoplasm as well as of the microsymbiont (Berry and Sunell 1990). In spite of the apical meristem, the growth of individual lobes is limited; additional branch lobes are formed as lateral primordia in the vascular pericycle of the preceding nodule lobe.

It is not understood which features decide whether a particular cortical cell will be infected by the microsymbiont. In most symbioses, infected cortical cells are interspersed with uninfected cells. In nodules formed on the roots of *Datisca* or *Coriaria* species, the pattern of infected cells is unusual in that they form a continuous patch, kidney-shaped in cross section, not interspersed with uninfected cells, on one side of the acentric stele (Fig. 5C; Newcomb and Pankhurst 1982; Hafeez et al. 1984; see also chapter by Persson and Huss-Danell, in this volume).

3.1
How to Deal with the Oxygen Dilemma of Nitrogen Fixation

As mentioned in the Introduction, symbiotic nitrogen fixation requires aerobic respiration to fulfill its energy demands, while the nitrogenase enzyme complex has to be protected from oxygen. So on the one hand, oxygen access to nodules, which may present a problem in poorly aerated soils, has to be secured. On the other hand, the access of oxygen to the intracellular nitrogen-fixing bacteria has to be tightly controlled. In most symbioses, *Frankia* forms vesicleswith multilayered envelopes, thus contributing to the protection of nitrogenase from O_2. For *Alnus* sp. it has been shown that, like in the free-living state, the number of lipid monolayers of vesicles in nodules is correlated with the external oxygen tension, supporting the assumption that *Frankia* contributes to oxygen protection in nodules (Silvester et al. 1988a).

Actinorhizal nodules are surrounded by a superficial periderm that is more or less impermeable to gas (Silvester et al. 1990). Actinorhizal plants that grow in wet or waterlogged soils have developed mechanisms for gas transport to the nodules. In the cases of *Casuarina*, *Gymnostoma*, *Myrica*, and *Comptonia* species that often grow in wetlands, oxygen is provided to nodules via air spaces in so-called nodule roots (Silvester et al. 1990). Nodule roots are formed at the tips of nodule lobes and grow upward; their length is negatively correlated with the aeration of the root substrate (Fig. 5D; Tjepkema 1978; Silvester et al. 1988b).

In nodules of *Casuarina* sp. *Frankia* does not form vesicles (Berg and McDowell 1987), but the plant provides an oxygen diffusion barrier via an unusually hydrophobic type of lignification of the walls of infected cells (Berg and McDowell 1988) that leads to a microaerobic environment in these cells, and discontinuity in the intercellular air spaces (Zeng et al. 1989). The similarity with the oxygen protection systems of legume nodules is further underlined by the fact that high amounts of a symbiotic hemoglobin are formed in the infected cells of *Casuarina* nodules (Fleming et al. 1987; Jacobsen-Lyon et al. 1995). While hemoglobins were also found in the infected cells of other actinorhizal nodules (*Alnus* sp.; Suharjo and Tjepkema 1995; Sasakura et al. 2006; *Myrica gale*; Pathirana and Tjepkema 1995; Heckmann et al. 2006; *Datisca glomerata*; Pawlowski et al. 2007), these were class I hemoglobins (*Alnus*, *Myrica*) or truncated hemoglobins from plants (*Datisca*) which are not implicated in O_2 transport, but in NO detoxification (summarized by Pawlowski 2008). For *D. glomerata*, evidence suggests that a truncated hemoglobin from *Frankia* plays a role in oxygen transport within *Frankia* vesicles (Pawlowski et al. 2007).

Nodules of *Alnus* species cannot form nodule roots, but when growing in well-drained soils are well aerated since their periderm is interrupted by lenticels and their cortex contains large intercellular spaces (Wheeler et al. 1979). When growing in waterlogged soil, the lenticels can become hypertrophied (Batzli and Dawson 1999). Nevertheless, they do not present the only aeration system of *Alnus* nodules. Previously it was thought that here, the problem of gas access was solved mainly via thermo-osmotically mediated gas transport from the aerial parts to the roots and nodules (Schröder 1989). However, recent results have shown that it is more likely that stem photosynthesis, using internally sourced CO_2 from respiration and the transpiration stream, plays a role in root aeration in young trees (Armstrong and Armstrong 2005).

Nodules of *Coriaria* and *Datisca*, in which the infected cells are only found on one side of the acentric stele, are surrounded by a dense periderm with suberized cell walls (Hafeez et al. 1984; Mirza et al. 1994). Aeration of *Coriaria* sp. nodules is achieved by a single elongate lenticel (Silvester et al. 1990), while nodules of *Datisca glomerata*, which grows in moist riparian places like streambeds or seasonal drainages, can form nodule roots under waterlogged conditions and lenticels under well-drained conditions. For nodules of both *Datisca* sp. and *Coriaria* sp., the only actinorhizal plant where nitrogen-fixation activity can adapt to changing oxygen tension similar to nitrogen fixation in legume nodules, an unusual concentration of mitochondria at the base of the symbiotic vesicleshas been reported (Hafeez et al. 1984; Mirza et al. 1994). This phenomenon has been linked to oxygen scavenging which is supposed to leave the vesicles, the envelopes of which do not consist of many lipid layers, in a microaerobic atmosphere (Silvester et al. 1999).

Altogether, mechanisms developed for O_2 access and protection of nitrogenase from O_2 in actinorhizal nodules are diverse, reflecting the broad host range of *Frankia*. Both plant and microsymbiont can contribute.

4
Symbiotic Signaling

The signal exchange between bacterium and host plant that leads to uptake of bacteria into the plant root, and nodule formation, has been examined in detail in legume–rhizobia symbioses (see chapter by Untergasser et al., in this volume). Basically, specific lipochitooligosaccharides, the so-called Nod factors, induce several reactions in the plant, including (1) root hair deformation; (2) preparation of cells for infection thread growth; (3) cortical cell divisions, i.e., the formation of a nodule primordium; and (4) on some plants, Nod factors are sufficient to induce the formation of nodule primordia or whole bacteria-free nodules (*Medicago sativa*; Truchet et al. 1991; *Glycine soja*; Stokkermans and Peters 1994). Given the phylogenetic relationship of both symbioses, it is expected that *Frankia* and rhizobia use similar means to influence plant morphogenesis. The *Frankia* Nod factor equivalents have not been characterized yet, which is partially due to the fact that in contrast to legume symbioses, no convenient bioassay is available. Root hair deformation on actinorhizal plants can also be induced by several nonsymbiotic soil bacteria (Knowlton et al. 1980). However, a partial purification of the root hair deforming factor from the supernatant of *Frankia* cultures led to the conclusion that no lipochitooligosaccharides are involved in the infection of actinorhizal plants (Cérémonie et al. 1999). Since the genomic sequences of three *Frankia* strains are available (Normand et al. 2007), it could be shown that they do not contain homologues of all common *nod* genes encoding the enzymes involved in the common backbone of rhizobial Nod factors.

Analysis of symbiotic legume mutants has led to the identification of several components of the plant signal transduction chain responsible for the perception of, and response to, rhizobial Nod factors (see chapter by Untergasser et al., in this volume). A summary is shown in Fig. 6. The same signal transduction pathway is also involved in the establishment of arbuscular mycorrhizal (AM) symbioses in response to yet unidentified fungal signal factors, the so-called Myc factors. AM symbioses are evolutionarily much older than nitrogen-fixing root nodule symbioses; fossil evidence for AM symbioses dates them back at least 400 million years (Remy et al. 1994) while root nodule symbioses evolved 50–100 million years ago. It is assumed that during the evolution of root nodule symbioses, mechanisms established for AM symbioses were recruited (Kistner and Parniske 2002). Both signal transduction

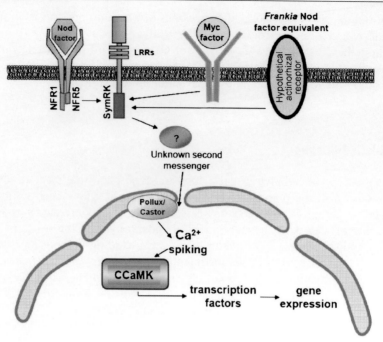

Fig. 6 Microbial factor signal transduction in root symbioses. The names of *Lotus japonicus* proteins are used. The common parts of the pathway are the plasma membrane receptor kinase SymRK (DMI2), the ion channels in the nuclear envelope, Castor and Pollux (DMI1), and the nuclear calcium/calmodulin dependent protein kinase (CCaMK, DMI3). Binding of Nod factors to a heterodimer of LysM receptor kinases NFR1 and NFR2, or binding of Myc factors or *Frankia* Nod factor equivalents to hypothetical receptors, leads to signaling to SymRK. An unknown second messenger activates Castor and Pollux, leading to nuclear Ca^{2+} spiking. The amplitude and frequency of the Ca^{2+} spikes is measured by CCaMK which in turn signals to transcription factors, leading to gene expression. The activated genes differ depending on whether Nod factors or Myc factors started the signaling cascade

pathways (Fig. 6) share three common factors: a plasma membrane receptor kinase with three leucine-rich repeats (LRRs) in its N-terminal extracellular domain (SymRK/DMI2/NORK; Endre et al. 2002; Stracke et al. 2002), an ion channel from the nuclear envelope (Castor/Pollux/DMI1; Ané et al. 2004; Imaizumi-Anraku et al. 2005), and a nuclear calcium/calmodulin dependent protein kinase (CCaMK/DMI3; Levy et al. 2004).

Recent results have shown that this common part of the Nod and Myc factor signal transduction pathway is also exploited by *Frankia* signaling. *SymRK* homologues were identified in actinorhizal plants from two different phylogenetic groups, *D. glomerata* and *C. glauca*. Suppression of the transcription of *DgsymRK* and *CgsymRK*, respectively, using RNAi constructs in transgenic hairy roots of composite plants led to the inhibition of nodule induction by compatible *Frankia* strains in both species, and to the inability to enter an AM symbiosis with *Glomus intraradices* (Gherbi et al. 2008; Markmann et al. 2008). The complementation of a legume (*Lotus japonicus*) *symRK* mutant was attempted using *symRK* genes from *L. japonicus*, *D. glomerata*, *C. glauca*, and rice or tomato (Gherbi et al. 2008; Markmann et al. 2008). Complementation was always successful for the AM symbiosis, i.e., the *symRK* orthologues from all plant species under examination could complement the AM-deficient phenotype of *symRK* mutants of *L. japonicus* (Gherbi et al. 2008; Markmann et al. 2008). However, in contrast to *symRK* orthologues from symbiotic plants (*L. japonicus*, *C. glauca*, *D. glomerata*), *symRK* from rice or tomato could not complement the nodulation phenotype. In contrast to the SymRK proteins from symbiotic plants, SymRK proteins from rice or tomato did not contain three LRRs in their extracellular domain, but only two (rice SymRK, like SymRKs from other monocotyledonous plants, was also lacking a large part of the N-terminal extracellular region). The only *symRK* orthologue from a non-nodule forming plant that could complement the *L. japonicus symRK* mutant, *Tropaeolum majus symRK*, encoded a protein containing three LRRs in its extracellular domain. *T. majus* is a member of the Brassicaceae family, belonging to the Eurosid II clade, and the closest relative of Eurosid I of the non-nodule forming plants examined. Hence, the acquirement of the third LRR repeat has been suggested to be the basic feature that allowed the development of a root nodule symbiosis with nitrogen-fixing bacteria (Markmann et al. 2008).

Altogether, symbiotic signaling by *Frankia* Nod factor equivalents, rhizobial Nod factors, and fungal Myc factors uses a common pathway. However, it remains unclear whether *Frankia* Nod factor equivalents are recognized by homologues of LysM receptor kinases that recognize rhizobial Nod factors (Limpens et al. 2003; Radutoiu et al. 2003) and are responsible for host specificity on the plant side.

Two strands of evidence seem to indicate that the induction of actinorhizal nodules, or the entry of microbes into actinorhizal roots, is less strictly controlled than that of legume nodules. First, actinorhizal nodules can contain not only different *Frankia* strains but also, in their outer cortex, non-nitrogen-fixing related actinomycetes that cannot reinfect the plant on their own (Mirza et al. 1992; Ramirez-Saad et al. 1998). In some cases, nitrogen-fixing *Frankia* strains isolated from field-collected actinorhizal nodules could not reinfect the host plants from which they were isolated, but were able to infect other actinorhizal plants (see, e.g., Torrey 1990). So it seems as if effective *Frankia* strains can help ineffective strains to enter a host plant root. There is even evidence that non-*Frankia* soil bacteria can support nodulation by inducing root hair deformation on *Alnus* sp. (Knowlton et al. 1980; Berry and Torrey 1983). Second, a saprophytic soil fungus, *Penicillium nodositatum*, can use the actinorhizal intracellular infection pathway of *Alnus* sp. to induce the formation parasitic myconodules resembling single-lobed actinorhizal nodules and containing cortical cells filled by branching infection threadscontaining fungal hyphae (Pommer 1956; Sequerra et al. 1994; 1995). This fact shows that symbiotic signaling by *Frankia* can be mimicked by a nonsymbiotic fungus. Hence, in contrast with legume symbioses, no part of the *Frankia* cell wall is required to suppress plant defense during infection (Scheidle et al. 2005; Tellström et al. 2007). Altogether, it seems that the entry of bacteria into the root is less tightly controlled in actinorhizal plants than in legumes.

The overlap in the transduction of bacterial signals from rhizobia and *Frankia* raises the question whether the downstream events in response to rhizobial Nod factors and *Frankia* Nod factor equivalents are similar as well. In legumes, Nod factor signaling leads to the accumulation of cytokinin, which in turn leads to the formation of the nodule primordium (Murray et al. 2007; Tirichine et al. 2007). On the legume *Medicago sativa*, bacteria-free nodules (so-called pseudonodules) can be induced by auxin transport inhibitors (Hirsch et al. 1989) or bacteria secreting cytokinin (Cooper and Long 1994). The reaction of legume roots to rhizobial Nod factors also involves the inhibition of auxin transport (Mathesius et al. 1998), and autoregulatory nodulation mutants of legumes tend to be affected in auxintransport during nodulation (van Noorden et al. 2006; Prayitno et al. 2006). Also for actinorhizal plants, induction of bacteria-free nodules ("pseudoactinorhiza") by cytokinin(Rodriguez-Barrueco and de Castro 1973) and by auxin transport inhibitors has been described (Santi 2003). Hence, it is likely that more similarities exist in the nodule induction mechanisms of rhizobia and *Frankia* strains.

5
Autoregulation of Nodule Number

In any plant–microbial symbiosis, the host has to be able to regulate the proportion of symbiotic tissue with regard to photosynthate partitioning. This can be achieved by controlling the number of new infections, and by controlling the development of existing infections/nodules.

In both legumes and actinorhizal plants, nodule induction is controlled by the availability of nitrogen in the soil. In the presence of soil nitrogen, roots are not susceptible to infection (nitrogen inhibition; Carroll et al. 1985; Parsons et al. 1993). A second control mechanism is the so-called autoregulation of nodulation (Caetano-Anollés and Gresshoff 1991; Wall 2000) that controls the number of infections and also suppresses early nodule development. The molecular basis for autoregulation/nitrogen inhibition in legumes is complex and connected with the regulation of lateral root formation (Wopereis et al. 2000; Krusell et al. 2002; Nishimura et al. 2002; Penmetsa et al. 2003; Schnabel et al. 2005). Furthermore, mechanisms seem to differ in different legume species (see, e.g., Schmidt et al. 1999). Many of the legume analyses involved plant mutants where autoregulation was affected; no such studies could be performed on actinorhizal plants with their longer generation times. Using grafting experiment with mutants, and analyzing the mutated genes, it could be shown that autoregulation of nodulation is mediated by a shoot-derived factor, and that receptor-like kinases are involved (Krusell et al. 2002; Nishimura et al. 2002). In *Lotus japonicus*, a single system based on the *har1* gene limits the extent of cortical cell division centers, thus regulating lateral root as well as nodule formation (Wopereis et al. 2000).

Nitrogen-fixing nodules induced by *Frankia* form as discrete clusters, which is explained as an initial burst of nodule induction followed by a feedback inhibition that reduces the susceptibility of the growing root to further infections (Wall and Huss-Danell 1997; Valverde and Wall 1999b; Chaia and Raffaele 2000). Independent of the infection mechanism, nodulation always takes place in the elongation region behind the root tip (Burggraaf et al. 1984; Dobritsa and Novik 1992; Wall and Huss-Danell 1997; Valverde and Wall 1999b; Wall et al. 2003). This feedback inhibition of infection and nodule development in actinorhizal plants is analogous to autoregulation in legume-rhizobia symbioses in that it starts before symbiotic nitrogen fixation commences (Caetano-Anollés and Gresshoff 1991; Valverde and Wall 1999b; Wall 2000).

In double-inoculation experiments with both intra- and intercellularly infected actinorhizal plants, the development of younger nodule primordia was arrested at a stage before *Frankia* entered these primordia. When existing nodule lobes were removed, previously suppressed nodule primordia developed into mature nodules (Wall and Huss-Danell 1997; Valverde and Wall 1999b). During intracellular infection via root hairs, infection thread-like structurescontaining *Frankia* hyphae are already present in some root hairs and even prenodules by the time primordium induction takes place, comparable to the situation during autoregulation in the legume *L. japonicus* (Wopereis et al. 2000). Regarding intercellularly infected actinorhizal plant species, the delayed outgrowth of primodia after the removal of nodules indicates that the plant is able to control the propagation of *Frankia* in the root apoplast.

Once bacterial nitrogen fixation has commenced, N inhibition regulates nodule number and total nodule biomass in legumes (MacConnell and Bond 1957) as well as actinorhizal plants (Arnone et al. 1994; Kohls and Baker 1989; Thomas and Berry 1989). N inhibition can also be the result of exogenous N application (Baker and Parsons 1997). Detailed studies with split root systems have demonstrated that N inhibition is both localized and systemic in actinorhizal symbioses (Pizelle 1965; 1966; Gentili and Huss-Danell 2002; 2003).

Mechanisms for N inhibition due to nitrogen-fixing nodules differ between intracellularly infected plant genera such as *Alnus* (Wall and Huss-Danell

1997) and intercellularly infected genera such as *Discaria* (Valverde and Wall 1999b). If N inhibition is temporarily suspended in nodulated plants, i.e., when nodulated roots are placed under an argon atmosphere without N_2, new nodules would develop in intracellularly infected hosts, but in intercellularly infected plants instead the biomass of already existing nodules would increase without new nodule formation (Wall et al. 2003). Similar effects on nodule biomass, but not nodule number, were observed when nodulated *D. trinervis* plants were kept in the dark for 6 days (Valverde and Wall 2003). However, if nitrogen-fixing nodules were completely removed from the roots, new infections would occur and new nodules would be formed on roots of intracellularly infected (Wall and Huss-Danell 1997) as well as intercellularly infected plants (Valverde and Wall 1999b).

N inhibition of nodulation depends on the external N/P ratio. Like N levels, P levels in soil affect nodulation directly, not via an unspecific effect on plant growth (Israel 1993; Reddell et al. 1997; Valverde et al. 2002; Gentili and Huss-Danell 2002). Inhibition of nodulation by high levels of inorganic N in the soil can be counteracted by high inorganic P levels. This P effect has been demonstrated for intracellularly as well as intercellularly infected plants (Yang 1995; Yang et al. 1997; Wall et al. 2000; Valverde et al. 2002; Gentili and Huss-Danell 2002; 2003). With increasing P levels, the ratio of nodule biomass to whole-plant biomass increases. Like the response to the temporary suspension of N inhibition by nodule nitrogen fixation, the strategy to achieve this result seems to differ among actinorhizal plants depending on the infection pathway. While in intracellularly infected *A. incana*increased P levels led to increased nodule numbers per plant (Wall et al. 2000), in the intercellularly infected *Discaria trinervis* increased P levels resulted in an increase of the size of individual nodules, while nodule numbers remained unaffected (Valverde et al. 2002). Current knowledge on N inhibition, P effects, and autoregulation is summarized in Fig. 7.

In the legume genus *Medicago*, the effect of ethylene inhibitors is similar to that of mutations in the autoregulation system. The hypernodulating *Medicago truncatula* mutant *sickle* is ethylene-insensitive (Penmetsa and Cook 1997), and exposure to the inhibitors of ethylene synthesis (aminoethoxyvinyl glycine, AVG) leads to increased nodulation (Peters and Christ-Estes 1989) and counteracts the nitrogen inhibition of nodulation (Ligero et al. 1991) in *M. sativa*. However, not all legumes show the same reaction to ethylene—e.g., in soybean the regulation of nodulation is independent of ethylene signaling (Schmidt et al. 1999). Studies on the intercellularly infected actinorhizal species *Discaria trinervis* revealed that nodulation was increased by treatment with AVG or with the ethyleneperception inhibitor Ag^+ to some extent, and nodulation was reduced somewhat by treatment with ethylene donors; there was no significant effect on ethylene inhibitors or donors on the autoregulation of nodulation (Valverde and Wall 2005). Altogether, ethylene

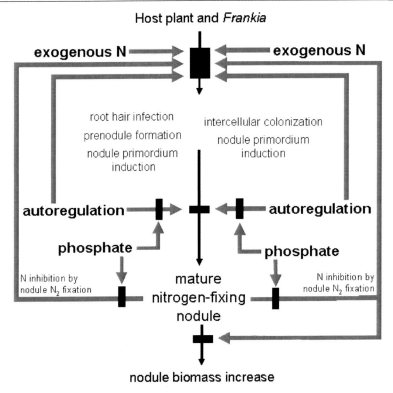

Fig. 7 Regulation of nodulation in intracellularly vs intercellularly infected actinorhizal plants (based on Wall 2000). Differences between intracellularly infected plants (*left side*) and intercellularly infected plants (*right side*) are indicated. It should be mentioned that extra root hair infection can be aborted before prenodule formation and nodule primordium induction by localized deposition of callose (Berry et al. 1986; Berry and Mc-Cully 1990), suggesting that one level of nodulation control operates at the point between root hair infection to prenodule induction

seemed to be involved in modulating the susceptibility of the basal portion of *D. trinervis* seedling roots to nodulation.

6
Concluding Remarks and Future Perspectives

The inability to transform *Frankia*, and the woody nature of the host plants, has impeded research on actinorhizal symbioses so far. While *Frankia* strains still cannot be transformed, the availability of genome sequences can now be used to assess and test their metabolic potential. Analyses of the root and nodule transcriptome of several actinorhizal plant species are in progress and will help to identify genes whose products are involved in the infection process, and to elucidate pathways involved in nodule metabolism. The lack of plant mutants still represents a handicap. The only actinorhizal plant species that can be developed as a genetic system is *Datisca glomerata* with its generation time of 6 months; unfortunately, so far no protocol for stable transformation could be developed for this species, and it is infected by a nonculturable *Frankia* strain (Fig. 1).

One of the most urgent objectives in actinorhizal research is now to identify the Nod factorequivalents of *Frankia* strains. Once their structure is understood, the basis of host specificity can be examined.

References

Ané JM, Kiss GB, Riely BK, Penmetsa RV, Oldroyd GE, Ayax C, Lévy J, Debellé F, Baek JM, Kalo P, Rosenberg C, Roe BA, Long SR, Dénarié J, Cook DR (2004) *Medicago truncatula DMI1* required for bacterial and fungal symbioses in legumes. Science 303:1364–1367

Armstrong W, Armstrong J (2005) Stem photosynthesis not pressurized ventilation is responsible for light-enhanced oxygen supply to submerged roots of alder (*Alnus glutinosa*). Ann Bot 96:591–612

Arnone JA, Kohls SJ, Baker DD (1994) Nitrate effects on nodulation and nitrogenase activity of actinorhizal *Casuarina* studied in split root systems. Soil Biol Biochem 26:599–606

Baker A, Parsons R (1997) Evidence for N feedback regulation of N_2 fixation in *Alnus glutinosa* L. J Exp Bot 48:67–73

Baker DD (1987) Relationships among pure cultured strains of *Frankia* based on host specificity. Physiol Plant 70:245–248

Baker DD, Mullin BC (1992) Actinorhizal symbioses. In: Stacey G, Burris RH, Evans HJ (eds) Biological nitrogen fixation. Chapman & Hall, New York, pp 259–292

Batzli JM, Dawson JO (1999) Development of flood-induced lenticels in red alder nodules prior to the restoration of nitrogenase activity. Can J Bot 77:1373–1377

Benson DR, Clawson ML (2000) Evolution of the actinorhizal plant symbiosis. In: Triplett EW (ed) Prokaryotic nitrogen fixation: a model system for analysis of a biological process. Horizon Scientific, Symondham, pp 207–224

Benson DR, Silvester WB (1993) Biology of *Frankia* strains, actinomycete symbionts of actinorhizal plants. Microbiol Rev 57:293–319

Benson DR, Vanden Heuvel BD, Potter D (2004) Actinorhizal symbioses: diversity and biogeography. In: Gillings M (ed) Plant microbiology. BIOS Scientific, Oxford, pp 97–127

Berg RH (1999) *Frankia* forms infection threads. Can J Bot 77:1327–1333

Berg RH, McDowell L (1987) Endophyte differentiation in *Casuarina* actinorhizae. Protoplasma 136:104–117

Berg RH, McDowell L (1988) Cytochemistry of the wall of infected cells in *Casuarina* actinorhizae. Can J Bot 66:2038–2047

Berry A, Harriott O, Moreau R, Osman S, Benson D, Jones A (1993) Hopanoid lipids compose the *Frankia* vesicle envelope, presumptive barrier of oxygen diffusion to nitrogenase. Proc Natl Acad Sci USA 90:6091–6094

Berry AM, McCully ME (1990) Callose-containing deposits in relation to root-hair infections of *Alnus rubra* by *Frankia*. Can J Bot 68:798–802

Berry AM, Sunell LA (1990) The infection process and nodule development. In: Schwintzer CR, Tjepkema JD (eds) The biology of *Frankia* and actinorhizal plants. Academic, New York, pp 61–81

Berry AM, Torrey JG (1983) Root hair deformation in the infection process of *Alnus rubra*. Can J Bot 61:2863–2876

Berry AM, McIntyre L, McCully ME (1986) Fine structure of root hair infection leading to nodulation in the *Frankia–Alnus* symbiosis. Can J Bot 64:292–305

Berry AM, Rasmussen U, Bateman K, Huss-Danell K, Lindwall S, Bergman B (2002) Arabinogalactan proteins are expressed at the symbiotic interface in root nodules of *Alnus* spp. New Phytol 155:469–479

Bolaños L, Brewin NJ, Bonilla I (1996) Effects of boron on *Rhizobium*–legume cell-surface interactions and nodule development. Plant Physiol 110:1249–1256

Bolaños L, Redondo-Nieto M, Bonilla I, Wall LG (2002) Boron requirement for growth, nitrogen fixation and nodulation of *Frankia* BCU110501. Physiol Plant 115:563–570

Bosco M, Fernandez MP, Simonet P, Materassi R, Normand P (1992) Evidence that some *Frankia* sp. strains are able to cross boundaries between *Alnus* and *Elaeagnus* host specificity groups. Appl Environ Microbiol 58:1569–1576

Burggraaf AJP, Van der Linden J, Tak T (1983) Studies on the localization of infectible cells on *Alnus glutinosa* roots. Plant Soil 74:175–188

Caetano-Anollès G, Gresshoff PM (1991) Plant genetic control of nodulation. Annu Rev Microbiol 45:345–382

Callaham D, Torrey JG (1977) Prenodule formation and primary nodule development in roots of *Comptonia* (Myricaceae). Can J Bot 55:2306–2318

Callaham D, Newcomb W, Torrey JG, Peterson RL (1979) Root hair infection in actinomycete-induced root nodule initiation in *Casuarina, Myrica,* and *Comptonia*. Bot Gaz (Suppl) 140:S1–S9

Carroll BJ, McNeil DL, Gresshoff PM (1985) A supernodulation and nitrate-tolerant symbiotic (nts) soybean mutant. Plant Physiol 78:34–40

Cérémonie H, Debelle F, Fernandez MP (1999) Structural and functional comparison of *Frankia* root hair deforming factor and rhizobia Nod factor. Can J Bot 77:1293–1301

Chaia E, Raffaele E (2000) Spatial patterns of root branching and actinorhizal nodulation in *Discaria trinervis* seedlings. Symbiosis 29:329–341

Clawson ML, Benson DR (1999) Natural diversity of *Frankia* strains in actinorhizal root nodules from promiscuous hosts in the family Myricaceae. Appl Environ Microbiol 65:4521–4527

Clawson ML, Carú M, Benson DR (1998) Diversity of *Frankia* strains in root nodules of plants from the families Elaeagnaceae and Rhamnaceae. Appl Environ Microbiol 64:3539–3543

Cooper JB, Long SR (1994) Morphogenetic rescue of *Rhizobium meliloti* nodulation mutants by trans-zeatin secretion. Plant Cell 6:215–225

Dobritsa SV, Novik SN (1992) Feedback regulation of nodule formation in *Hippophaë rhamnoides*. Plant Soil 144:45–50

Doyle JJ (1998) Phylogenetic perspectives on nodulation: evolving views of plants and symbiotic bacteria. Trends Plant Sci 3:473–478

Endre G, Kereszt A, Kevei Z, Mihacea S, Kaló P, Kiss GB (2002) A receptor kinase gene regulating symbiotic nodule development. Nature 417:962–966

Fleischer A, Titel C, Ehwald R (1998) The boron requirement and cell wall properties of growing and stationary suspension-cultured *Chenopodium album* L. cells. Plant Physiol 117:1401–1410

Fleming AI, Wittenberg JB, Wittenberg BA, Dudman WF, Appleby CA (1987) The purification, characterization and ligand-binding kinetics of hemoglobins from root nodules of the nonleguminous *Casuarina glauca—Frankia* symbiosis. Biochim Biophys Acta 911:209-220

Fontaine MS, Lancelle SA, Torrey JG (1984) Initiation and ontogeny of vesicles in cultured *Frankia* sp. strain HFPArI3. J Bacteriol 160:921-927

Frühling M, Hohnjec N, Schröder G, Küster H, Pühler A, Perlick AM (2000) Genomic organization and expression properties of the *VfENOD5* gene from broad bean (*Vicia faba* L.). Plant Sci 155:169-178

Gauthier D, Jaffre T, Prin Y (1999) Occurrence of both *Casuarina*-infective and *Elaeagnus*-infective *Frankia* strains within actinorhizae of *Casuarina collina*, endemic to New Caledonia. Eur J Soil Biol 35:9-15

Gentili F, Huss-Danell K (2002) Phosphorus modifies the effects of nitrogen on nodulation in split-root systems of *Hippophaë rhamnoides*. New Phytol 153:53-61

Gentili F, Huss-Danell K (2003) Local and systemic effects of phosphorus and nitrogen on nodulation and nodule function in *Alnus incana*. J Exp Bot 54:2757-2767

Geurts R, Fedorova E, Bisseling T (2005) Nod factor signaling genes and their function in the early stages of *Rhizobium* infection. Curr Opin Plant Biol 8:346-352

Gherbi H, Duhoux E, Franche C, Pawlowski K, Berry AM, Bogusz D (1997) Cloning of a full-length symbiotic hemoglobin cDNA and in situ localization of the corresponding mRNA in *Casuarina glauca* root nodule. Physiol Plant 99:608-616

Gherbi H, Markmann K, Svistoonoff S, Estevan J, Autran D, Giczey G, Auguy F, Péret B, Laplaze L, Franche C, Parniske M, Bogusz D (2008) SymRK defines a common genetic basis for plant root endosymbioses with arbuscular mycorrhiza fungi, rhizobia, and *Frankia* bacteria. Proc Natl Acad Sci USA 105:4928-4932

Greene EA, Erard M, Dedieu A, Barker DG (1998) MtENOD16 and 20 are members of a family of phytocyanin-related early nodulins. Plant Mol Biol 36:775-783

Hafeez F, Akkermans ADL, Chaudhary AH (1984) Observations on the ultrastructure of *Frankia* sp. in root nodules of *Datisca cannabina* L. Plant Soil 79:383-402

Heckmann AB, Hebelstrup KH, Larsen K, Micaelo NM, Jensen EØ (2006) A single hemoglobin gene in *Myrica gale* retains both symbiotic and non-symbiotic specificity. Plant Mol Biol 61:769-779

Hirsch AM, Bhuvaneswari TV, Torrey JG, Bisseling T (1989) Early nodulin genes are induced in alfalfa root outgrowths elicited by auxin transport inhibitors. Proc Natl Acad Sci USA 86:1244-1248

Huang J-B, Zhao Z-Y, Chen G-X, Liu H-C (1985) Host range of *Frankia* endophytes. Plant Soil 87:61-65

Huguet V, Mergeay M, Cervantes E, Fernandez MP (2004) Diversity of *Frankia* strains associated to *Myrica gale* in Western Europe: impact of host plant (*Myrica* vs. *Alnus*) and of edaphic factors. Environ Microbiol 6:1032-1041

Huguet V, Gouy M, Normand P, Zimpfer JF, Fernandez MP (2005) Molecular phylogeny of Myricaceae: a reexamination of host-symbiont specificity. Mol Phylogenet Evol 34:557-685

Huss-Danell K, Bergman B (1990) Nitrogenase in *Frankia* from root nodules of *Alnus incana* (L.) Moench: immunolocalization of the Fe- and MoFe proteins during vesicle differentiation. New Phytol 116:443-455

Imaizumi-Anraku H, Takeda N, Charpentier M, Perry J, Miwa H, Umehara Y, Kouchi H, Murakami Y, Mulder L, Vickers K, Pike J, Downie JA, Wang T, Sato S, Asamizu E, Tabata S, Yoshikawa M, Murooka Y, Wu GJ, Kawaguchi M, Kawasaki S, Parniske M,

Hayashi M (2005) Plastid proteins crucial for symbiotic fungal and bacterial entry into plant roots. Nature 433:527–531
Israel DW (1993) Symbiotic dinitrogen fixation and host-plant growth during development of and recovery from phosphorus deficiency. Physiol Plant 88:294–300
Jacobsen-Lyon K, Jensen EØ, Jorgensen JE, Marcker KA, Peacock WJ, Dennis ES (1995) Symbiotic and nonsymbiotic hemoglobin genes of *Casuarina glauca*. Plant Cell 7:213–223
Kistner D, Parniske M (2002) Evolution of signal transduction in intracellular symbiosis. Trends Plant Sci 7:511–518
Knowlton S, Berry AM, Torrey JG (1980) Evidence that associated soil bacteria may influence root hair infection of actinorhizal plants by *Frankia*. Can J Microbiol 26:971–977
Kohls SJ, Baker DD (1989) Effects of substrate nitrate concentration on symbiotic nodule formation in actinorhizal plants. Plant Soil 118:171–179
Krusell L, Madsen LH, Sato S, Aubert G, Genua A, Szczyglowski K, Duc G, Kaneko T, Tabata S, de Bruijn F, Pajuelo E, Sandal N, Stougaard J (2002) Shoot control of root development and nodulation is mediated by a receptor-like kinase. Nature 420:422–426
Lalonde M (1979) Techniques and observations of the nitrogen fixing *Alnus* root nodule symbiosis. In: Subba Rao NS (ed) Recent advances in biological nitrogen fixation. Oxford University Press and IBH, New Delhi, pp 421–434
Laplaze L, Duhoux E, Franche C, Frutz T, Svistoonoff S, Bisseling T, Bogusz D, Pawlowski K (2000a) *Casuarina glauca* prenodule cells display the same differentiation as the corresponding nodule cells. Mol Plant Microbe Interact 13:107–112
Laplaze L, Ribeiro A, Franche C, Duhoux E, Auguy F, Bogusz D, Pawlowski K (2000b) Characterization of a *Casuarina glauca* nodule-specific subtilisin-like protease gene, a homolog of *Alnus glutinosa ag12*. Mol Plant Microbe Interact 13:113–117
Lévy J, Bres C, Geurts R, Chalhoub B, Kulikova O, Duc G, Journet EP, Ané JM, Lauber E, Bisseling T, Dénarié J, Rosenberg C, Debellé F (2004) A putative Ca^{2+} and calmodulin-dependent protein kinase required for bacterial and fungal symbioses. Science 303:1361–1364
Ligero F, Caba JM, Lluch C, Olivares J (1991) Nitrate inhibition of nodulation can be overcome by the ethylene inhibitor aminoethoxyvinylglycine. Plant Physiol 97:1221–1225
Limpens E, Franken C, Smit P, Willemse J, Bisseling T, Geurts R (2003) LysM domain receptor kinases regulating rhizobial Nod factor-induced infection. Science 302:630–633
Liu Q, Berry AM (1991a) The infection process and nodule initiation in the *Frankia–Ceanothus* root nodule symbiosis: a structural and histochemical study. Protoplasma 163:82–92
Liu Q, Berry AM (1991b) Localization and characterization of pectic polysaccharides in roots and root nodules of *Ceanothus* spp. during intercellular infection by *Frankia*. Protoplasma 164:93–101
Lumini E, Bosco M, Fernandez MP (1996) PCR-RFLP and total DNA homology revealed three related genomic species among broad-host-range *Frankia* strains. FEMS Microbiol Ecol 21:303–311
Maggia L, Bousquet J (1994) Molecular phylogeny of the actinorhizal Hamamelidae and relationships with host promiscuity toward *Frankia*. Mol Ecol 3:459–467
Markmann K, Giczey G, Parniske M (2008) Functional adaptation of a plant receptor-kinase paved the way for the evolution of intracellular root symbioses with bacteria. PLoS Biol 6:e68
Mathesius U, Schlaman HRM, Spaink HP, Sautter C, Rolfe BG, Djordjevic M (1998) Auxin transport inhibition precedes root nodule formation in white clover roots and is regulated by flavonoids and derivatives of chitin oligosaccharides. Plant J 14:23–34

McConnell JT, Bond G (1957) A comparison of the effects of combined nitrogen on nodulation in non-legumes and legumes. Plant Soil 8:378–386

Miller IM, Baker DD (1986) The initiation development and structure of root nodules in *Elaeagnus angustifolia* (Elaeagnaceae). Protoplasma 128:107–119

Mirza MS, Hahn D, Akkermans ADL (1992) Isolation and characterization of *Frankia* strains from *Coriaria nepalensis*. Syst Appl Microbiol 15:289–295

Mirza S, Pawlowski K, Hafeez FY, Chaudhary AH, Akkermans ADL (1994) Ultrastructure of the endophyte and localization of *nifH* transcripts in root nodules of *Coriaria nepalensis* Wall. by *in situ* hybridization. New Phytol 126:131–136

Murray JD, Karas BJ, Sato S, Tabata S, Amyot L, Szczyglowski K (2007) A cytokinin perception mutant colonized by *Rhizobium* in the absence of nodule organogenesis. Science 315:101–104

Navarro E, Jaffre T, Gauthier D, Gourbiere F, Rinaudo G, Simonet P, Normand P (1999) Distribution of *Gymnostoma* spp. microsymbiotic *Frankia* strains in New Caledonia is related to soil type and to host-plant species. Mol Ecol 8:1781–1788

Navarro E, Nalin R, Gauthier D, Normand P (1997) The nodular microsymbionts of *Gymnostoma* spp. are *Elaeagnus*-infective *Frankia* strains. Appl Environ Microbiol 63:1610–1616

Newcomb W, Pankhurst CE (1982) Fine structure of actinorhizal nodules of *Coriaria arborea* (Coriariaceae). NZJ Bot 20:93–103

Nishimura R, Hayashi M, Wu G-J, Kouchi H, Imaizumi-Anraku H, Murakami Y, Kawasaki S, Akao S, Ohmori M, Nagasawa M, Harada K, Kawaguchi M (2002) HAR1 mediates systemic regulation of symbiotic organ development. Nature 420:426–429

Normand P, Lapierre P, Tisa LS, Gogarten JP, Alloisio N, Bagnarol E, Bassi CA, Berry AM, Bickhart DM, Choisne N, Couloux A, Cournoyer B, Cruveiller S, Daubin V, Demange N, Francino MP, Goltsman E, Huang Y, Kopp OR, Labarre L, Lapidus A, Lavire C, Marechal J, Martinez M, Mastronunzio JE, Mullin BC, Niemann J, Pujic P, Rawnsley T, Rouy Z, Schenowitz C, Sellstedt A, Tavares F, Tomkins JP, Vallenet D, Valverde C, Wall LG, Wang Y, Medigue C, Benson DR (2007) Genome characteristics of facultatively symbiotic *Frankia* sp. strains reflect host range and host plant biogeography. Genome Res 17:7–15

Parniske M (2000) Intracellular accommodation of microbes by plants: a common developmental program for symbiosis and disease? Curr Opin Plant Biol 3:320–328

Parsons R, Silvester WB, Harris S, Gruijters WTM, Bullivant S (1987) *Frankia* vesicles provide inducible and absolute oxygen protection for nitrogenase. Plant Physiol 83:728–731

Parsons R, Stanforth A, Raven JA, Sprent JI (1993) Nodule growth and activity may be regulated by a feedback mechanism involving phloem nitrogen. Plant Cell Environ 16:125–136

Pathirana SM, Tjepkema JD (1995) Purification of hemoglobin from the actinorhizal root nodules of *Myrica gale* L. Plant Physiol 107:827–831

Pawlowski K (2008) Nodules and oxygen. Plant Biotechnol 25:291–298

Pawlowski K, Twigg P, Dobritsa S, Guan C, Mullin BC (1997) A nodule-specific gene family from *Alnus glutinosa* encodes glycine- and histidine-rich proteins expressed in the early stages of actinorhizal nodule development. Mol Plant Microbe Interact 10:656–664

Pawlowski K, Swensen S, Guan C, Hadri A-E, Berry AM, Bisseling T (2003) Distinct patterns of symbiosis-related gene expression in actinorhizal nodules from different plant families. Mol Plant Microbe Interact 16:796–807

Pawlowski K, Jacobsen KR, Alloisio N, Ford Denison R, Klein M, Tjepkema JD, Winzer T, Sirrenberg A, Guan C, Berry AM (2007) Truncated hemoglobins in actinorhizal nodules of *Datisca glomerata*. Plant Biol 9:776–785

Penmetsa RV, Cook DR (1997) A legume ethylene-insensitive mutant hyperinfected by its rhizobial symbiont. Science 275:527–530

Peters NK, Crist-Estes DK (1989) Nodule formation is stimulated by the ethylene inhibitor aminoethoxyvinylglycine. Plant Physiol 91:690–693

Pizelle G (1965) L'azote minéral et la nodulation de l'aune glutineux (*Alnus glutinosa*). Observations sur des plantes cultivées avec systémes racinaires compartimentés. Bulletin de l'Ecole Nationale Supérieure Agronomique 7:55–63

Pizelle G (1966) L'azote minéral et la nodulation de l'aune glutineux (*Alnus glutinosa*). II. Observations sur l'action inhibitrice de l'azote minéral a l'égard de la nodulation. Ann Inst Pasteur (Paris) 111:259–264

Pommer EH (1956) Beiträge zur Anatomie und Biologie der Wurzelknöllchen von *Alnus glutinosa* (L.) Gaertn. Flora 143:603–634

Prayitno J, Rolfe BG, Mathesius U (2006) The ethylene insensitive *sickle* mutant of *Medicago truncatula* shows altered auxin transport regulation during nodulation. Plant Physiol 142:168–180

Racette S, Torrey JG (1989) Root nodule initiation in *Gymnostoma* (Casuarinaceae) and *Shephardia* (Elaeagnaceae) induced by *Frankia* strain HFPGpI1. Can J Bot 67:2873–2879

Radutoiu S, Madsen LH, Madsen EB, Felle HH, Umehara Y, Grønlund M, Sato S, Nakamura Y, Tabata S, Sandal N, Stougaard J (2003) Plant recognition of symbiotic bacteria requires two LysM receptor-like kinases. Nature 425:585–592

Ramirez-Saad H, Janse JD, Akkermans ADL (1998) Root nodules of *Ceanothus caeruleus* contain both the N_2-fixing *Frankia* endophyte and a phylogenetically related Nod$^-$/Fix$^-$ actinomycete. Can J Microbiol 44:140–148

Reddell P, Yun Y, Shipton WA (1997) Do *Casuarina cunninghamiana* seedlings dependent on symbiotic N_2 fixation have higher phosphorus requirements than those supplied with adequate fertilizer nitrogen? Plant Soil 189:213–219

Redondo-Nieto M, Rivilla R, El-Hamdaoui A, Bonilla I, Bolaños L (2001) Boron deficiency affects early infection events in the pea–*Rhizobium* symbiotic interaction. Aust J Plant Physiol 28:819–823

Remy W, Taylor TN, Hass H, Kerp H (1994) Four hundred-million-year-old vesicular arbuscular mycorrhizae. Proc Natl Acad Sci USA 91:11841–11843

Ribeiro A, Akkermans AD, van Kammen A, Bisseling T, Pawlowski K (1995) A nodule-specific gene encoding a subtilisin-like protease is expressed in early stages of actinorhizal nodule development. Plant Cell 7:785–794

Rodriguez-Barrueco C, de Castro FB (1973) Cytokinin-induced pseudonodules on *Alnus glutinosa*. Physiol Plant 29:227–280

Santi C (2003) Approche moléculaire de la mise en place du nodule actinorhiyien chez les arbres tropicaux de la famille des Casuarinaceae. PhD dissertation, Université Montpellier II

Sasakawa H, Hiyoshi T, Sugiyama T (1988) Immuno-gold localization of nitrogenase in root nodules of *Elaeagnus pungens* Thunb. Plant Cell Physiol 29:1147–1152

Sasakura F, Uchiumi T, Shimoda Y, Suzuki A, Takenouchi K, Higashi S, Abe M (2006) A class 1 hemoglobin gene from *Alnus firma* functions in symbiotic and nonsymbiotic tissues to detoxify nitric oxide. Mol Plant Microbe Interact 19:441–450

Scheidle H, Gross A, Niehaus K (2005) The lipid A substructure of the *Sinorhizobium meliloti* lipopolysaccharides is sufficient to suppress the oxidative burst in host plants. New Phytol 165:559–565

Scheres B, Van Engelen F, Van der Knap E, Van de Wiel C, Van Kammen A, Bisseling T (1990) Sequential induction of nodulin gene expression in the developing pea nodule. Plant Cell 2:687–700

Schmidt JS, Harper JE, Hoffman TK, Bent AF (1999) Regulation of soybean nodulation independent of ethylene signaling. Plant Physiol 119:951–960

Schnabel E, Journet EP, De Carvalho-Niebel F, Duc G, Frugoli J (2005) The *Medicago truncatula SUNN* gene encodes a CLV1-like leucine-rich repeat receptor kinase that regulates nodule number and root length. Plant Mol Biol 58:809–822

Schröder P (1989) Aeration of the root system in *Alnus glutinosa* L. Gaertn. Ann Sci Forest 46:310–314

Schwintzer CR, Berry AM, Disney LD (1982) Seasonal patterns of root nodule growth, endophyte morphology, nitrogenase activity, and shoot development in *Myrica gale*. Can J Bot 60:746–757

Sequerra J, Capellano A, Faure-Raynard M, Moiroud A (1994) Root hair infection process and myconodule formation of *Alnus incana* by *Penicillium nodositatum*. Can J Bot 72:955–962

Sequerra J, Capellano A, Gianinazzi-Pearson V, Moiroud A (1995) Ultrastructure of cortical root cells of *Alnus incana* infected by *Penicillium nodositatum*. New Phytol 130:545–555

Silvester WB, Silvester JK, Torrey JG (1988a) Adaptation of nitrogenase to varying oxygen tension and the role of the vesicle in root nodules of *Alnus incana* ssp. *rugosa*. Can J Bot 66:1772–1779

Silvester WB, Whitbeck J, Silvester JK, Torrey JG (1988b) Growth, nodule morphology and nitrogenase activity of *Myrica gale* grown with roots at various oxygen levels. Can J Bot 66:1762–1771

Silvester WB, Harris SL, Tjepkema JD (1990) Oxygen regulation and hemoglobin. In: Schwintzer CR, Tjepkema JD (eds) The biology of *Frankia* and actinorhizal plants. Academic, New York, pp 157–176

Silvester WB, Langenstein B, Berg RH (1999) Do mitochondria provide the oxygen diffusion barrier in root nodules of *Coriaria* and *Datisca*? Can J Bot 77:1358–1366

Soltis DE, Soltis PS, Morgan DR, Swensen SM, Mullin BC, Dowd JM, Martin PG (1995) Chloroplast gene sequence data suggest a single origin of the predisposition for symbiotic nitrogen fixation in angiosperms. Proc Natl Acad Sci USA 92:2647–2651

Stokkermans TJW, Peters NK (1994) *Bradyrhizobium elkanii* lipooligosaccharide signals induce complete nodule structures on *Glycine soja* Siebold et Zucc. Planta 193:413–420

Stracke S, Kistner C, Yoshida S, Mulder L, Sato S, Kaneko T, Tabata S, Sandal N, Stougaard J, Szczyglowski K, Parniske M (2002) A plant receptor-like kinase required for both bacterial and fungal symbiosis. Nature 417:959–9562

Suharjo UKJ, Tjepkema JD (1995) Occurence of hemoglobin in the nitrogen-fixing root nodules of *Alnus glutinosa*. Physiol Plant 95:247–252

Swensen SM (1996) The evolution of actinorhizal symbioses—evidence for multiple origins of the symbiotic association. Am J Bot 83:1503–1512

Swensen SM, Mullin BC (1997) Phylogenetic relationships among actinorhizal plants. The impact of molecular systematics and implications for the evolution of actinorhizal symbioses. Physiol Plant 99:565–573

Tellström V, Usadel B, Thimm O, Stitt M, Küster H, Niehaus K (2007) The lipopolysaccharide of *Sinorhizobium meliloti* suppresses defense-associated gene expression in cell cultures of the host plant *Medicago truncatula*. Plant Physiol 143:825–837

Thomas KA, Berry AM (1989) Effects of continuous nitrogen application and nitrogen preconditioning on nodulation and growth of *Ceanothus griseus* var. *horizontalis*. Plant Soil 118:181–187

Tirichine L, Sandal N, Madsen LH, Radutoiu S, Albrektsen AS, Sato S, Asamizu E, Tabata S, Stougaard J (2007) A gain-of-function mutation in a cytokinin receptor triggers spontaneous root nodule organogenesis. Science 315:104–107

Tjepkema JD (1978) The role of oxygen diffusion from the shoots and the nodule roots in nitrogen fixation by root nodules of *Myrica gale*. Can J Bot 56:1365–1371

Torrey JG (1990) Cross-inoculation groups within *Frankia*. In: Schwintzer CR, Tjepkema JD (eds) The biology of *Frankia* and actinorhizal plants. Academic, New York, pp 83–106

Truchet G, Roche P, Lerouge P, Vasse J, Camut S, De Billy F, Promé JC, Dénarié J (1991) Sulfated lipo-oligosaccharide signals of *Rhizobium meliloti* elicit root nodule organogenesis in alfalfa. Nature 351:670–673

Valverde C, Wall LG (1999a) Time course of nodule development in the *Discaria trinervis* (Rhamnaceae)–*Frankia* symbiosis. New Phytol 141:345–354

Valverde C, Wall LG (1999b) Regulation of nodulation in *Discaria trinervis* (Rhamnaceae)–*Frankia* symbiosis. Can J Bot 77:1302–1310

Valverde C, Wall LG (2003) The regulation of nodulation, nitrogen fixation and ammonium assimilation under a carbohydrate shortage stress in the *Discaria trinervis*–*Frankia* symbiosis. Plant Soil 254:155–165

Valverde C, Wall LG (2005) Ethylene modulates the susceptibility of the root for nodulation in actinorhizal *Discaria trinervis*. Physiol Plant 124:121–131

Valverde C, Ferrari A, Wall LG (2002) Phosphorus and the regulation of nodulation in the actinorhizal symbiosis between *Discaria trinervis* (Rhamnaceae) and *Frankia* BCU110501. New Phytol 153:43–52

Van Brussel AAN, Bakhuizen R, Van Spronsen PC, Spaink HP, Tak T, Lugtenberg BJJ (1992) Induction of pre-infection thread structures in the leguminous host plant by mitogenic lipo-oligosaccharides of *Rhizobium*. Science 257:70–71

Van Noorden GE, Ross JJ, Reid JB, Rolfe BG, Mathesius U (2006) Defective long-distance auxin transport regulation in the *Medicago truncatula* super numeric nodules mutant. Plant Physiol 140:1494–506

Vanden Heuvel BD, Benson DR, Bortiri E, Potter D (2004) Low genetic diversity among *Frankia* spp. strains nodulating sympatric populations of actinorhizal species of Rosaceae, *Ceanothus* (Rhamnaceae) and *Datisca glomerata* (Datiscaceae) west of the Sierra Nevada (California). Can J Microbiol 50:989–1000

Wall LG (2000) The actinorhizal symbiosis. J Plant Growth Regul 19:167–182

Wall LG, Huss-Danell K (1997) Regulation of nodulation in *Alnus-Frankia* symbiosis. Physiol Plant 99:594–600

Wall LG, Hellsten A, Huss-Danell K (2000) Nitrogen, phosphorus, and the ratio between them affect nodulation in *Alnus incana* and *Trifolium pratense*. Symbiosis 29:91–105

Wall LG, Valverde C, Huss-Danell K (2003) Regulation of nodulation in the absence of N_2 is different in actinorhizal plants with different infection pathways. J Exp Bot 54:1253–1258

Wheeler CT, Gordon JC, Ching TM (1979) Oxygen relations of the root nodules of *Alnus rubra* Bong. New Phytol 82:449–457

Wheeler CT, Watts SH, Hillman JR (1983) Changes in carbohydrates and nitrogenous compounds in the root nodules of *Alnus glutinosa* in relation to dormancy. New Phytol 95:209–218

Wopereis J, Pajuelo E, Dazzo FB, Jiang Q, Gresshoff PM, De Bruijn FJ, Stougaard J, Szczyglowski K (2000) Short root mutant of *Lotus japonicus* with a dramatically altered symbiotic phenotype. Plant J 23:97–114

Yang W-C, de Blank C, Meskiene I, Hirt H, Bakker J, van Kammen A, Franssen H, Bisseling T (1994) *Rhizobium* Nod factors reactivate the cell cycle during infection and nodule primordium formation, but the cycle is only completed in primordium formation. Plant Cell 6:1415–1426

Yang Y (1995) The effect of phosphorus on nodule formation and function in the *Casuarina–Frankia* symbiosis. Plant Soil 176:161–169

Yang Y, Shipton WA, Reddel P (1997) Effects of phosphorus supply on in vitro growth and phosphatase activity of *Frankia* isolates from *Casuarina*. Plant Soil 189:75–79

Zhang Z, Torrey JG (1985) Studies of an effective strain of *Frankia* from *Allocasuarina lehmanniana* of the Casuarinaceae. Plant Soil 97:1–16

Zhang Z, Lopez MF, Torrey JG (1984) A comparison of cultural characteristics and infectivity of *Frankia* isolates from root nodules of *Casuarina* species. Plant Soil 78:79–90

Physiology of Actinorhizal Nodules

Tomas Persson[1] (✉) · Kerstin Huss-Danell[2]

[1]Department of Botany, Stockholm University, 106 91 Stockholm, Sweden
Tomas.Persson@botan.su.se

[2]Department of Agricultural Research for Northern Sweden,
Swedish University of Agricultural Sciences (SLU), 901 83 Umeå, Sweden

1	Introduction	156
2	The Actinorhizal Nodule	158
2.1	Nodule Anatomy	158
2.2	The *Frankia*–Host Cell Interface	159
3	Primary Metabolism in Non-Symbiotic *Frankia*	162
3.1	Carbon Metabolism in Non-Symbiotic *Frankia*	162
3.2	Nitrogen Metabolism in Non-Symbiotic *Frankia*	163
4	Nodule Primary Metabolism	163
4.1	Carbon Metabolism in Nodules	163
4.2	Nitrogen Metabolism in Nodules	165
4.2.1	N_2 Fixation	166
4.2.2	Ammonium Assimilation in Nodules	166
4.2.3	Nitrogen Transport Forms in Actinorhizal Plants	167
4.2.4	N_2 Fixation and the Effects of Carbon and Nitrogen Availability	169
5	Hydrogen Metabolism	170
6	The Oxygen Dilemma of N_2 Fixation – How Actinorhizal Plants Deal with It	170
7	Future Prospects	171
References		172

Abstract Some plants live in symbiosis with N_2-fixing bacteria. *Frankia*, a genus of soil actinomycetes, can infect roots and induce root nodules on so-called actinorhizal plants. These are about 200 species of essentially woody angiosperms belonging to eight families. The root nodule is a unique organ where the plant anatomy and metabolism has adapted to facilitate the hosting of *Frankia*. In exchange for carbon compounds *Frankia* provides plant-available nitrogen rendering the host independent of soil nitrogen. When free-living, *Frankia* is capable of fixing N_2 for its own growth, and an important characteristic are specialized cells, so-called vesicles, where dinitrogenase can function in an aerobic environment. Among actinorhizal genera there is a broad range of anatomical and biochemical adaptations in *Frankia* and the host plant to enable both N_2 fixation and aerobic metabolism in the root nodules. Plant and bacterial metabolism form a complex intertwined network with bacterial and plant metabolites shuttled across membranes.

N_2 fixation is energy costly and nodules represent strong carbon sinks. The plant provides the nodules with photosynthates mostly in the form of sucrose. Dicarboxylates are imported to *Frankia* to support bacterial development and N_2 fixation. Ammonium is assumed to be exported from *Frankia* to the cytoplasm of the infected cells and converted to appropriate amino acids and amides for transport to the nitrogen sinks of the plant. The N_2 fixation activity in *Frankia* is controlled by external factors such as the plant's photosynthetic activity, which affects the carbon supply to the nodule, and nitrogen availability in the plant.

Abbreviations

2-OG	2-Oxoglutarate
AAT	Amino acid transferase
Ala	Alanine
Alc	Allantoic acid
Arg	Arginine
AS	Asparagine synthetase
As	Argininosuccinate
Asn	Asparagine
Asp	Aspartic acid
Cit	Citrulline
CMP	Carbamoyl phosphate
CMPS	Carbamoyl phosphate synthase
Frc	Fructose
GDH	Glutamate dehydrogenase
Gln	Glutamine
Glc	Glucose
Gly	Glycine
GOGAT	Glutamate oxo-glutarate aminotransferase
GS	Glutamine synthetase
Mal	Malate
MDH	Malate dehydrogenase
OA	Oxalacetate
OCT	Ornithine carbamoyl transferase
Orn	Ornithine
PEP	Phosphoenolpyruvate
PEPC	Phosphoenolpyruvate carboxylase
Pyr	Pyruvate
SS	Sucrose synthase
Suc	Succinate
TCA	Tricarboxylic acid

1
Introduction

Some plants are capable of growing on nitrogen poor soil. This is due, not to adaptation to live with a low level of nitrogen but rather to help from N_2-fixing bacteria. *Frankia* is a genus of saprophytic filamentous soil bacteria capable of providing plant-available nitrogen to actinorhizal plants by fixing atmospheric nitrogen gas. This symbiosis between host and *Frankia* is localized to specialized root organs known as nodules. Mature actinorhizal nodules usually make up 1–10% of the total biomass of their host plant (Huss-Danell 1990).

Frankia is a multicellular bacterium, an actinomycete, which grows with hyphae in symbiosis as well as in culture, hereafter referred to as free-living. *Frankia* produces spores in culture and in some symbioses. In nearly all host genera and as free-living under nitrogen-poor conditions, *Frankia* produces vesicles where N_2 fixation takes place. When free-living, *Frankia* extracts nutrients from its surrounding medium and, under nitrogen deprivation, fixes N_2 from the air for its own growth. In symbiosis *Frankia* receives carbon compounds and other nutrients from the plant in return for nitrogenous products. This complex exchange of metabolites requires modifications in both plant and bacterial metabolic profiles.

The N_2 fixation, catalyzed by dinitrogenase, produces ammonia. Reduced nitrogen is then exported from *Frankia* to the plant cytosol for assimilation into amino acids, and these are transported to nitrogen sinks via the xylem to support plant growth and development.

Actinorhizal symbioses (Table 1) involve plants from eight different plant families, belonging to the Fagales (Betulaceae, Casuarinaceae and Myricaceae), the Rosales (Rosaceae, Elaeagnaceae and Rhamnaceae) and the Cucurbitales (Coriariaceae and Datiscaceae; see also Normand and Fernandez, this volume). All actinorhizal plants are perennial, and with the exception of *Datisca* they are all woody. Nodule metabolism is therefore season-dependent, with preparation for winter, a dormancy phase in winter and a growth and development phase during spring and summer. Infection pathways by *Frankia*, nodule formation, nodule morphology, nodule anatomy and nodule physiology vary greatly among actinorhizal plants from different families (Huss-Danell 1997).

Host-symbiont coevolution has led to great diversity among *Frankia* strains. Different clades vary dramatically in their capacity for saprophytic growth (Normand et al. 2006).

Actinorhizal symbioses comprise much of the diversity of nature's solutions to the dilemma of N_2-fixation and are thus an important source of information to learn from when attempting to understand how symbiotic N_2-fixing symbioses function and when attempting to create new symbioses. From a practical point of view, actinorhizal symbioses have many applications such as production of wood and bioenergy, as ornamental plants and in soil reclamation (Schwintzer and Tjepkema 1990; Russo 2005).

Understanding the unique metabolic profile of a nodule is crucial for future improvement of nitrogen assimilation. To determine different options for nodule metabolism, it is necessary to elucidate how nitrogen derived from N_2 fixation is incorporated into nitrogenous solutes for xylem transport, and which carbon compounds are supplied by the host plant, in different symbiotic systems. Various aspects of actinorhizal physiology and metabolism have been partially covered in earlier summaries by Benson and Silvester (1993); Huss-Danell (1990, 1997) and Valverde and Huss-Danell (2008). Actinorhizal nodule physiology is the focus of this work.

2
The Actinorhizal Nodule

Early interactions, infection and nodulation in actinorhizal symbioses were recently summarized (Wall and Berry 2008) and will not be considered here. Mature actinorhizal nodules are actually clusters of coralloid structures, consisting of multiple branched lobes (Fig. 1) (Newcomb and Wood 1987; Berry and Sunell 1990), with each lobe representing a modified lateral root. In the following, we usually refer to "nodule lobe" when discussing nodules.

2.1
Nodule Anatomy

A nodule lobe (Fig. 1) differs from a root in that it lacks a root cap and root hairs. Like roots, nodule lobes have a central vascular system, and an expanded cortex. *Frankia* is located in the nodule cortex. The nodule periderm is thought to block water uptake from the soil (Bond 1956), leaving the nodule dependent on the vascular system for its water supply. The nodule periderm reduces gas exchange with the environment. The periderm is however interrupted by one (*Coriaria*) or several (*Alnus*) lenticels (Batzli and Dawson 1999; Silvester and Harris 1989; Silvester et al. 1990). Alternatively, some nodules (*Casuarina, Gymnostoma, Myrica, Comptonia*) are aerated by agraviotropically growing nodule roots (Fletcher 1955; Torrey 1976; Newcomb and Wood 1987) – the lobe meristem changes its activity and gives rise to a usually upward-growing root with large air spaces in the cortex facilitating air transport from better aerated soil to the nodule (Tjepkema 1983). *Datisca* nodules are aerated by both lenticels and nodule roots (Pawlowski et al. 2007). Dinitrogenase is oxygen labile and its synthesis is inhibited by oxygen, but on the other hand the energy demand of the N_2 fixation reaction must be supported by ATP derived from respiratory activity, leading to the so-called oxygen dilemma. To cope with this, the plant and/or *Frankia* has developed features to provide a microanaerobic environment for dinitrogenase within actively respiring cells (see Sect. 6).

In culture, *Frankia* shows three structures, hyphae, sporangia and vesicles (Newcomb and Wood 1987). *Frankia* grows with septate hyphae (filaments) about 0.5 µm in diameter. Hyphae can differentiate into vegetative sporangia containing spores of about the same diameter as hyphae. Moreover, at low nitrogen availability hyphae can differentiate into more or less spherical and septate vesicles, up to about 5 µm in diameter and being the site of N_2 fixation. *Frankia* are delimited by a cell membrane, a bacterial cell wall and a layered structure, the so-called envelope consisting of lipid laminae. The envelope is dominated by hopanoid lipids and is particularly well developed with many layers around vesicles (Berry et al. 1993).

Symbiotic vesicles (Newcomb and Wood 1987; Berg 1999) are a prominent structure in nodules, except for *Casuarina* and *Allocasuarina*. Depending on the host family, vesicles have different localization in the infected cell, different shapes and are either septate or non-septate (Newcomb and Wood 1987; Huss-Danell 1997; Berg 1999).

The cellular make-up of the actinorhizal nodule (Fig. 1) is generated by activity in the nodule apical meristem, leading to a developmental gradient of infected cells in the cortex, from young infected cortical cells in the infection zone that are gradually being filled with *Frankia* hyphae, to mature infected cells in the N_2-fixation zone, to senescent infected cells, where *Frankia* is degraded. Infected cortical cells are, with the exception of *Datisca* and *Coriaria*, interspersed with uninfected cells. The relative arrangement of cells in the nodule cortex varies in different actinorhizal symbioses (reviewed by Silvester et al. 2008). In *Datisca* (Fig. 1B) and *Coriaria*, the continuous patch of infected cells is kidney-shaped in cross-section and located on one side of the acentric vascular system, while in all other genera (exemplified by *Alnus* Fig. 1A) the groups of infected cells are interspersed in the cortex surrounding the vascular system.

2.2
The *Frankia*–Host Cell Interface

In root nodules there are several barriers between host cytoplasm and *Frankia* cytoplasm (Fig. 2). The host cytoplasm is enclosed by the host plasma membrane (Newcomb and Wood 1987; Berry and Sunell 1990), also called the perisymbiont membrane (Pawlowski and Sprent 2008). Outside the plasma membrane the host surrounds *Frankia* with a perisymbiont matrix (Berry et al. 2002; Pawlowski and Sprent 2008) which is contiguous with the host cell wall. This matrix has also been termed "invaginated host cell wall" (Berry and Sunell 1990) or interfacial matrix (Berg 1999) or has commonly been called "capsule" (e.g. Newcomb and Wood 1987). Arabinogalactans have been localized to the perisymbiont matrix (Berry et al. 2002), which has also been shown to be rich in pectin, cellulose, hemicellulose and proteins (Lalonde and Knowles 1975; Berg 1990; Liu and Berry 1991).

Frankia cytoplasm is surrounded by a cell membrane and the *Frankia* cell wall (Fig. 2). Outside the bacterial cell wall *Frankia* forms the so-called envelope, or lipid laminae, which is a multi-layered structure rich in hopanoids, a bacterial steroid lipid (Berry et al. 1993). The envelope is particularly de-

Fig. 1 Root nodule(s) of **A** *Alnus acuminata* and **B** *Datisca glomerata* together with schematic horizontal and longitudinal sections of their nodule lobes. The vascular tissue is centrally positioned in *Alnus* while acentric in *Datisca*. In *D. glomerata* a nodule root can grow out of the lobe apex, a nodule root is visible in the photograph (*arrow*). The outermost shaded structure represents periderm, *internal shaded green areas* represent infected region and *black* denotes vascular tissue. le-lenticel; m-meristem; 1-infection zone; 2-mature, N_2-fixing zone; 3-senescing zone. In *D. glomerata* the infected cells are in a continuous patch while in *Alnus* the infected cells are intermixed with uninfected cells. Photos taken by **A** Luis Wall and **B** Tomas Persson

veloped with a large number of layers around the symbiotic vesicles while there are only a few layers around hyphae.

The complex layering at the *Frankia*-plant interface requires intricate metabolic and communication pathways between the two partners. In actinorhizal systems only a dicarboxylate transporter located at the perisymbiont membrane in *Alnus glutinosa* has been described so far (Jeong et al. 2004). In comparison, several transporters have been found to reside in the perisymbiont membrane in legumes, including transporters for zinc (Moreau et al. 2002), iron (Moreau et al. 1995, 1998; LeVier et al. 1996), sulfate (Krusell et al. 2005), ammonium/ammonia (Tyerman et al. 1995; Niemietz and Tyerman 2000) and dicarboxylate transporters (Udvardi et al. 1988). Calcium channels and pumps (Tyerman et al. 1995; Niemietz and Tyerman 2000), cation and anion channels/transporters (Udvardi et al. 1991; Roberts and Tyerman 2002) and aquaporin/ion channels (Fortin et al. 1987; Miao et al. 1992; Weaver et al. 1994; Rivers et al. 1997; Dean et al. 1999) have also been described. Similar transporters are likely to reside in the perisymbiont membrane in actinorhizal nodules but have yet to be demonstrated. It is important to note that corresponding transporters must also be present in the plasma membrane of *Frankia*.

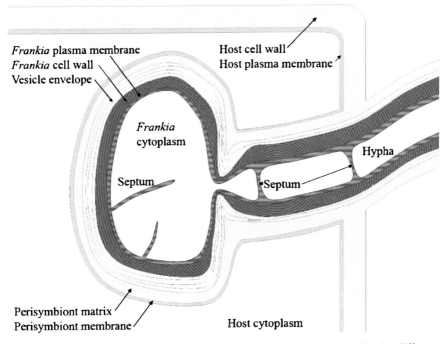

Fig. 2 Schematic drawing of part of an *Alnus* nodule cell with a *Frankia* hypha differentiated into a symbiotic vesicle. The *Frankia* cytoplasm is surrounded by *Frankia* plasma membrane, *Frankia* cell wall and *Frankia* envelope. Note that the host plasma membrane is continuous with the membrane surrounding *Frankia*, the perisymbiont membrane. The perisymbiont matrix (older term: capsule) is contiguous with the plant cell wall and has a chemical composition with similarities to that of a primary plant cell wall. Note also that the envelope is thicker, has many more layers, around the symbiotic vesicle than around the hypha. Based on Berry and Sunell (1990); Berry et al. (1993); Berg (1999), the drawing is not to scale

3
Primary Metabolism in Non-Symbiotic *Frankia*

The metabolic profiles of free-living and symbiotic *Frankia* differ and must be studied independently.

Using sequence analysis of the 16S rRNA gene, nif-genes and other genes, a *Frankia* phylogenetic tree, consisting of three clusters, was constructed (Benson and Clawson 2000). The clusters (referred to as cluster 1, 2 and 3) have different, but often overlapping, physiological properties. All cluster 1 *Frankia* strains, infecting plants from the order Fagales (Betulaceae, Casuarinaceae and Myricaceae), are culturable, while cluster 3 strains, infecting Cucurbitables (Coriariaceae and Datiscaceae), are still unculturable. Cluster 2 *Frankia* strains in symbiosis with plants from the order Rosales (Rosaceae) and the genus *Ceanothus* (among the Rhamnaceae) are still unculturable, while symbionts of actinorhizal plants within Rhamnaceae and Elaeagnaceae are culturable. Hence, the metabolism of several *Frankia* strains can still not be analyzed in free-living state. Our understanding of *Frankia* metabolism is however quickly enhanced, thanks to the rapidly expanding genomics toolbox.

3.1
Carbon Metabolism in Non-Symbiotic *Frankia*

Carbon metabolism in free-living *Frankia* has been studied exclusively in culture. The culturable *Frankia* strains require no growth factor and can thus be grown on minimal media (Benson and Silvester 1993).

Frankia strains can use a variety of carbon sources, including short chain fatty acids such as propionate and acetate, TCA cycle intermediates, pyruvate and some sugars (Benson and Silvester 1993). Propionate can serve as a carbon source for all culturable strains, with the exception of the isolate EAN1pec (Tisa et al. 1983) nodulating Elaeagnaceae and potentially others. Because propionate uptake can be inhibited by carbonyl cyanide *m*-chlorophenylhydrazone (CCCP), sodium azide, *p*-hydroxymercuribenzoate and *N*-ethylmalemide, it was suggested that uptake is actively mediated by a specific permease (Stowers et al. 1986). Amino acids such as Asp and Gln promote growth of *Frankia* strain CcI1 (Zhang and Benson 1992). Some *Frankia* strains have been found to grow on minimal media having only Tween 20 and 80 as the sole carbon source (Lechevalier and Lechevalier 1990). Interestingly, several strains have been found to degrade cellulose (Safo-Sampah and Torrey 1988). Free-living *Frankia* can store carbon in the form of trehalose and glycogen (Lopez et al. 1983, 1984).

3.2
Nitrogen Metabolism in Non-Symbiotic *Frankia*

A variety of organic and inorganic nitrogen sources are reported to promote growth of free-living *Frankia*, including amino acids, urea, nitrate, ammonia and N_2 (Benson and Silvester 1993). Ammonium in the medium is assimilated by the bacteria via the GS/GOGAT pathway producing Gln and Glu (Benson and Schultz 1990). *Frankia* sp. strain CpI1 grown on media with low concentrations of ammonium show active uptake. In contrast, when grown on elevated concentrations of ammonium (> 0.5 mM) it enters the bacteria passively in unprotonated form (Mazzucco and Benson 1984).

Frankia grown aerobically in culture has shown the ability to fix N_2 at a sufficient rate for its own growth (Tjepkema et al. 1980; Gauthier et al. 1981; Murry et al 1984; Noridge and Benson 1984; Meesters et al. 1987). These studies have also shown that active dinitrogenase is localized in *Frankia* vesicles.

4
Nodule Primary Metabolism

Root nodules are metabolically very active. The nodule is fed carbon sources, largely as sucrose, via the phloem; this input is used as an energy source by the symbiont in N_2 fixation and for the nodule for its own development. Moreover, the plant is thought to provide *Frankia* amino acids (Fig. 3) and must provide *Frankia* with nutrients. The nitrogen fixed and exported by *Frankia* is converted to plant available nitrogenous compounds in the nodule and transported to nitrogen sinks via the xylem. Because of the intricate relation between the two organisms, a complex integrated metabolic system is required for proper nodule development and function. Moreover, the nodule primary metabolism requires active and communicating pathways over the symbiotic interface.

4.1
Carbon Metabolism in Nodules

Nodules are strong sinks for carbon, as is evidenced by the fact that uninfected cells in most types of actinorhizal nodules contain many plastids rich in starch grains (Newcomb and Wood 1987; Pawlowski 2002). The unloading of sucrose from the phloem in root nodules is followed by catabolism by either cytosolic sucrose synthase or by cytosolic, apoplastic or vacuolar invertase (van Ghelue et al. 1996; Lalonde et al. 2004). A part of the available carbon in nodules is converted to carbon compounds supplied to *Frankia*.

In symbiosis with *Alnus*, *Frankia* exhibits aerobic metabolism (Ching et al. 1983; Vikman et al. 1992). So-called vesicle clusters isolated mainly from *Al*-

nus nodules (Akkermans et al. 1977; van Straten et al. 1977; Lundquist and Huss-Danell 1991a; Berry et al. 2002) have been a useful tool in the study of *Frankia* metabolism in symbiosis, not only carbon metabolism but also studies of dinitrogenase (Lundquist and Huss-Danell 1991a,b), hydrogenase (Benson et al. 1980; Sellstedt 1989), GS (Lundquist and Huss-Danell 1992), lipid composition of vesicle envelope (Kleemann et al. 1994) and arabinogalactan proteins in the symbiotic interface (Berry et al. 2002). However, when evaluating experiments with vesicle clusters one should be aware of the fact that vesicle clusters are not only symbiotic vesicles with a piece of their subtending hyphae, but that *Frankia* is also surrounded by (parts of) the perisymbiont matrix and the perisymbiont membrane. When adding compounds to vesicle clusters in respiration studies negative results show either that *Frankia* can not use the compound or that the compound could not be taken up under the experimental conditions when cell integrity and concentration gradients in the infected cells are disrupted.

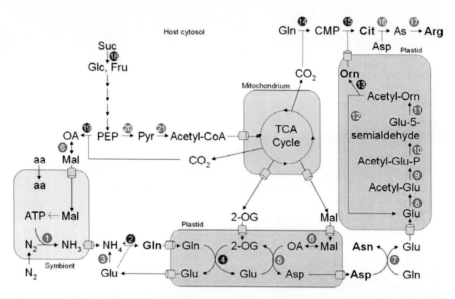

Fig. 3 Suggested general nitrogen assimilation scheme in nodules of plants with N_2-fixing symbionts. The figure shows routes that would produce the main nitrogen compounds Asn, Arg, Gln, Orn and/or Cit (highlighted in *bold*) found in the xylem of these plants. Enzymes highlighted in *red* are found active in nodules of *Alnus*. aa is an abbreviation for amino acids. 1–dinitrogenase; 2–GS; 3–GDH; 4–GOGAT; 5–AAT; 6–MDH; 7–AS; 8–glutamate *N*-acetyl transferase; 9–*N*-acetylglutamate 5-phosphotransferase; 10–*N*-acetylglutamyl-5-phosphate:NAD(P) + oxidoreductase; 11–*N*2-acetylornithine:2-oxoglutarate aminotransferase; 12–acetylornithine aminotransferase; 13–ornithine carbamoylphosphate; 14–CMPS; 15–OCT; 16–argininosuccinate synthase; 17–argininosuccinate lyase; 18–SS or invertase; 19–PEPC; 20–pyruvate kinase; 21–pyruvate dehydrogenase complex

Supplying vesicle clusters with Suc, maltose, trehalose, Glc, Frc, Glc-6-phosphate or 6-phosphogluconate in combination with NAD increases respiration (reviewed by Valverde and Huss-Danell 2008). Enzymatic activity in vesicle clusters and increased vesicle cluster respiration by the combination of malate, Glu and NAD indicate that dicarboxylates are likely to be the carbon source supplied to *Frankia* by its actinorhizal host (reviewed by Huss-Danell 1997). Malate and possibly other carbon sources feed the TCA cycle, glyoxylate cycle and gluconeogenesis pathway (Benson and Schultz 1990), producing ATP. Malate is, moreover, the carbon source for amino acid synthesis in *Alnus glutinosa* nodules (McClure et al. 1983). Other sources of carbon have been suggested and may work exclusively or in parallel with malate or in other actinorhizal systems. Additional and important support for the idea that malate or other dicarboxylates are exported to *Frankia* comes from the finding that a nodule-specific dicarboxylate transporter has been characterized and localized to the symbiotic interface in *A. glutinosa* (Jeong et al. 2004). At least in this symbiosis, *Frankia* can thus be supplied with dicarboxylates as rhizobia are in legume symbioses (Udvardi et al. 1988; Jeong et al. 2004; Prell and Poole 2005).

Frankia can store carbon and energy in the form of glycogen and trehalose in both free-living and symbiotic states (Benson and Eveleigh 1979; Lopez et al. 1983). Glycogen granules have been seen in *Frankia* using transmission electron microscopy of nodules of *Chamaebatia*, *Myrica*, *Comptonia*, *Coraria*, and *Elaeagnus* although not in *Discaria* and *Dryas* (Newcomb and Wood 1987).

Like roots, nodules have a non-light dependent CO_2-fixing capacity (Huss-Danell 1990; McClure et al. 1983) seen as a back-up process in case of depletion of carbon compounds for ammonia assimilation (Valverde and Huss-Danell 2008). Carbon dioxide fixation has been recorded in *Alnus glutinosa* (Schubert et al. 1981; McClure et al. 1983), *A. hirusta*, *A. incana*, *A. rubra*, *A. sinuata*, *Casuarina montana*, *Colletia cruciata* and *Datisca cannabina* (Huss-Danell 1990). Using pulse-chase experiments with $^{14}CO_2$ in nodules of *A. glutinosa* McClure et al. (1983) established that CO_2 was converted to organic acids by PEPC (Fig. 3).

4.2
Nitrogen Metabolism in Nodules

Although *Frankia* can fix N_2 both when free-living and in symbiosis, there are differences in the further metabolism of nitrogen. While free-living *Frankia* can assimilate its fixed nitrogen (see Sect. 3.2), this process is likely done in host cells in root nodules. The continuing assimilation process is then species or genus dependent among the diverse actinorhizal plants.

4.2.1
N₂ Fixation

N_2 reduction in actinorhizal plants takes place in symbiotic vesicles that differentiate from hyphal tips of the symbiotic *Frankia*. This rather early observation (Mian and Bond 1978) has been verified by in situ localization, immunogold localization and immunochemical detection (Sasakawa et al. 1988; Huss-Danell and Bergman 1990; Pawlowski et al. 1995; Valverde and Wall 2003b). No vesicles are found in *Casuarina* or *Allocasuarina*, however; dinitrogenase activity was instead localized to the hyphae (Laplaze et al. 2000). Berg (1999) considered the intracellular non-septate hyphae in *Casuarina* to serve as the symbiotic vesicle equivalent (Berg 1999).

The product of the reduction of N_2 is ammonia, which in legume nodules is protonated to form ammonium in the acidic peribacteroid space (Udvardi and Day 1997; Day et al. 2001).

Dinitrogenase:

$$N_2 + 12\text{–}24\text{ATP} + 8\,e^- + 8H^+ \xrightarrow{Mg^{2+}} 2NH_3 + 12\text{–}24\text{ADP} + 12\text{–}24P_i + H_2$$

An ammonium transporter has been localized to the peribacteroid membrane of *Lotus japonicus* nodules (Tyerman et al. 1995). The pH of the perisymbiont matrix in actinorhizal nodules has not been investigated but a similar system is not unlikely.

4.2.2
Ammonium Assimilation in Nodules

On the basis of enzyme and NMR studies (Lundberg and Lundquist 2004) there is no doubt that exogenously applied ammonium is assimilated in nodules by the GS/GOGAT pathway.

GS:

$$NH_4^+ + \text{Glu} + \text{ATP} \xrightarrow{Mg^{2+}} \text{Gln} + \text{ADP} + P_i$$

GOGAT:

$$\text{Gln} + 2\text{-oxo-glutarate} + \text{NAD(P)H} + H^+ \rightarrow 2\text{Glu} + \text{NAD(P)}$$

For ammonium from N_2 fixation it appears that GS activity and protein abundance are low in vesicle cluster extracts and neither GDH nor GOGAT activity could be detected (Blom et al. 1981; Akkermans et al. 1983; Lundquist and Huss-Danell 1992). A situation where the plant or bacterium represses *Frankia* GS/GOGAT activity would be similar to the situation in pea nodules where plant-derived amino acids have been suggested to feed the bacteria, thus implying nitrogen shuttling in both directions across the symbiotic interface (see Fig. 3; Lodwig et al. 2003). In the general model for nitrogen fixation and assimilation N_2 is reduced by dinitrogenase in symbiotic vesicles into ammonia (Huss-Danell and Bergman 1990; Prin et al. 1993; Pawlowski et al. 1995), which is thought to be exported to the cytosol in the form of ammonium like in legume symbioses (O'Gara and Shanmugam 1976; Guan et al. 1996; Baker and Parsons 1997; Udvardi and Day 1997; Day et al. 2001; Pawlowski 2002). In the plant cytosol ammonium is assimilated via the GS/GOGAT cycle, as suggested by immunological studies in *Alnus incana* (Lundquist and Huss-Danell 1992) and evidenced by high transcription of a gene encoding a cytosolic GS in the infected cells of *A. glutinosa* (Guan et al. 1997).

In contrast to the GS expression pattern in *Alnus* nodules, the pattern of GS expression in *Datisca glomerata* and possibly the phylogenetically related actinorhizal genus *Coriaria* seem to differ (Berry et al. 2004). An active expression of GS is found in vesicles and in uninfected cortical cells, not in the infected cells (Pawlowski et al. 1996). Analysis of the concentrations of nitrogenous solutes in nodules suggests that in this symbiosis *Frankia* excretes, not ammonia, but another assimilation product, probably Arg, which is transported to the uninfected cortical tissue where it is reduced to NH_4^+, entering the GS/GOGAT cycle for further assimilation (Berry et al. 2004).

4.2.3
Nitrogen Transport Forms in Actinorhizal Plants

The major transported nitrogen compound in legumes and actinorhizal plants often comprises up to 70% of total nitrogen compounds (Valverde and Wall 2003a). On the basis of the type of nitrogenous compounds that dominate in the xylem, actinorhizal plants fall into three groups: plants that transport mainly Cit, plants that transport mainly Gln and/or Asn and plants with a more mixed composition of major xylem compounds (Table 1). When measuring ATP costs of synthesis of the exported nitrogen compounds, it was concluded that Alc, Arg, Asn, Cit and Gly are energetically the most efficient compounds (Schubert 1986). Production of these compounds for xylem transport follows different metabolic pathways as outlined in Fig. 3. In *Alnus glutinosa* and in certain legumes GS transcripts have been found in the pericycle of the nodule vascular system (Guan et al. 1996; Temple et al. 1995; Forde et al. 1989). This indicates presence of free ammonium from infected cells, or the result of a degradation and reassimilation of nitrogenous compounds during transport. Moreover, Glu concentration seems to be higher in xylem tissue than in nodules (Blom et al. 1981). For legumes Lea and Miflin (1980) suggested a catabolism/assimilation cycle in which nodule assimilated products may be degraded in the pericycle and reassimilated by the GS/GOGAT cycle.

The host provides the carbon backbone for nitrogen assimilation in the form of malate in alfalfa (Vance 2000) and presumably also in actinorhizal

Table 1 Nitrogen compounds and their transport in actinorhizal plants

Family	Plant species	Major xylem N compounds	Refs.
Betulaceae	Alnus crispa	Cit	Schubert 1986
	A. glutinosa	Cit	Gardner and Leaf 1960
	A. incana	Cit	Miettinen and Virtanen 1952
			Wheeler and Bond 1970
	A. inokumai	Gln, Asn, Glu, Cit	Hafeez et al. 1984
	A. nitida	Cit	Schubert 1986
	A. rubra	Cit	
Casuarinaceae	Casuarina cunninghamiana	Cit, Asn, Arg	Sellstedt and Atkins 1991
	C. equisetifolia	Asn	Bollard 1957
		Cit	Walsh et al. 1984
		Cit, Asn, Orn	Sellstedt and Atkins 1991
	C. glauca	Asn, Cit, Orn	Sellstedt and Atkins 1991
Coriariaceae	Coriaria ruscifolia	Gln, Asn, Cit	Bollard 1957
		Glu, Gln, Arg, Asn	Wheeler and Bond 1970
			Hafeez et al. 1984
Datiscaceae	Datisca glomerata	Gln, Asn	Schubert 1986
		Glu, Gln, Arg, Asn	Wheeler and Bond 1970
			Hafeez et al. 1984
		Gln, Glu	Berry et al. 2004
Eleagnaceae	Elaeagnus angustifolia	Asn	Schubert 1986
	E. umbellata	Asn	Schubert 1986
	Hippophaë rhamnoides	Asn	Wheeler and Bond 1970
Myricaceae	Comptonia spp.	Asn	Schubert 1986
	Myrica cerifera	Asn, Gln	Schubert 1986
	M. gale	Asn	Leaf et al. 1959
Rhamnaceae	Ceanothus americana	Asn	Schubert 1986
	Discaria trinervis	Asn	Valverde 2000
Rosaceae	No information found		

plants (Jeong et al. 2004). Biosynthesis of the amide Asn in root nodules starts in the plastids with the conversion of Gln by the combined action of GOGAT and AAT to Asp. Asparagine synthetase then catalyzes the conversion of Asp to Asn in the cytosol.

Synthesis of Cit and Arg takes place via enzymes of the urea cycle and Glu is the precursor of both. Initiating Cit biosynthesis, Gln and CO_2 are used as substrates by carbamoyl phosphate synthetase to form carbamoyl phosphate (CMP) and Glu. Ornithine carbamoyl transferase then forms Cit from the two products CMP and Orn. Orn is derived from Glu via the plastidic Orn-pathway. Using ATP, Asp is added onto the Cit by argininosuccinate synthase. Fumarate is removed from argininosuccinate by argininosuccinase forming Arg.

These pathways have been partly resolved in actinorhizal nodules (e.g. Lundberg and Lundquist 2004). Most studies have been conducted with the Cit transporting *Alnus glutinosa* and *A. incana* as a model for metabolism (see Fig. 3), but *Discaria trinervis, Datisca glomerata, Elaeagnus umbellata, Myrica gale, Casuarina glauca* and *C. cunninghamiana* have also been used. In *D. trinervis*, an example of Asn transporters (Table 1), activity of GS, AAT, MDH, GDH, GOGAT and AS were found, indicating an active amide synthesizing pathway. Collected xylem sap from nodulated roots was found to have Asn as the main exported compound, while Asp, Glu, Ala and Ser were detected in lower concentrations (Valverde and Wall 2003a).

4.2.4
N_2 Fixation and the Effects of Carbon and Nitrogen Availability

There are many external factors that may influence the sink strength of nodules in actinorhizal plants. Bacterial N_2-fixing activity in nodules is controlled by the plant as it provides carbon as an energy source. It was proposed by Patrick (1990) that partitioning of photosynthates in plants is mostly controlled by transport processes in the sink adjusting turgor to control phloem unloading. The nodule organ is unusual as it has considerable fluxes of solutes both in (carbon and nitrogen solutes) and out (nitrogen solutes). Walsh et al. (1998) concluded that the rate of nitrogen export was not linked to carbon import in legume nodules. The conclusions made indicate that carbon unloading is influenced, not by turgor in the nodule but rather by the energy needs in the root system. Tricot et al. (1997) linked nodule development to growth rate of the particular root segment carrying the nodule. Numerous experiments show a correlation between photosynthesis and symbiotic N_2 fixation. By using both elevated CO_2 (Soussana and Hartwig 1996; Schortemeyer et al. 1999) and increased light intensity (Hardy and Havelka 1976; Bethlenfalvay and Phillips 1977) a correlation between higher photosynthesis and elevated nitrogen fixation has been established. Similarly reduced photosynthesis has been shown to result in lower N_2 fixation rates both in legumes and actinorhizal nodules (Feigenbaum and Mengel 1979; Huss-Danell and Sellstedt 1985; Fujita et al. 1988; Huss-Danell 1990; Vikman et al. 1990; Lundquist and Huss-Danell 1991a,b; Huss-Danell et al. 1992; Lundquist et al. 2003). There is evidence for regulation of nodule number by carbon availability in legumes (Atkins et al. 1989) as well as intrinsic autoregulation in actinorhizal plants through shoot signaling (Wall and Huss-Danell 1997; Valverde and Wall 2003b). In a series of experiments (summarized in Huss-Danell 1997) where *Alnus incana* was exposed to severe stress, including carbon deprivation, dinitrogenase activity decreased when measured both in vivo (intact plants) and in vitro (vesicle clusters). Further, these reduced activities were correlated with loss of both the Fe-protein and the MoFe-protein in dinitrogenase. When the stress treatments stopped dinitrogenase activities and proteins recovered at a rate correlated to infection of new nodule cells by *Frankia* and differentiation of new symbiotic vesicles and synthesis of dinitrogenase. Another aspect of carbon starvation was that dinitrogenase became more sensitive to oxygen inactivation.

Also nitrogen is important for the regulation of nodule growth and activity. Nitrate inhibition of nodule growth in legumes results from a decreased supply of photoassimilates to the nodule leading to a discontinuation of cell growth and expansion (Fujikake et al. 2003). Almost half of all nitrogen in legume nodules is supplied by the plant via the phloem (Layzell et al. 1973; Pate et al. 1981; Schubert 1986), and Parsons et al. (1993) suggested a nitrogen feedback mechanism during nodule development. Nodules respond by increasing growth and dinitrogenase activity during low nitrogen concentrations in the phloem and decreasing activity during high nitrogen availability (Parsons et al. 1993). It is however not easy to completely distinguish between effects of carbon and nitrogen availability on N_2 fixation as carbon and nitrogen are closely related in the plant metabolism, in particular amino acid metabolism.

5
Hydrogen Metabolism

H_2 is formed as a byproduct when dinitrogenase uses N_2 as a substrate (Benson et al. 1979; Roelofsen and Akkermans 1979). Uptake hydrogenases oxidize H_2, forming protons and electrons (Roelofsen and Akkermans 1989; Benson et al. 1980). The electrons are used for energy production in the respiratory chain through oxidative phosphorylation (Robson 2001). Moreover, the activity of uptake hydrogenase prevents an accumulation of H_2, which is a competitive inhibitor of N_2 reduction by dinitrogenase (Rasche and Arp 1989). Uptake hydrogenase (Hup+) has been demonstrated in nearly 20 *Frankia* strains isolated from ten different host plants (Leul et al. 2005) but there are *Frankia* strains lacking uptake hydrogenase activity (Hup–; Sellstedt and Huss-Danell 1984). Higher dinitrogenase activity and growth in *Alnus incana* was found when inoculated with Hup+ *Frankia* compared to a Hup– *Frankia* (Sellstedt et al. 1986). However, the *Frankia* strains differed genetically not only in Hup, which makes the comparison more complicated.

6
The Oxygen Dilemma of N_2 Fixation – How Actinorhizal Plants Deal with It

N_2 fixation requires a large amount of ATP, i.e. respiratory activity, while dinitrogenase is irreversibly denatured by oxygen. This is the oxygen dilemma of N_2 fixation: the bacteria have to protect dinitrogenase from oxygen while maintaining a high level of respiration.

In legume nodules, where the bacteria cannot contribute to the protection of dinitrogenase from oxygen, this is solved in a standard way. An oxygen diffusion barrier, interrupted only at the tip of indeterminate nodules, encloses the infected tissue, while a high concentration of a plant oxygen transport protein, leghemoglobin, is maintained in the cytosol of the infected cells (Hunt and Layzell 1993).

In actinorhizal nodules, *Frankia* can contribute to oxygen protection by having its vesicles surrounded by a hopanoid-rich lipid envelope thought to hinder oxygen entry into the vesicle (Berry et al. 1993). A range of strategies for oxygen protection have been developed (Huss-Danell 1997): one example is in *Casuarina* nodules where *Frankia* does not form vesicles but the nodules are rather rich in hemoglobin (Silvester et al. 1990; Tjepkema and Asa 1987), another solution is in *Alnus* nodules where *Frankia* forms spherical vesicles and the nodules have low levels of reactive heme compounds (Silvester et al. 1988). In *A. incana* the vesicle envelope thickness responds, by adding or retracting layers, to ambient oxygen pressure (Kleemann et al. 1994). These mechanisms are described in more detail by Normand and Fernandez, in this volume.

The oxygen dilemma is not only a matter of static structures, but can be more or less well solved depending on physiological conditions in nodules. This was seen in *Alnus incana* where dinitrogenase activity in carbon-starved plants was more sensitive to oxygen inactivation than in control plants (Lundquist et al. 2003).

7
Future Prospects

To further the field of actinorhizal research methods for constructing transgenic *Frankia* must be developed. Moreover, there is a need for sequence data from more strains of *Frankia*.

As most actinorhizal plants are trees, the development of transgenic trees with a short generation time is most promising. *Casuarina* is currently the only actinorhizal tree for which genetic transfer methods exist. Moreover, *Casuarina glauca* has a small genome size similar to *Arabidopsis thaliana*. Methods for developing transgenic *Alnus* would be highly valuable. Sequencing one or both of these host genera would be of great interest. Unfortunately, *Alnus glutinosa* has a comparatively large genome. We suggest that there is a choice of study objects, where there is focus on one or few genera for a more complete understanding. Contrasting that argument we appreciate the diversity of actinorhizal plants and the fact that they offer a broad range of physiological solutions to how an N_2-fixing symbiosis functions.

Acknowledgements We thank Katharina Pawlowski, Luis Wall and Jenna Persson for critical reading of the manuscript.

References

Akkermans ADL, Roelofsen W, Blom J, Huss-Danell K, Harkink R (1983) Utilization of carbon and nitrogen compounds by *Frankia* in synthetic media and in root nodules of *Alnus glutinosa*, *Hippophae rhamnoides*, and *Datisca cannabina*. Can J Bot 61:2793–2800

Andreev IM, Dubrovo PN, Krylova VV, Izmailov SF (1999) Functional identification of ATP-driven Ca^{2+} pump in the peribacteroid membrane of broad bean root nodules. FEBS Lett 447:49–52

Atkins CA, Pate JS, Sanford PJ, Dakora FD, Matthews I (1989) Nitrogen nutrition of nodules in relation to "N-hunger" in cowpea (*Vigna unguiculata* L. Walp). Plant Physiol 90:1644–1649

Baker A, Parsons R (1997) Rapid assimilation of recently fixed N_2 in root nodules of *Myrica gale*. Physiol Plant 99:640–647

Batzli JM, Dawson JO (1999) Development of flood-induced lenticels in red alder nodules prior to the restoration of nitrogenase activity. Can J Bot 77:1373–1377

Benson DR, Arp DJ, Burris RH (1980) Hydrogenase in actinorhizal root nodules and root nodule homogenates. J Bacteriol 142:138–144

Benson DR, Clawson ML (2000) Evolution of the actinorhizal plant symbioses. In: Triplett EW (ed) Prokaryotic nitrogen fixation: A model system for analysis of biological process. Horizon Scientific Press, Wymondham, UK, pp 207–224

Benson DR, Eveleigh DE (1979) Ultrastructure of the nitrogen-fixing symbiont of *Myrica pennsylvanica* L. (Bayberry) root nodules. Bot Gaz 140:S15–S21

Benson DR, Schultz N (1990) Physiology and biochemistry of *Frankia* in culture. In: Schwintzer CR, Tjepkema JD (eds) The Biology of *Frankia* and Actinorhizal Plants. Academic Press, San Diego, CA, pp 107–127

Benson DR, Silvester WB (1993) Biology of *Frankia* strains, actinomycete symbionts of actinorhizal plants. Microbiol Rev 57:293–319

Berg RH (1990) Cellulose and xylans in the interface capsule in symbiotic cells of actinorhizae. Protoplasma 159:35–43

Berg RH, McDowell L (1987) Endophyte differentiation in *Casuarina* actinorhizae. Protoplasma 136:104–117

Berry AM, Harriottv OT, Moreau RA, Osman SF, Benson DR, Jones AD (1993) Hopanoid lipids compose the *Frankia* vesicle envelope, presumptive barrier of oxygen diffusion to nitrogenase. Proc Natl Acad Sci USA 90:6091–6094

Berry AM, Murphy TM, Okubara PA, Jacobsen KR, Swensen SM, Pawlowski K (2004) Novel expression pattern of cytosolic glutamine synthetase in nitrogen-fixing root nodules of the actinorhizal host, *Datisca glomerata*. Plant Physiol 135:1849–1862

Berry AM, Rasmussen U, Bateman K, Huss-Danell K, Lindwall S, Bergman B (2002) Arabinogalactan proteins at the symbiotic interface in root nodules of *Alnus* spp. New Phytol 155:469–479

Berry AM, Sunell LA (1990) The infection process and nodule development. In: Schwintzer CR, Tjepkema JD (eds) The Biology of *Frankia* and Actinorhizal Plants. Academic Press, San Diego, CA, pp 61–81

Bethlenfalvay GJ, Phillips DA (1977) Effect of light intensity on efficiency of carbon dioxide and nitrogen reduction in *Pisum sativum* L. Plant Physiol 60:868–871

Blom J, Roelofsen W, Akkermans ADL (1981) Assimilation of nitrogen in root nodules of alder (*Alnus glutinosa*). New Phytol 89:321–326

Bond G (1956) Some aspects of translocation in root nodule plants. J Exp Bot 7:387–394

Bryan JK (1990) Advances in the biochemistry of amino acid biosynthesis. In: Miflin BJ (ed) The Biochemistry of Plants, Vol. 16. Academic Press Inc, San Diego, CA, pp 161–195

Day DA, Poole PS, Tyerman SD, Rosendahl L (2001) Ammonia and amino acid transport across symbiotic membranes in nitrogen-fixing legume nodules. Cell Mol Life Sci 58:61–71

Dean RM, Rivers RL, Zeidel ML, Roberts DM (1999) Purification and functional reconstitution of soybean nodulin 26: An aquaporin with water and glycerol transport properties. Biochemistry 38:347–353

Feigenbaum S, Mengel K (1979) The effect of reduced light intensity and sub-optimal potassium supply on N_2 fixation and N turnover in *Rhizobium*-infected lucerne. Physiol Plant 45:245–249

Fletcher WW (1955) The development and structure of the root nodules of *Myrica gale* L. with special reference to the nature of the endophyte. Ann Bot 19:501–513

Forde BG, Day HM, Turton JF, Shen WJ, Cullimore JV, Oliver JE (1989) Two glutamine synthetase genes from *Phaseolus vulgaris* L. display contrasting developmental and spatial patterns of expression in transgenic *Lotus corniculatus* plants. Plant Cell 1:391–401

Fortin MG, Morrison NA, Verma DPS (1987) Nodulin-26, a peribacteroid membrane nodulin is expressed independently of the development of the peribacteroid compartment. Nucl Acids Res 15:813–824

Fujikake H, Yamazaki A, Ohtake N, Sueyoshi K, Matsuhashi S, Ito T, Mizuniwa C, Kume T, Hashimoto S, Ishioka NS, Watanabe S, Osa A, Sekine T, Uchida H, Tsuji A, Ohyama T (2003) Quick and reversible inhibition of soybean root nodule growth by nitrate involves a decrease in sucrose supply to nodules. J Exp Bot 54:1379–1388

Fujita K, Masuda T, Ogata S (1988) Dinitrogen fixation, ureide concentration in xylem exudate and translocation of photosynthates in soybean as influenced by pod removal and defoliation. Soil Sci Plant Nutr 34:265–275

Gentili F, Huss-Danell K (2003) Local and systemic effects of phosphorus and nitrogen on nodulation and nodule function in *Alnus incana*. J Exp Bot 54:2757–2767

Gardner IC, Leaf G (1960) Translocation of citrulline in *Alnus* glutinosa. Plant Physiol 35:948–950

Gauthier D, Diem HG, Dommergues Y (1981) In vitro nitrogen fixation by two actinomycete strains isolated from *Casuarina* nodules. Appl Environ Microbiol 41:306–308

Gest H, Kamen MD, Bregoff HM (1950) Studies on the metabolism of photosynthetic bacteria. V. Photoproduction of hydrogen and nitrogen fixation by *Rhodospirillum rubrum*. J Biol Chem 182:153–170

Van Ghelue M, Ribeiro A, Solheim B, Akkermans ADL, Bisseling T, Pawlowski K (1996) Sucrose synthase and enolase expression in actinorhizal nodules of *Alnus glutinosa*: comparison with legume nodules. Mol Gen Genet 250:437–446

Guan C, Ribeiro A, Akkermans ADL, Jing Y, Van Kammen A, Bisseling T, Pawlowski K (1996) Nitrogen metabolism in actinorhizal nodules of *Alnus glutinosa*: expression of glutamine synthetase and acetylornithine transaminase. Plant Mol Biol 32:1177–1184

Hardy RWF, Havelka UD (1976) Photosynthate as a major factor limiting nitrogen fixation by field-grown legumes with emphasis on soybeans. In: Nutman PS (ed) Symbiotic nitrogen fixation. Cambridge University Press, Cambridge, UK, pp 421–439

Harper JE, Cowingan KA, Barbera AC, Abd-Alla MH (1997) Hypernodulation of soybean, mung bean and hyacinth bean is controlled by a common shoot signal. Crop Sci 37:1242–1246

Hunt S, Layzell DB (1993) Gas-exchange of legume nodules and the regulation of nitrogenase activity. Annu Rev Plant Physiol Plant Mol Biol 44:483–511

Huss-Danell K (1990) The physiology of actinorhizal nodules. In: Schwintzer CR, Tjepkema JD (eds) The Biology of Frankia and Actinorhizal Plants. Academic Press, Inc., New York, NY, pp 129–156

Huss-Danell K (1991) Influence of host (Alnus and Myrica) genotype on infectivity, N_2 fixation, spore formation and hydrogenase activity in Frankia. New Phytol 119:121–127

Huss-Danell K (1997) Actinorhizal symbioses and their N_2 fixation. New Phytol 136:375–405

Huss-Danell K, Bergman B (1990) Nitrogenase in Frankia from root nodules of Alnus incana (L.) Moench: immunolocalization of the Fe- and MoFe-proteins during vesicle differentiation. New Phytol 116:443–455

Huss-Danell K, Gentili F, Valverde C, Wall L, Wiklund A (2002) Phosphorus is important in nodulation of actinorhizal plants and legumes. In: Finan T, O'Brian M, Layzell D, Vessey K, Newton WE (eds) Nitrogen fixation: global perspectives. CAB International, Wallingford, pp 163–166

Huss-Danell K, Lundquist P-O, Ohlsson H (1992) Distribution of biomass and nitrogen among plant parts and soil nitrogen in a young Alnus incana stand. Can J Bot 70:1545–1549

Huss-Danell K, Sellstedt A (1985) Nitrogenase activity in response to darkening and defoliation of Alnus incana. J Exp Bot 36:1352–1358

Jeong J, Suh S, Guan C, Tsay Y-F, Moran N, Oh CJ, An CS, Demchenko KN, Pawlowski K, Lee Y (2004) A nodule-specific dicarboxylate transporter from alder is a member of the peptide transporter family. Plant Physiol 134:969–978

Kaiser BN, Moreau S, Castelli J, Thomson R, Lambert A, Bogliolo S, Puppo A, Day DA (2003) The soybean NRAMP homologue, GmDMT1, is a symbiotic divalent metal transporter capable of ferrous iron transport. Plant J 35:295–304

Kleemann G, Alskog G, Berry AM, Huss-Danell K (1994) Lipid composition and nitrogenase activity of symbiotic Frankia (Alnus incana) in response to different oxygen concentrations. Protoplasma 183:107–115

Krusell L, Krause K, Ott T, Desbrosses G, Kraemer U, Sato S, Nakamura Y, Tabata S, James E, Sandal N, Stougaard J, Kawaguchi M, Miyamoto A, Suganuma N, Udvardi M (2005) The sulfate transporter, SST1 is crucial for symbiotic nitrogen fixation in Lotus japonicus root nodules. Plant Cell 17:1–12

Lalonde M, Knowles R (1975) Ultrastructure, composition, and biogenesis of the encapsulation material surrounding the endophyte in Alnus crispa var. mollis root nodules. Can J Bot 53:1951–1971

Lalonde S, Wipf D, Frommer WB (2004) Transport mechanisms for organic forms of carbon and nitrogen between source and sink. Annu Rev Plant Biol 55:341–372

Laplaze L, Ribeiro A, Franche C, Duhoux E, Auguy F, Bogusz D, Pawlowski K (2000) Characterization of a Casuarina glauca nodule-specific subtilisin-like protease gene, a homolog of Alnus glutinosa ag12. Mol Plant-Microbe Interact 13:113–117

Layzell DB, Rainbird RM, Atkins CA, Pate JS (1979) Economy of photosynthate use in nitrogen-fixing legume nodules. Plant Physiol 64:888–891

Lea PJ, Miflin BJ (1980) Transport and metabolism of asparagine and other nitrogen compounds within the plant. In: Miflin BJ (ed) The Biochemistry of Plants, Vol. 5. Academic Press, New York, NY, pp 569–608

Leaf G, Gardner IC, Bond G (1959) Observations on the composition and metabolism of the nitrogen-fixing root nodules of Myrica. Biochem J 72:662–667

Lechevalier MP (1984) The taxonomy of the genus Frankia. Plant Soil 78:1–6

Lechevalier MP, Lechevalier HA (1990) Systematics, isolation and culture of *Frankia*. In: Schwintzer CR, Tjepkema JD (eds) The Biology of *Frankia* and Actinorhizal Plants. Academic Press, San Diego, CA, pp 35–60

Leul M, Mohapatra A, Sellstedt A (2005) Biodiversity of hydrogenases in *Frankia*. Curr Microbiol 50:17–23

LeVier K, Day DA, Guerinot ML (1996) Iron uptake by symbiosomes from soybean root nodules. Plant Physiol 111:893–900

Liu Q, Berry AM (1991) Localization and characterization of pectic polysaccharides in roots and root nodules of *Ceanothus* spp. during intercellular infection by *Frankia*. Protoplasma 163:93–101

Lodwig EM, Hosie AHF, Bourdés A, Findlay K, Allaway D, Karunakaran R, Downie JA, Poole PS (2003) Amino-acid cycling drives nitrogen fixation in the *Rhizobium*-legume symbiosis. Nature 422:722–726

Lopez MF, Fontaine MS, Torrey JG (1984) Levels of trehalose and glycogen in *Frankia* sp. HFPArI3 (*Actinomycetales*). Can J Microbiol 30:746–752

Lopez MF, Whaling CS, Torrey JG (1983) The polar lipids and free sugars of *Frankia* in culture. Can J Bot 61:2834–2842

Lundberg P, Lundquist P-O (2004) Primary metabolism in N_2-fixing *Alnus incana-Frankia* symbiotic root nodules studied with ^{15}N and ^{31}P nuclear magnetic resonance spectroscopy. Planta 219:661–672

Lundquist P-O, Huss-Danell K (1991a) Nitrogenase activity and amounts of nitrogenase proteins in a *Frankia-Alnus incana* symbiosis subjected to darkness. Plant Physiol 95:808–813

Lundquist P-O, Huss-Danell K (1991b) Response of nitrogenase to altered carbon supply in a *Frankia-Alnus incana* symbiosis. Physiol Plant 83:331–338

Lundquist P-O, Huss-Danell K (1992) Immunological studies of glutamine synthetase in *Frankia-Alnus incana* symbioses. FEMS Microbiol Lett 91:141–146

Lundquist P-O, Näsholm T, Huss-Danell K (2003) Nitrogenase activity and root nodule metabolism in response to O_2 and short-term N_2 deprivation in dark-treated *Frankia-Alnus incana* plants. Physiol Plant 119:244–252

Mazzucco C, Benson DR (1984) ^{14}C-Methylammonium transport by *Frankia* sp. strain CpI1. J Bacteriol 160:636–641

McClure PR, Coker GT, Schubert KR (1983) CO_2 fixation in roots and nodules of *Alnus glutinosa*: Role of PEP carboxylase and carbamyl phosphate synthetase in dark CO_2 fixation, citrulline synthesis, and N_2 fixation. Plant Physiol 71:652–657

Meesters TM, VanVliet WM, Akkermans ADL (1987) Nitrogenase is restricted to the vesicles in *Frankia* strain EAN1$_{pec}$. Physiol Plant 70:267–271

Mian S, Bond G (1978) The onset of nitrogen fixation in young alder plants and its relation to differentiation in the nodular endophyte. New Phytol 80:187–192

Miao GH, Hong Z, Verma DPS (1992) Topology and phosphorylation of soybean nodulin-26, an intrinsic protein of the peribacteroid membrane. J Cell Biol 118:481–490

Moreau S, Day DA, Puppo A (1998) Ferrous iron is transported across the peribacteroid membrane of soybean nodules. Planta 207:83–87

Moreau S, Meyer JM, Puppo A (1995) Uptake of iron by symbiosomes and bacteroids from soybean nodules. FEBS Lett 361:225–228

Moreau S, Thomson RM, Kaiser BN, Trevaskis B, Guerinot ML, Udvardi MK, Puppo A, Day DA (2002) GmZIP1 encodes a symbiosis-specific zinc transporter in soybean. J Biol Chem 277:4738–4746

Murry MA, Fontaine MS, Tjepkema JD (1984) Oxygen protection of nitrogenase in *Frankia* sp. HFPArI3. Arch Microbiol 139:162–166

Newcomb W, Wood SM (1987) Morphogenesis and fine structure of *Frankia* (Actinomycetales): the microsymbiont of nitrogen-fixing actinorhizal root nodules. Int Rev Cytol 109:1-88

Niemietz CM, Tyerman SD (2000) Channel-mediated permeation of ammonia gas through the peribacteroid membrane of soybean nodules. FEBS Lett 465:110-114

Noridge NA, Benson DR (1986) Isolation and nitrogen-fixing activity of *Frankia* sp. strain CpI1 vesicles. J Bacteriol 166:301-305

Normand P, Lapierre P, Tisa LS, Gogarten JP, Alloisio N, Bagnarol E, Bassi CA, Berry AM, Bickhart DM, Choisne N, Couloux A, Cournoyer B, Cruveiller S, Daubin V, Demange N, Francino MP, Goltsman E, Huang Y, Kopp OR, Labarre L, Lapidus A, Lavire C, Marechal J, Martinez M, Mastronunzio JE, Mullin BC, Niemann J, Pujic P, Rawnsley T, Rouy Z, Schenowitz C, Sellstedt A, Tavares F, Tomkins JP, Vallenet D, Valverde C, Wall LG, Wang Y, Medigue C, Benson DR (2006) Genome characteristics of facultatively symbiotic *Frankia* sp. strains reflect host range and host plant biogeography. Genome Res 17:7-15

O'Gara F, Shanmugam KT (1976) Regulation of nitrogen fixation by rhizobia: export of fixed nitrogen as ammonium ion. Biochim Biophys Acta 437:313-321

Pate JS, Atkins CA, Hamel K, McNeil DL, Layzell DB (1979) Transport of organic solutes in phloem and xylem of a nodulated legume. Plant Physiol 63:1082-1088

Pate JS, Atkins CA, Rainbird RM (1981) Theoretical and experimental costing of nitrogen fixation and related processes in nodules of legumes. In: Gibson AH, Newton WE (eds) Current Perspectives in Nitrogen Fixation. Aust Acad Sci, Canberra, Australia, pp 105-116

Patrick JW (1990) Sieve element unloading: cellular pathway, mechanism and control. Physiol Plant 78:298-308

Pawlowski K (2002) Actinorhizal symbioses. In: Leigh GJ (ed) Nitrogen Fixation at the Millennium. Elsevier Science, Pergamon Press, Amsterdam, pp 167-189

Pawlowski K, Akkermans ADL, Van Kammen A, Bisseling T (1995) Expression of *Frankia nif* genes in nodules of *Alnus glutinosa*. Plant Soil 170:371-376

Pawlowski K, Jacobsen KR, Alloisio N, Denison RF, Klein M, Winzer T, Sirrenberg A, Guan C, Berry AM (2007) Truncated hemoglobins in actinorhizal nodules of *Datisca glomerata*. Plant Biol 9:776-785

Pawlowski K, Ribeiro A, Guan C, Van Kammen A, Berry AM, Bisseling T (1996) Actinorhizal nodules from different plant families. In: Stacey G, Mullin BC, Gresshoff PM (eds) Biology of Plant-Microbe Interactions, Vol. 1. IS-MPMI, St. Paul, MN, pp 417-422

Pawlowski K, Sprent JI (2008) Comparison between actinorhizal and legume Symbiosis. In: Pawlowski K, Newton WE (ed) Nitrogen-fixing Actinorhizal Symbioses. Springer, Dordrecht, The Netherlands, pp 261-288

Prell J, Poole P (2006) Metabolic changes of rhizobia in legume nodules. Trend Microbiol 14:161-168

Prin Y, Mallein-Garin F, Simonet P (1993) Identification and localization of *Frankia* strains in *Alnus* nodules by in situ hybridization of *nifH* mRNA with strain-specific oligonucleotide probes. J Exp Bot 44:815-820

Rasche ME, Arp DJ (1989) Hydrogen inhibition of nitrogen reduction by nitrogenase in isolated soybean nodule bacteroids. Plant Physiol 91:663-668

Rivers RL, Dean RM, Chandy G, Hall JE, Roberts DM, Zeidel ML (1997) Functional analysis of nodulin 26, an aquaporin in soybean root nodule symbiosomes. J Biol Chem 272:16256-16261

Roberts DM, Tyerman SD (2002) Voltage-dependent cation channels permeable to NH_4^+, K^+, and Ca^{2+} in the symbiosome membrane of the model legume *Lotus japonicus*. Plant Physiol 128:370-378

Robson R (2001) Biodiversity of hydrogenases. In: Cammack R, Frey F, Robson R (eds) Hydrogen as Fuel: Learning from Nature. Taylor and Francis Inc, New York, NY, pp 9–32

Russo RO (2005) Nitrogen-fixing trees with actinorhiza in forestry and agroforestry. In: Werner D, Newton WE (eds) Nitrogen Fixation in Agriculture, Forestry, Ecology and the Environment. Springer, Dordrecht, The Netherlands, pp 143–171

Safo-Sampah S, Torrey JG (1988) Polysaccharide-hydrolyzing enzymes of *Frankia* (Actinomycetales). Plant Soil 112:89–97

Sasakawa H, Hiyoshi T, Sugiyama T (1988) Immunogold localization of nitrogenase in root nodules of *Elaeagnus pungens* Thunb. Plant Cell Physiol 29:1147–1152

Schortemeyer J, Atkin OK, McFarlane N, Evans JR (1999) The impact of elevated atmospheric CO_2 and nitrate supply on growth, biomass allocation, nitrogen partitioning and N_2 fixation of *Acacia melanoxylon*. Aust J Plant Physiol 26:737–747

Schubert KR (1986) Products of biological nitrogen fixation in higher plants: synthesis, transport, and metabolism. Annu Rev Plant Physiol 37:539–574

Schwintzer CR, Tjepkema JD (eds) (1990) The Biology of *Frankia* and Actinorhizal Plants. Academic Press, San Diego, USA

Sellstedt A (1989) Occurrence and activity of hydrogenase in symbiotic *Frankia* from field-collected *Alnus incana*. Physiol Plant 75:304–308

Sellstedt A, Atkins CA (1991) Composition of amino compounds transported in xylem of *Casuarina* sp. J Exp Bot 42:1493–1497

Sellstedt A, Huss-Danell K (1984) Growth, nitrogen fixation and relative efficiency of nitrogenise in *Alnus incana* grown in different cultivation systems. Plant Soil 78:147–158

Silvester WB, Harris SL (1989) Nodule structure and nitrogenase activity of *Coriaria arborea* in response to varying oxygen partial pressure. Plant Soil 118:97–110

Silvester WB, Harris SL, Tjepkema JD (1990) Oxygen regulation and hemoglobin. In: Schwintzer CR, Tjepkema JD (ed) The Biology of *Frankia* and Actinorhizal Plants. Academic Press, San Diego, CA, pp 157–176

Soussana JF, Hartwig UA (1996) The effect of elevated CO_2 on symbiotic N_2 fixation: a link between the carbon and nitrogen cycles in grassland ecosystems. Plant Soil 187:321–332

Stowers MD, Kulkarni RK, Steele DB (1986) Intermediary carbon metabolism in *Frankia*. Arch Microbiol 143:78–84

Temple SJ, Heard J, Ganter G, Dunn K, Sengupta-Gopalan C (1995) Characterization of a nodule-enhanced glutamine synthetase from alfalfa: nucleotide sequence, in situ localization, and transcript analysis. Mol Plant-Microbe Interact 8:218–227

Tisa L, McBride M, Ensign JC (1983) Studies of growth of *Frankia* isolates in relation to infectivity and nitrogen fixation (acetylene reduction). Can J Bot 61:2768–2773

Tjepkema JD (1983) Oxygen concentration within the nitrogen-fixing root nodules of *Myrica gale* L. Am J Bot 70:59–63

Tjepkema JD, Asa DJ (1987) Total and CO-reactive heme content of actinorhizal nodules and the roots of some non-nodulated plants. Plant Soil 100:225–236

Tjepkema JD, Ormerod W, Torrey JG (1980) Vesicle formation and acetylene reduction activity in *Frankia* sp. CpI1 cultured in defined nutrient media. Nature 287:633–635

Torrey JG (1976) Initiation and development of root nodules of *Casuarina* (Casuarinaceae). Am J Bot 63:335–344

Tricot F, Crozat Y, Pellerin S (1997) Root growth and nodule establishment on pea (*Pisum sativum* L.). J Exp Bot 48:1935–1941

Tyerman SD, Whitehead LF, Day DA (1995) A channel-like transporter for NH_4^+ on the symbiotic interface of N_2-fixing plants. Nature 378:629–632

Udvardi MK, Day DA (1997) Metabolite transport across symbiotic membranes of legume nodules. Annu Rev Plant Physiol Plant Mol Biol 48:493-523

Udvardi MK, Lister DL, Day DA (1991) ATPase activity and anion transport across the peribacteroid membrane of isolated soybean symbiosomes. Arch Microbiol 156:362-366

Udvardi MK, Price GD, Gresshoff PM, Day DA (1988) A dicarboxylate transporter on the peribacteroid membrane of soybean nodules. FEBS Lett 231:36-40

Valverde C (2000) Regulación de la nodulación radicular en la simbiosis *Discaria trinervis-Frankia*. PhD Thesis, Facultad de Ciencias Exactas, Universidad Nacional de La Plata, La Plata, Argentina

Valverde C, Huss-Danell K (2008) Carbon and Nitrogen Metabolism in Actinorhizal Nodules. In: Pawlowski K, Newton WE (eds) Nitrogen-fixing Actinorhizal Symbioses. Springer, Dordrecht, The Netherlands, pp 167-198

Valverde C, Wall LG (2003a) Ammonium assimilation in *Discaria trinervis* root nodules. Regulation of enzyme activities and protein levels by the availability of macronutrients (N, P and C). Plant Soil 254:139-153

Valverde C, Wall LG (2003b) The regulation of nodulation, nitrogen fixation and assimilation under a carbohydrate shortage stress in the *Discaria trinervis -Frankia* symbiosis. Plant Soil 254:155-165

Vikman P-Å (1992) The symbiotic vesicle is a major site for respiration in *Frankia* from *Alnus incana* root nodules. Can J Microbiol 38:779-784

Vikman P-Å, Lundquist P-O, Huss-Danell K (1990) Respiratory capacity, nitrogenase activity and structural changes of *Frankia*, in symbiosis with *Alnus incana*, in response to prolonged darkness. Planta 182:617-625

Wall LG, Berry AM (2008) Early interactions, infection and nodulation in actinorhizal symbioses. In: Pawlowski K, Newton WE (eds) Nitrogen-fixing Actinorhizal Symbioses. Springer, Dordrecht, The Netherlands, pp 147-166

Wall LG, Huss-Danell K (1997) Regulation of nodulation in *Alnus-Frankia* symbiosis. Physiol Plant 99:594-600

Walsh KB, Thorpe MR, Minchin PEH (1998) Photoassimilate partitioning in nodulated soybean III. The effect of changes in nodule activity shows that carbon supply to the nodule is not linked to nodule nitrogen metabolism. J Exp Bot 49:1827-1834

Weaver CD, Shomer NH, Louis CF, Roberts DM (1994) Nodulin 26, a nodule-specific symbiosome membrane protein from soybean, is an ion channel. J Biol Chem 269:17858-17862

Wheeler CT, Bond G (1970) The amino acids of non-legume root nodules. Phytochem 9:705-708

Zhang X, Benson DR (1992) Utilization of amino acids by *Frankia* sp. strain CpI1. Arch Microbiol 158:256-261

Part III
Cyanobacterial Symbioses

Physiological Adaptations in Nitrogen-fixing *Nostoc*–Plant Symbiotic Associations

John C. Meeks

Section of Microbiology, University of California, Davis, CA 95616, USA
jcmeeks@ucdavis.edu

1	Introduction	182
2	Specificity of the Associations	184
2.1	Plant Partners	184
2.2	Cyanobacterial Partners	185
3	Physiological Adaptation in the Associations	186
3.1	Establishment of a Symbiotic Association	187
3.1.1	Induction of Hormogonium Differentiation	187
3.1.2	Control of Hormogonium Movement	188
3.1.3	Colonization and Repression of Hormogonium Differentiation	189
3.2	Development of a Functional Association	189
3.2.1	Growth and Metabolism	190
3.2.2	Heterocyst Differentiation	195
4	Conclusions and Future Perspectives	198
4.1	Specificity	198
4.2	Physiological Adaptation in the Associations	199
4.3	Perspectives	199
References		200

Abstract *Nostoc* species establish nitrogen-fixing symbiotic associations with representatives of the four main lineages of terrestrial plants: bryophyte hornworts and liverworts, the pteridophyte fern *Azolla*, gymnosperm cycads, and the angiosperm genus *Gunnera*. However, the plant partners represent only narrowly selected groups within these lineages. The plant partner benefits by the acquisition of fixed nitrogen, but the benefits to the *Nostoc* partner are unclear. Thus, the associations are considered a commensal form of symbiosis. A working hypothesis of this chapter is that these associations evolved as the lineage of the plant partner emerged. Inherent in this hypothesis is that the plant partners may have evolved different regulatory signals and targets in control of the *Nostoc* partner. The physiological interactions between the two partners can be modeled as a two-step process. First is the establishment of an association and involves the differentiation and behavior of motile hormogonium filaments, the infective units. Hormogonium formation is essential, but not singularly sufficient for establishment of an association. Second is the development of a functional nitrogen-fixing association involving the differentiation and behavior of heterocysts, the functional units. Heterocysts are the cellular sites of nitrogen fixation, protecting nitrogenase from inactivation by oxygen. The symbiotic growth state of *Nostoc* spp. is characterized by a reduced rate of growth, depressed carbon dioxide and ammonium assimilation, transition to a heterotrophic metabolic mode, an elevated heterocyst frequency, and an enhanced rate of nitrogen fixation. In all but one association (cycads), dinitrogen-derived ammonium is made available to the plant partner. Physiological measurements indicate that different reactions of *Nostoc* photosynthetic carbon dioxide and ammonium assimilation are modulated by the various plant partners. These results appear to support the working hypothesis and indicate that different mechanisms may be operational, allowing for manipulation of different strategies in engineering new plant partners for symbiotic nitrogen fixation.

1
Introduction

Some filamentous cyanobacterial species or strains of the genus *Nostoc*, order Nostocales, have the unusual property of fixing nitrogen in free-living and plant-associated symbiotic growth states (Meeks 1998). The terrestrial symbiotic plant partners include the spore-producing non-vascular bryophytes, the vascularized spore-producing pteridophyte aquatic fern family Azollaceae, the seed producing gymnosperm order Cycadales, and the angiosperm family Gunneraceae (Adams 2000; Bergman et al. 1996; Meeks 1998; Rai et al. 2000). *Nostoc* species also establish lichenized (Paulsrud et al. 1998) and non-lichenized (Mollenhauer et al. 1996) associations with fungi, but these will not be considered here. A question to be asked is, are there physiological adaptations of the *Nostoc* species that may reflect on the mechanisms of symbiotic interaction with the diverse plant partners?

Cyanobacteria are characterized by the phenotypic traits of oxygen-evolving photosynthesis and a prokaryotic cell structure, which present a dilemma with regard to nitrogen fixation. The nitrogenase enzyme complex is highly sensitive to inactivation by oxygen; as a consequence, cyanobacteria have evolved two strategies to protect nitrogenase from excess oxygen. The behavioral strategy is temporal segregation of photosynthesis to the daylight hours and fixation of nitrogen during the night when oxygen is no longer produced and aerobic respiration can lower the intracellular oxygen tension (Fay 1992; Gallon 1992). This strategy is practiced by both unicellular and non-differentiating filamentous cyanobacteria. Nitrogen-fixing associations with these cyanobacteria have not been documented.

The morphological strategy is to confine oxygenic photosynthesis and nitrogen fixation into separate cells as compartments, such that the two processes can occur simultaneously in the filaments, leading to greater growth efficiency. Under conditions of combined nitrogen limitation, vegetative cells differentiate into microoxic cells, heterocysts, which are specialized for nitrogen fixation. Heterocyst differentiation is terminal and occurs only in, and in part defines, two filamentous cyanobacteria orders: the Nostocales and the Stigonematales (Castenholz 2001). In the free-living growth state, heterocysts appear singly in a non-random spacing pattern in the filament and constitute 5–10% of the cells. The microoxic state of heterocysts is achieved by a cessation of the oxygen-evolving reaction of photosynthesis, the deposition of a gas and solute impermeable extra wall layer of polysaccharide and glycolipid, and an increased rate of aerobic respiration to consume any oxygen that may enter at the cell pole connections to adjacent vegetative cells (Wolk et al. 1994). The loss of complete photosynthesis dictates a shift from an autotrophic to a heterotrophic metabolic mode in heterocysts. Heterocysts and vegetative cells, therefore, establish a reciprocal source-sink relationship; vegetative cells provide heterocysts with reductant in the form of carbohydrate and heterocysts provide vegetative cells with reduced nitrogen in the form of glutamine (Wolk et al. 1994). This metabolic relationship is characteristic of a multicellular organism, of which the heterocyst-forming cyanobacteria clearly represent. The differentiation of heterocysts and the cellular protection of nitrogenase from oxygen inactivation may be a basis of the recruitment of *Nostoc* species into symbiotic associations with plants, in which the plants do not need to provide an additional mechanism for oxygen protection. Three aspects of the heterocyst spacing pattern, which are relevant also to symbiotic growth, are typically considered: establishment of pattern, maintenance of pattern, and disruption of pattern (Meeks and Elhai 2002).

The cyanobacteria are an ancient and phylogenetically cohesive group of prokaryotes dating back to at least 2500, and perhaps 3500, million years ago (Ma) (Schopf 2000). However, ancestors to the extant Nostocales are thought to be amongst the latest to have emerged in the cyanobacterial radiation, with the heterocyst-formers estimated at around 1500 Ma (Tomitani et al. 2006). Based on the fossil record, the terrestrial plants emerged sequentially starting with the bryophytes at 480–510 Ma, pteridophytes at 420 Ma, but with *Azolla* fossils appearing around 120 Ma, Cycadales at 250 Ma and Gunneraceae about 70–90 Ma (Raven 2002). A fundamental question is whether the cyanobacterial symbiotic associations evolved near to the times when the different plant lineages emerged, or did the associations form within a similar period of time after emergence of the Gunneraceae? There is no fossil specimen depicting any cyanobacterial symbiont within a plant partner. The significance of the question is rooted in the need to understand mechanisms of interactions between the partners in any attempts to exploit cyanobacterial symbiotic nitrogen fixation in agricultural applications. If the associations developed sequentially, different mechanisms could have evolved in stabilization of each association in the establishment of a nitrogen-fixing factory for the plant partner. Different mechanism would present alternative strategies in engineering plant partners for symbiotic association. Two approaches to initial insight into the mechanisms of symbiotic interaction are to define the specificity of partners in various associations and to determine the physiological characteristics of the partners during establishment of the associations. Essentially all of the experimental effort in the latter approach has been directed toward the *Nostoc* partner, with very little information accruing about the plant partners.

2
Specificity of the Associations

Specificity between partners has implications on the variety of mechanisms that may need to be involved in stabilization of the symbiotic interaction. Broadly specific associations are considered to be flexible and, perhaps, less highly evolved.

2.1
Plant Partners

Although the plant partners in cyanobacterial associations span the phylogenetic spectrum of terrestrial plants, they are representative of only a narrow range of partners within each group (Adams 2000; Bergman et al. 2002; Meeks 1998; Rai et al. 2000; see also Rai et al. 2002). Four out of six of the extant bryophyte hornwort genera have a cyanobacterial partner, but only two liverworts form endophytic associations, and moss associations are rare and of unresolved nutritional benefit. The pteridophyte fern associations are confined to the single genus, *Azolla*, in the family Azollaceae. All examined cycads have a symbiotic *Nostoc* partner, but cyanobacterial associations are lacking in all other gymnosperms. A similar distribution holds for the angiosperms, with the Gunneraceae being the only representative. Except for the cycads, the other plant partners share a common ecological distribution of growth in high water content habitats. Cycads have a dry and a rain forest, as well as a grassland, distribution (Costa and Lindblad 2002). Excepting *Azolla*, which can form extensive and uncontrollable surface blooms in eutrophic lakes and rivers (Carrapico et al. 1996), none of the associations represent the dominant vegetation in any ecosystem.

In all associations, the cyanobacterial partner is confined to a specific location in the plant tissue, commonly referred to as a symbiotic cavity or gland. The cavity is formed in all plants irrespective of the presence of a compatible cyanobacterium. In hornworts and liverworts, the cyanobacteria are in a slime-filled cavity that opens on the ventral surface of the gametophyte thallus. The cyanobacteria are present in a cavity in the dorsal leaves of floating *Azolla*. Consistent with increasing structural complexity, the cyanobacteria are found in specialized secondary roots in cycads, called coralloid roots, and are confined to a zone between the inner and outer cortex. Only in the Gunneraceae are the cyanobacteria intracellular, within special mucus secreting stem gland cells at the base of each petiole.

Thus, there is a clear degree of plant specificity, the genetic basis of which is unknown. The most common ecological parameter is a moist habitat, but even that is not exclusive. The most common physiological feature is production of mucus or slime by the plant partner. One gets the impression of an evolutionary experiment that provided a limited competitive advantage for the plant partner, but that did not lead to a dominant role in any ecosystem.

2.2
Cyanobacterial Partners

In the most part, cyanobacterial partners have been identified based on the morphological characteristics of free-living representatives following isolation and culture. The almost exclusive isolate from all associations is one or more species or strains of *Nostoc*. Occasionally, related *Calothrix* sp. (Nostocales) and *Chlorogloeopsis* sp. (Stigonematales) have been cultured from hornwort and cycad associations (Adams 2000). One limitation here, of course, is the requirement for growth under laboratory conditions. The primary symbiotic cyanobacterium in *Azolla* appears to be recalcitrant to culture in the laboratory (Tang et al. 1990). Cyanobacteria, identified as either *Nostoc* or *Anabaena*, have been cultured from *Azolla* associations, but these are thought to be secondary symbionts that contribute little or no nutritional advantage to the plant partner (Peters and Meeks 1989). A major phenotypic difference between the genera *Nostoc* and *Anabaena* in culture is that some *Anabaena* vegetative filaments are motile by gliding, while all *Nostoc* vegetative filaments are sessile. *Nostoc* spp. gain gliding motility by the differentiation of vegetative filaments into morphologically distinct, heterocyst-free, filaments called hormogonia (Rippka et al. 1979). Based on morphological characteristics, the primary uncultured symbiont of *Azolla* has historically been assigned to the genus *Anabaena*, although hormogonium-like filaments have been observed near the plant apical meristem (Peters and Meeks 1989).

Speciation in cyanobacteria is highly problematic (Castenholz 2001). Many of the morphological features used in classical taxonomy of field samples, such a cell length and width, extent of sheath and/or slime production, pigmentation, and cellular differentiation are very plastic, depending on growth conditions and unknown environmental signals. Thus, specificity has been tested experimentally by cross-infection studies. The associations are readily reconstituted in the laboratory with the bryophyte (Enderlin and Meeks 1983; Adams 2000) and *Gunnera* (Johansson and Bergman 1994) partners, but less so with cycads (Ow et al. 1999). Reconstitution experiments indicate broad specificity of *Nostoc* isolates from all associations, including lichens, with bryophyte (Enderlin and Meeks 1983; West and Adams 1997) or *Gunnera* (Johansson and Bergman 1994) partners. This is particularly true of an isolate from the cycad *Macrozamia* sp., identified as *Nostoc punctiforme* strain PCC 73102 (synonym strain ATCC 29133) and defined as the type strain of *Nostoc* Cluster 1 (Rippka and Herdman 1992). *N. punctiforme* strain PCC 73102 reconstitutes associations with the hornwort *Anthoceros punctatus* (Enderlin and Meeks 1983; Meeks 2003) and *Gunnera* spp. (Johansson and Bergman 1994). *N. punctiforme* is amenable to genetic manipulation (Cohen et al. 1994; Summers et al. 1995) and the genome of strain ATCC 29133 has been completely sequenced (Meeks 2005b; Meeks et al. 2001). *N. punctiforme* has emerged as a model organism for studies of symbiotic interaction (Meeks 2003) and *Nostoc* cellular developmental alternatives (Meeks et al. 2002).

More definitive studies to identify the cyanobacterial symbionts in plant associations have utilized molecular genetic approaches of restriction fragment length polymorphism and sequence analysis of conserved genes and/or intergenic regions (Costa et al. 2001; Rasmussen and Svenning 2001; West and Adams 1997). Due to limitations in size of the database and speciation problems in general, these studies do not lead to absolute identities, but do allow for estimates of diversity. The results of such studies indicate considerable diversity in the *Nostoc* symbionts of bryophytes, both geographically and in a single gametophyte thallus, in cycad coralloid roots from both greenhouse and natural field samples, and naturally grown *Gunnera* species (Rasmussen and Nilsson 2002).

While it appears that differentiation of hormogonia is necessary for symbiotic interaction, it is not singularly sufficient. Many hormogonia-forming *Nostoc* strains are not competent to establish a symbiotic association in the laboratory (Enderlin and Meeks 1983; Johansson and Bergman 1994). Other factors, such as chemotactic attraction (see below) may be required. An unresolved question is why other hormogonia-forming genera are not more commonly found in symbiotic association. Perhaps only strains in the genus *Nostoc* have the genetic capacity to establish associations with plants, and the other documented genera of symbiotic isolates were either only casually associated or infected by chance and did not contribute to the function of an association.

3
Physiological Adaptation in the Associations

The symbiotic *Nostoc*–plant associations have been characterized as a largely (but not exclusive) unidirectional flow of signals from plant to cyanobacterium (Meeks 1998). This characterization is based on the extensive physiological and morphological changes that occur in the symbiotic growth state of *Nostoc*, compared to the relatively minor changes that appear in the plant partner. Moreover, the selective advantage of *Nostoc* spp. in symbiotic association is not obvious. Thus, we have characterized these associations as a commensal form of symbiosis, where the plant partner clearly benefits by the provision of fixed nitrogen and the *Nostoc* partner neither benefits nor is harmed (Meeks 2005a; Meeks and Elhai 2002).

The symbiotic competence of a *Nostoc* strain is distinguished by responses to plant control over two developmental states, those of hormogonia and heterocysts, as well as over growth and metabolism. The interactions mostly likely occur as a continuum, but they can be modeled as a two-stage process. The first stage is infection involving the differentiation and behavior of hormogonia, which serve as the infective units. The second stage is development of a functional association and involves growth, and the differentiation and behavior of heterocysts.

3.1
Establishment of a Symbiotic Association

The establishment of a symbiotic association can be subdivided into the three substages of induction of hormogonium differentiation, control of the direction of hormogonium gliding, and infection of the symbiotic cavity followed by repression of hormogonium differentiation.

3.1.1
Induction of Hormogonium Differentiation

In any physical interaction, such as symbiotic association, the partners must come together, either by random chance or by directed movement. In the *Nostoc*–plant associations, *Nostoc* is the motile partner through the differentiation of hormogonium filaments. Hormogonia are motile by a gliding mechanism, which requires contact with a substratum; the ultimate substratum is the plant tissue. A multiplicity of physicochemical factors induces the differentiation of hormogonia (Meeks and Elhai 2002; Tandeau de Marsac 1994). Hormogonia are a non-growth state, thus, their development proceeds as a cycle (Campbell and Meeks 1989). Upon induction, there is a cessation of net macromolecular synthesis, including DNA replication (Herdman and Rippka 1988), but cell division continues for a period resulting in smaller and differently shaped cells in the filaments. The filaments fragment at the vegetative cell–heterocyst connections, resulting in a loss of the capacity for nitrogen fixation. Gliding motility is initiated by 18–24 h after induction and the filaments remain motile for another 48–60 h; this interval defines the infection window. The filaments then become sessile, and they re-enter the vegetative cell cycle of growth and division; biomass components increase in an undefined sequence and heterocysts differentiate, first at the ends of the filaments.

To initiate a symbiotic association, the plant partners release a chemical signal that induces the differentiation of hormogonia. The signal has been collectively called a hormogonium-inducing factor (HIF). HIF is produced by gametophyte tissue of bryophytes (referred to as exudate) (Campbell and Meeks 1989; Knight and Adams 1966), the coralloid roots of the cycad *Zamia* sp. (Ow et al. 1999) and is in the mucilage of the stem glands of *Gunnera* spp. (Johansson and Bergman 1992). Even roots of non-symbiotic wheat seedlings release substances that induce hormogonia in *Nostoc* sp. (Gantar et al. 1993). Production and/or release of HIF are enhanced by nitrogen starvation of hornwort tissue (Campbell and Meeks 1989). The chemical identity of HIF is not known. The factor(s) from *A. punctatus* and *Gunnera* spp. is a small molecule of between 0.5 and 12 kDa and is inactivated by heat. The activity in *A. punctatus* is complexed by polyvinylpyrrolidone (Campbell and Meeks 1989), while that of *Gunnera* spp. is inactivated by protease treatment (Rasmussen et al. 1994). The HIF from wheat appears to be larger than 12 kDa (Gantar et al. 1993). These various properties imply different chemical identities for the HIF activity, but this will not be known until the factor(s) from each plant partner is purified and characterized in detail.

The targets of HIF in *Nostoc* spp. are also unknown. We have now determined that the formation of hormogonia from vegetative filaments involves the differential transcription of more than 1820 genes, 52% of which are up-regulated (Campbell et al. 2007). While these transcriptome experiments have not yet identified HIF targets, they set the stage for working backwards from regulated genes to the regulators and signal transduction systems.

In the *Azolla* associations, the *Anabaena/Nostoc* is retained throughout the life cycle and passed to subsequent generations in the spore. Thus, this association has likely not been reconstituted with an environmental source of *Anabaena/Nostoc* for some time; it is also recalcitrant to reconstitution with any symbiotically competent *Nostoc* isolate (Peters and Meeks 1989). The *Anabaena/Nostoc* population in the fern apical meristem consists of undifferentiated filaments, morphologically similar to hormogonia (Peters and Meeks 1989). Whether formation of these undifferentiated filaments is regulated is unknown. Since the *Anabaena/Nostoc* filaments are pulled into the symbiotic cavity, via *Azolla* hair cells, by growth and morphogenesis (Peters and Meeks 1989), the filaments need not be motile.

3.1.2
Control of Hormogonium Movement

It is reasonable to assume that chemotaxis of *Nostoc* spp. to a plant partner is important in efficient formation of an association by the low population sizes of *Nostoc* in a habitat. Recent experiments have established that hormogonia of *Nostoc* spp. are chemotactic to the same exudate from bryophytes (Knight and Adams 1996) and mucilage from *Gunnera* spp. (Nilsson et al. 2006) that induce the differentiation of hormogonia. Hormogonia are known to be phototactic (Tandeau de Marsac 1994). Thus, a chemotactic response that is dominant over phototaxis could be instrumental in colonization of *Gunnera* stem gland cells where the hormogonia must migrate away from light into the mucilage filled channels (Johansson and Bergman 1992).

More rigorous evidence of the role of chemotaxis in establishment of an association needs to be generated by multiple approaches, including genetic analyses. For example, identification and subsequent removal or complexing of the attractant(s) should lead to failure to infect the plant partner. Moreover, the genome of *N. punctiforme* contains five loci of genes encoding chemotaxis-like proteins that could constitute complete signal transduction systems, one or more of which could be responsible for a chemotactic response to plant signals (Meeks 2005a; Meeks et al. 2001). Specific genes in four of the loci are transcribed primarily or exclusively in hormogonia (Campbell et al. 2007).

3.1.3
Colonization and Repression of Hormogonium Differentiation

Motile hormogonia have been observed to enter the symbiotic cavities of the liverwort *Blasia pustilla* (Adams 2000; Kimura and Nakano 1990). Similar observations have not been recorded for hornworts or cycads. A detailed light and electron microscopic study has well defined the colonization process in *Gunnera* spp., from hormogonium induction, migration in the channel, to intracellular localization into achlorophyllous gland cells (Johansson and Bergman 1992).

In at least the hornwort association, symbiotically associated tissue produces HIF (Campbell and Meeks 1989). Thus, *Nostoc* filaments in the symbiotic cavity, which is open to the environment, continue to be exposed to the signal to differentiate hormogonia. Continued hormogonium formation is potentially lethal by extinction and counter-productive to differentiation of heterocysts and fixation of nitrogen. *A. punctatus* produces a hormogonium repressing factor (HRF) that is dominant over HIF (Cohen and Meeks 1996). Physiological and genetic analyses imply that the HRF consists of more than one component (Campbell et al. 2003; Cohen and Yamasaki 2000). The factors target a genomic locus in *N. punctiforme* that appears to synthesize a metabolite that functions as a repressor of hormogonium differentiation (Campbell et al. 2003).

3.2
Development of a Functional Association

The symbiotic growth state of a *Nostoc* sp. is broadly characterized by a reduced relative growth rate, a shift to a heterotrophic metabolic mode, an increased heterocyst frequency and the release of fixed nitrogen to the plant partner. In addition to the increased heterocyst frequency, the vegetative cells are markedly larger than those in free-living cultures and the cell–cell connections are very weak, such that manually excised symbiotic colonies, mounted on a microscope slide and dispersed by pressure on the cover slip, appear as short two to three cell filaments or unicells in the microscope (Meeks 1990). The life cycle of *Nostoc* species includes macroscopic globular forms (*Nostoc* balls) where the filaments are confined by a somewhat rigid sheath; the confined filaments lose their filamentous nature and give the appearance of unicells (aseriate). To some extent, the symbiotic growth morphology reflects this aseriate growth stage. In some *Nostoc* species, the aseriate growth morphology is enhanced by green light (Lazaroff 1973). Green light could influence growth morphology of *Nostoc* spp. in bryophyte and *Azolla* associations that are exposed to light, but should have little impact in the darkened cycad or *Gunnera* tissues.

3.2.1
Growth and Metabolism

Nostoc species, even in the natural habitat, can grow considerably faster than the plant partners, excepting, perhaps, for *Azolla*. Moreover, annual bryophytes grow much faster than the perennial cycads and *Gunnera*. Thus, in a stable association, growth of the *Nostoc* partner, such as *N. punctiforme*, must be slowed dramatically in concert with its plant partner. There are no indications as to whether growth is limited in response to metabolic changes or vice versa, or even if physical confinement to a minimal expanding symbiotic compartment is a limiting factor.

Growth control in the *Azolla* spp. association is unique and possibly unprecedented. The *Azolla* associations can double their biomass in about 2 days (Peters et al. 1980), which is comparable to laboratory culture of a recently isolated *Nostoc* symbiont (Enderlin and Meeks 1983). This observation would imply minimal growth control of the primary *Anabaena/Nostoc* symbiont under optimal *Azolla* growth conditions. The secondary symbionts, however, cannot be detected in cyanobacterial purifications from crushed whole plant tissues using molecular probes, while the primary symbiont is readily detectable (Meeks et al. 1988). This negative result indicates the secondary symbionts are present at orders of magnitude lower concentration than the primary symbiont. Conversely, doubling times of the isolated secondary symbionts in the laboratory are on the order of 48 h or less. When crushed *Azolla* preparations are plated for enrichment of nitrogen-fixing cyanobacteria, it takes an extraordinarily longer period of time for colonies to emerge compared to plating free-living cultures. These observations indicate highly stringent growth control over the secondary symbionts that takes a prolonged incubation period apart from the plant partner before they recover and can initiate growth. This growth control, however, appears not to be imposed on the primary symbiont and may be an important aspect of the evolutionary changes in this apparent obligate symbiont.

Metabolic studies have focused on carbon and nitrogen metabolism; they are summarized in Tables 1 and 2.

Photosynthetic Carbon Metabolism

Because there are few distinguishing characteristics, other than pigmentation, there have been no direct studies of photosynthetic activity of the *Nostoc* symbiont within the plant tissue. In *Azolla caroliniana*, indirect measurements indicate that the *Anabaena/Nostoc* partner contributes to less than 5% of the CO_2 fixed by the association, even though the cyanobacterium is estimated to contribute approximately 16% to the total chlorophyll and protein (Kaplan and Peters 1988). By utilization of *Nostoc* mutants resistant to photosynthetic inhibitors, Steinberg and Meeks (1991), determined that the *Nostoc* symbiont of *A. punctatus* does photosynthesize and could contribute, at most, 30% of the photosynthetically generated reductant required for nitrogen fixation in the association on a short term basis. These studies reinforce the largely heterotrophic nature of the *Nostoc* in symbiotic association.

Table 1 Photosynthetic characteristics of *Nostoc* in symbiotic association with terrestrial plants (derived from Meeks 1998)

Association	Photosynthetic CO_2 fixation in planta	Photosynthetic CO_2 fixation ex planta	Rubisco activity in vitro	Rubisco protein
Free-living *Nostoc* spp.	ND	128 nmol/min/mg protein	215–321 nmol/min/mg protein	52 mg/g cell protein
Hornwort, *Anthoceros punctatus*[1]	Positive, < 30%	12%	15%	100%
Fern, *Azolla* spp.[2]	Positive, < 5%	85%	ND	mRNA, < 10%
Cycad, *Cycas, Zamina, Macrozamia* spp.[3]	ND	ND	100%	ND
Angiosperm, *Gunnera* spp.[4]	ND	< 1%	100%	100%

Free-living *Nostoc* values were compiled from control experiments in the cited references. Symbiotic values are are given as a percentage of free-living values. Some of the symbiotic values were converted from qualitative data
References: 1. Steinberg and Meeks (1989, 1991), Meeks (1990). 2. Ray et al. (1979), Kaplan et al. (1986), Nierzwicki-Bauer and Haselkorn (1986). 3. Lindblad et al. (1987), Lindblad and Bergman (1986). 4. Soderbäck and Bergman (1992, 1993)
ND not determined

The capacity of the immediately isolated *Nostoc* symbiont to carry out complete photosynthetic CO_2 fixation ex planta varies from essentially no activity in *Nostoc* isolated from *Gunnera* and cycads, to 12% and 85% of the free-living rate in bryophyte and *Azolla* associations, respectively (Table 1). Examination of in vitro ribulose bisphosphate carboxylase/oxygenase (Rubisco) activity reveals a correlation with ex planta CO_2 fixation only in the *Nostoc* associated with bryophytes. In this association, however, the low Rubsico activity does not correlate with the high amount of Rubsico protein present. These results imply that photosynthetic CO_2 fixation is modulated, in a large part, by an irreversible inhibition of the catalytic activity of Rubisco in the *Nostoc* associated with bryophytes.

The in vitro Rubisco activity and protein content that are essentially the same as free-living cultures do not correlate with the ex planta lack of photosynthetic CO_2 fixation in the cycad and *Gunnera* associations (Table 1). In *G. tinctoria*, photosystem II activity of the *Nostoc* symbiont is down-regulated by modification of the D1 protein (Black and Osborne 2004). This is the only association in which photosynthetic electron transport in the *Nostoc* partner has been examined and more of these essential studies need to be done. The discrepancy in the cycad associations has not yet been addressed.

The results with *Anabaena/Nostoc* from *Azolla* are particularly confusing. Estimates of Rubisco gene expression indicate that only about 10% of the mRNA is present in the symbiont compared to typical free-living cultures (Nierzwicki-Bauer and Haslekorn 1986). This result is not consistent with the relatively high rates of photosynthetic CO_2 fixation ex planta. Unfortunately, the in vitro Rubisco catalytic activity and amount of protein have not been examined in the *Anabaena/Nostoc* symbiont to resolve the inconsistencies. In addition, as noted above, the *Anabaena/Nostoc* contributes little photosynthate to the association and appears to be an obligate symbiont, unable to grow apart from the plant partner. What then is the selective pressure to retain the potential for a high rate of photosynthesis if it cannot be utilized?

Since most of these studies have been done using different *Nostoc* isolates, it is problematic to draw broad conclusions regarding different mechanisms for modulating the photosynthetic activities of the symbiont. A more optimal situation would employ the same symbiont in association with both the hornwort *A. punctatus* and *G. tinctoria*, for example *N. punctiforme* PCC 73102, to determine if Rubisco and photosystem II activities are down-regulated only in the hornwort and *Gunnera* association, respectively.

Nitrogen Metabolism

Symbiotically associated *Nostoc* spp. fix nitrogen at a higher rate than free-living cultures, consistent with a higher heterocyst frequency (Table 2). Since the symbiotic *Nostoc* grow slower than free-living cultures, the nitrogen fixed in excess of that required for their growth is likely lost as a metabolic waste product and made available to the plant partner. A questions is, how much and in what form? In all but the cycad associations, ammonium is released by the symbiotic *Nostoc* spp. and the amounts vary from 40% to 90% of the fixed nitrogen (Table 2). Cyanobacteria assimilate NH_4^+, obtained from the environment or derived from nitrate/nitrite, dinitrogen, urea or organic nitrogen, by the sequential activities of glutamine synthetase (GS) and glutamate synthase; glutamate dehydrogenase has no significant role (Flores and Herrero 1994). Thus, studies have focused on modulation of GS (encoded by *glnA*) activity or synthesis as a cause of NH_4^+ release.

Table 2 Nitrogen fixation and assimilation characteristics of *Nostoc* in symbiotic association with terrestrial plants (derived from Meeks 1998)

Association	Heterocyst frequency (%)	Nitrogenase specific activity	Nitrogen release as % of fixed	Glutamine synthetase activity	Glutamine synthetase protein
Free-living *Nostoc* spp.	5–10	2.7–6.3 nmol/min/mg protein	< 10 as organic N	0.85–1.8 nmol/min/mg protein	6.8–7.6 mg/g cell protein
Hornwort, *Anthoceros punctatus*[1]	25–45	23.5	80 as NH_4^+	15%	100%
Fern, *Azolla* spp.[2]	26–30	6.2	40 as NH_4^+	30%	38%
Cycad, *Cycas*, *Zamina*, *Macrozamia* spp.[3]	17–46	26.7	unknown as organic N	100%	100%
Angiosperm, *Gunnera* spp.[4]	20–60	24.8	90 as NH_4^+	70%	100%

Free-living *Nostoc* values were compiled from control experiments in the cited references. Nitrogenase activity is acetylene reduction. Glutamine synthetase symbiotic activity and protein values are given as a percent of free-living values. Activity is of the transferase reaction. In some cases, the original data were normalized to units of chlorophyll *a*; these values were converted to protein assuming that chlorophyll makes up 4% of the total cellular protein
References: 1. Stewart and Rodgers (1977), Joseph and Meeks (1987), Steinberg and Meeks (1991). 2. Hill (1975), Ray et al. (1978), Meeks et al. (1987), Lee et al. (1988). 3. Lindblad et al. (1985), Lindblad and Bergman (1986), Pate et al. (1988). 4. Söderbäck et al. (1990), Bergman et al. (1992), Silvester et al. (1996)

Symbiotic *Nostoc* sp., immediately isolated from association with *A. punctatus*, has 20% of the capacity to assimilate exogenous NH_4^+ as does a free-living culture and releases approximately 80% of its fixed nitrogen as NH_4^+ in the intact association (Meeks et al. 1985). The specific in vitro activity of GS in the symbiotic *Nostoc* is about 15% that of its free-living culture, but the amount of GS protein is comparable in both growth states (Joseph and Meeks 1987). These results imply irreversible inhibition of GS catalytic activity, similar to Rubisco, and are superficially consistent with impaired GS as a cause of NH_4^+ release. Such a casual relationship can be examined by a simple comparison. The rate of symbiotic nitrogen fixation in this association is 12.5 nmol of NH_4^+/min/mg *Nostoc* protein and the in vitro specific biosynthetic activity of GS is 19.8 nmol of NH_4^+ min/mg protein (Meeks 2003). Thus, there would appear to be sufficient NH_4^+ assimilation activity to sequester all of the N_2-derived NH_4^+. This conclusion could be negated by two conditions: substrates for GS are not at saturating concentrations in the symbiotic *Nostoc*; and the N_2-derived NH_4^+ is assimilated only in the heterocyst and, if not assimilated, may directly diffuse out the heterocyst and into the symbiotic cavity. Metabolite concentrations are unknown in symbiotic or free-living *Nostoc* spp., therefore, the first possibility cannot be adequately analyzed. However, even operating at half-saturation values (ca. 10 nmol/min/mg protein), the catalytic activity is not consistent with the assimilation of only 2.5 nmol of NH_4^+ (20% of the fixed NH_4^+). Walsby (2007) calculated that the heterocyst envelope is impermeable to gasses, and therefore solutes, except at the pole junctions with adjacent vegetative cells. Thus, N_2-derived NH_4^+ must be transferred to adjacent vegetative cells before it could be released into the symbiotic cavity. Therefore, assumptions that reduced GS activity in heterocysts per se could be responsible for NH_4^+ release do not take into account the assimilation of NH_4^+ in vegetative cells. Of course, any release could be a combination of factors.

A situation similar to that of the hornwort, with some variation, holds for *Anabaena/Nostoc* associated with *Azolla* (Table 2). Here 40% of the N_2-derived NH_4^+ is released (Meeks et al. 1987), in vitro GS activity is 30%, GS protein is about 38% (Lee et al. 1988) and the *glnA* mRNA concentration is about 10% (Nierzwicki-Bauer and Haselkorn 1986) of free-living cultures. These results imply that GS activity is modulated by synthesis of the protein, rather than inhibition of catalytic activity. Whether the lower amount of GS mRNA is a consequence of regulated transcription or simply of a less efficient promoter structure is unknown. A comparison of catalytic rates similar to that in the hornwort association yields an even greater potential for assimilation of all of the N_2-derived NH_4^+, implying that modulation of GS activity and level may not be the complete story of NH_4^+ release in *Azolla* spp.

There is a notable disconnection between release of N_2-derived NH_4^+ and in vitro GS activity in the *Gunnera* association. In this association, *Nostoc* sp. in vitro GS activity and protein are 70–100% of free-living cultures (Bergman et al. 1992), while 90% of the N_2-derived NH_4^+ is released (Silvester et al. 1996). The most direct explanation for NH_4^+ release in this case would be depletion of substrates and reactants for GS activity. Conversely, the cycad associations are documented to release organic nitrogen to the plant partner (Pate et al. 1988). The *Nostoc* in association with cycads expressed essentially 100% of the in vitro GS activity and protein as a free-living culture (Table 2). The conclusions one can draw from these studies are that the mechanisms of release of fixed nitrogen in symbiotic associations may well vary with the plant partner, but there is insufficient information to currently propose models. As in CO_2 fixation, it would be of interest to compare the same *Nostoc* sp. in the three associations that can be reconstituted, in which GS activity is and is not irreversibly modulated.

3.2.2
Heterocyst Differentiation

Apart from the aseriate growth form, the most dramatic morphological change in symbiotically associated *Nostoc* spp. is the increase in frequency of heterocysts in the filaments (Table 2). There are two aspects of symbiotic heterocyst differentiation to be considered: what is the signal for induction of differentiation; and what is the spacing pattern and will the pattern(s) reflect on regulatory aspects of differentiation.

Signal for Differentiation

The environmental signal for heterocyst differentiation in free-living cultures is limitation for combined nitrogen (Meeks and Elhai 2002). The limitation signal is thought to be perceived by an elevated cellular level of 2-oxoglutarate (Muro-Pastor et al. 2001; Vazquez-Bermudez et al. 2002). 2-Oxoglutarate activates a transcriptional regulator, NtcA, that modulates transcription of genes encoding proteins for the acquisition of sources of nitrogen alternative to NH_4^+ (Herrero et al. 2001). If no source of combined nitrogen is available, activated NtcA, directly or indirectly, stimulates transcription of regulators of heterocyst differentiation and subsequent fixation of N_2. These include the positive regulators HetR and HetF and the negative regulators PatN and PatS (Fig. 1) (Golden and Yoon 2003; Herrero et al. 2004; Meeks and Elhai 2002; Zhang et al. 2006).

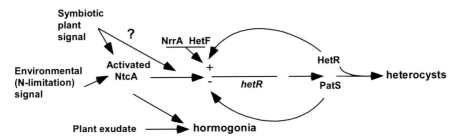

Fig. 1 Schematic of possible regulatory elements and pathways in free-living and plant partner induction of heterocyst and hormogonium differentiation

There are at least two lines of evidence that the same nitrogen limitation signal does not regulate symbiotic heterocyst differentiation. First, the symbiotic *Nostoc* colonies are immersed in a pool of N_2-derived NH_4^+ in the bryophyte, *Azolla* and *Gunnera* associations, yet heterocyst differentiation continues in these colonies. The concentration of NH_4^+ in the hornwort *A. punctatus* symbiotic cavity is calculated to be about 0.55 mM (Meeks 2003) and that in *Azolla* 0.8–6.0 mM (Canini et al. 1990). In free-living cultures, heterocyst differentiation is repressed by about 0.010 mM NH_4^+ (Meeks et al. 1982). Second, symbiotic *Nostoc* vegetative cells that differentiate into heterocysts do not show morphological signs of nitrogen limitation. Cyanobacteria store nitrogen as the multi-L-arginyl-poly(L-aspartic acid) non-protein polymer (cyanophycin), and as amino acids in phycobilisomes and carboxysomes (Rubisco) (Allen 1984). These polymers are catabolized in nitrogen-starved cells. All of these structures can be seen in symbiotic *Nostoc* vegetative cells, as well as cyanophycin and phycobilisomes in some heterocysts (Meeks 1990). These observations verify that the symbiotic *Nostoc* species assimilate sufficient N_2-derived NH_4^+ to supply their own metabolic needs.

The conclusion from these observations is that the plant partner produces an environment or chemical signal(s) that supersedes the nitrogen limitation signal. Genetic analyses of *N. punctiforme* in association with *A. punctatus* have established that the signal transduction pathway for symbiotic heterocyst differentiation includes functional HetR and HetF proteins, similar to the free-living state (Wong and Meeks 2002). Therefore, the symbiotic signal must enter the signal transduction pathway upstream from HetR/HetF (Fig. 1). NtcA and the response regulator NrrA are the only transcription factors currently known to operate upstream of HetR (Ehira and Ohmori 2006). Elevated transcription of *nrrA* alone does not lead to heterocyst differentiation, at least in filaments committed to hormogonium differentiation (Campbell et al. 2007). A mutation in *ntcA* in *N. punctiforme* leads to a hormogonium defective phenotype, which is also defective in symbiotic infection. Consequently, it cannot be determined whether a plant signal for symbiotic heterocyst differentiation bypasses NtcA. Characteristics of these *N. punctiforme* mutants in other plant associations have not been determined. The possibility of enhanced accumulation of 2-oxoglutarate in the heterotrophic symbiotic vegetative cells does not appear to be reasonable as an aberrant signal for nitrogen limitation. If this were to occur, the cells should respond as if they were nitrogen-limited and subsequently mobilize their nitrogen reserve polymers; symbiotic vegetative cells that differentiate into heterocysts do not appear to be nitrogen-limited.

Pattern of Spacing

The increase in heterocyst frequency leads to a disruption in the free-living spacing pattern of heterocysts in the filaments. The free-living pattern can be disrupted in two ways: (i) additional new heterocysts can appear in the vegetative cell interval between adjacent heterocysts, thereby shortening the interval (multiple singular heterocysts, Msh); or (ii) they may appear adjacent to existing heterocysts, yielding a cluster of heterocysts at specific sites in the filament (multiple contiguous heterocysts, Mch). A Mch pattern is commonly observed in mutants or genetic constructs during the establishment of pattern; *patS* (Yoon and Golden 1998, 2001) and *patU* (Meeks et al. 2002) deletions, and over expression of *hetR* (Buikema and Haselkorn 2001) and *hetF* (Wong and Meeks 2001), yield Mch. Mutation of *hetN* (Callahan and Buikema 2001) and *patB* (Jones et al. 2003) results in a Mch pattern, but the pattern appears in the second round of heterocyst differentiation, after the initial establishment of pattern. HetN and PatB are modeled as essential elements in the maintenance of pattern (Callahan and Buikema 2001; Jones et al. 2003).The Msh pattern of the *patN* mutant appears during establishment of pattern (Meeks et al. 2002); this is the only report of a Msh phenotype.

Due to the fragile nature of the filamentous growth pattern, the distorted size of vegetative cells and the frequent presence of phycobiliprotein-induced chlorophyll fluorescence in heterocysts, it is often difficult identify heterocysts and their spacing pattern in symbiosis (Meeks 1990, 1998). Nevertheless, both Msh and Mch patterns can be observed in light and electron micrographs (Meeks and Elhai 2002). The most persuasive light micrographs are of *Anabaena/Nostoc* in the *Azolla* association, taken by G. A. Peters, which depict a Msh pattern (Meeks and Elhai 2002). Examination of collages of electron micrographs of *Nostoc* in the *A. punctatus* association, taken within two months after reconstitution, yield an overwhelmingly Msh pattern as well (Meeks and Elhai 2002). However, the most definitive results come from analysis of experiments that compare the rates of nitrogen fixation and heterocyst frequency along the developmental gradient from plant meristematic to mature tissue in *Azolla* (Hill 1975), cycad (Lindblad et al. 1985) and *Gunnera* (Söderbäck et al. 1990). At the meristematic tip of the *Azolla* leaf, coralloid roots and *Gunnera* stems, the *Nostoc* filaments have a low to no frequency of heterocysts and nitrogen fixation is correspondingly low to absent. Progression away from the tip results in an essentially linear increase in both heterocyst frequency and rates of nitrogen fixation, and the heterocyst spacing reflects a Msh pattern. A distance is reached beyond which an inverse correlation appears between increasing heterocyst frequency and decreasing rates of nitrogen fixation. At this distance, and beyond, the Mch pattern becomes more prevalent. Heterocysts are terminally differentiated with a finite, but undefined, life span. The life span may well be longer in symbiosis compared to free-living growth and variable with respect to the life span of the plant partner.

We suggest that the symbiotic Mch pattern arises as a consequence of new heterocysts differentiating adjacent to a functionally dead heterocyst; confinement in the symbiotic cavity results in the non-functional heterocyst remaining in place rather than being detached and lost from the filament

(Meeks and Elhai 2002). If this analysis is correct, then the Msh pattern arises as a function of time and may be more a consequence of disruption of the maintenance, than of the establishment, of the spacing pattern. Maintenance of the pattern is the more difficult process to study in a randomly dividing population of filaments. Since alteration of the maintenance assigned genes *hetN* and *patB* result in Mch, the symbiotic pattern must be a consequence of the symbiotic manipulation of other, as yet unknown, genes and gene products involved in maintenance of the free-living pattern. Although *patN* is a possible candidate, its spacing pattern in free-living culture emerges during the establishment of pattern.

4
Conclusions and Future Perspectives

What conclusions can be drawn from the specificity and physiological adaptation results with respect to providing a foundation for future studies?

4.1
Specificity

The studies collectively indicate that a large (but how large is unknown) number of *Nostoc* species or strains establish symbiotic associations with phylogenetically and morphologically distinct plant partners. It is possible that some strains are specific for a plant partner, but most appear to be non-specific. Conversely, some free-living *Nostoc* strains do not form symbiotic associations, at least under benign laboratory culture conditions (Meeks 1990). Although symbiotic competence does not appear to be a universal characteristic of the genus *Nostoc*, there must be a genetic basis for expression of such competence. It is difficult, however, to imagine that each symbiotically competent *Nostoc* species or strain contains genetic information that allows it to interact uniquely with the various plant partners.

There is no reason to assume that bryophytes waited for at least 400 million years until emergence of the Gunneraceae to take advantage of the nitrogen fixation activity of heterocyst-forming cyanobacteria, unless other selective pressures are involved in establishment of the association. Because nitrogen-fixing cyanobacterial associations are actually narrowly distributed in the four major terrestrial plant lineages, it also does not seem reasonable that the genetic information of symbiotic competence was linearly transferred from hornworts through ferns and cycads to *Gunnera*. Therefore, one may logically deduce that each plant group learned how to manipulate properties of susceptible *Nostoc* spp. in the adjacent wet soil and that the regulatory signals targeting a specific *Nostoc* physiological characteristic need not be the same in each association.

4.2
Physiological Adaptation in the Associations

All plant partners appear to control the infection process by producing hormogonium-inducing factors and perhaps chemoattractants. However, details on the identities of the factors and attractants, and, thus, the variety of signal receptors and mechanistic responses are unknown. The act of infection appears to vary depending on the plant structures to be colonized. Once colonized, the symbiotic cavities in the *Azolla*, cycad, and *Gunnera* partners appear to be closed off from the external environment, including the presence of HIF. Therefore, perhaps only the bryophyte partners may need to produce hormogonium-repressing factors. If this is so, it is of interest that *N. punctiforme* ATCC 29133, isolated from the cycad *Macrozamia* sp, retains the genetic information to synthesize a hormogonium-repressing metabolite. This observation verifies the broad symbiotic competence of *N. punctiforme* and indicates the presence of an unknown selective pressure to retain that information.

Although growth control of *Nostoc* appears to be an essential aspect of the plant associations, nothing is known how it is achieved. Assimilatory phototrophic carbon and nitrogen metabolism by the *Nostoc* partner are depressed in all associations, but they appear to be depressed in proportion to a decreased growth rate. The different plant partners appear to target different reactions in at least the control of photosynthetic CO_2 fixation. Thus, plants may employ different mechanisms to achieve a similar physiological state of the same *Nostoc* in various associations. A causal relationship between growth and metabolism has not been established. Heterocyst differentiation is enhanced in the symbiotic growth state leading to higher rates of nitrogen fixation and release of fixed nitrogen as NH_4^+ in all but one association; the nitrogen fixation is primarily fueled by photosynthate from the plant partner. The enhanced heterocyst frequency yields a Msh spacing pattern in the most highly functional tissues of the plant partner. The Msh pattern may result from an alternation in the mechanisms that regulate the maintenance, rather than the establishment, of the heterocyst spacing pattern in the free-living growth state. Thus, knowledge of the maintenance mechanisms is essential in understanding plant-dependent enhancement of heterocyst differentiation.

4.3
Perspectives

Nitrogen-fixing *Nostoc* spp. offer the potential for genetic engineering of crop plants for symbiotic nitrogen fixation. *Nostoc* spp. carry their own oxygen protective mechanisms and, thus, inflict less of a burden on a plant partner to develop specialized structures in formation of a functional association. Considerable progress has been made in defining the physiological characteristic of the *Nostoc* spp. as they establish a functional association with their evolved plant partners. Ambiguities in physiological adaptation mechanisms could be resolved by using the same *Nostoc* isolate in various laboratory reconstituted associations. Advances in this area will also be facilitated by knowing the genome sequence of a genetically tractable symbiotic strain, *N. punctiforme* ATCC 29133. I hypothesize that the plant partners have, in fact, evolved different mechanisms of growth and metabolic control over the *Nostoc* symbiont to achieve a stable commensal relationship. Therefore, more effort needs to be expended in characterization of the plant partners. *Gunnera* species have now been genetically transformed (Wan-Ling Chiu, Virginia Commonwealth University, personal communication) and this could open exciting new avenues of research in identifying plant signals and in engineering new plant–*Nostoc* nitrogen-fixing associations.

Acknowledgements Work in the author's laboratory on the genomics of *N. punctiforme* ATCC 29133 is supported by the USA National Science Foundation, grant EF-0317104.

References

Adams DG (2000) Symbiotic interactions. In: Whitton BA, Potts M (eds) The ecology of cyanobacteria, their diversity in time and space. Kluwer, Dordrecht, pp 523–561

Allen MM (1984) Cyanobacterial cell inclusions. Ann Rev Microbiol 38:1–25

Bergman B (2002) The *Nostoc-Gunnera* symbiosis. In: Rai A, Bergman B, Rasmussen U (eds) Cyanobacteria in symbiosis. Kluwer, Dordrecht, pp 207–232

Bergman B, Matveyev A, Rasmussen U (1996) Chemical signaling in cyanobacterial–plant symbioses. Trends Plant Sci 1:191–197

Bergman B, Rai AN, Johansson C, Söderbäck E (1992) Cyanobacterial–plant symbiosis. Symbiosis 14:61–81

Black K, Osborne B (2004) An assessment of photosynthetic downregulation in cyanobacteria from the *Gunnera-Nostoc* symbiosis. New Phytol 162:125–132

Buikema WJ, Haselkorn R (2001) Expression of the *Anabaena hetR* gene from a copper-regulated promoter leads to heterocyst differentiation under repressing conditions. Proc Natl Acad Sci USA 98:2729–2734

Callahan SM, Buikema WJ (2001) The role of HetN in maintenance of the heterocyst pattern in *Anabaena* sp. PCC 7120. Mol Microbiol 40:941–950

Campbell EL, Meeks JC (1989) Characteristics of hormogonia formation by symbiotic *Nostoc* spp. in response to the presence of *Anthoceros punctatus* or its extracellular products. Appl Environ Microbiol 55:125–131

Campbell EL, Summers ML, Christman H, Martin ME, Meeks JC (2007) Global gene expression patterns of *Nostoc punctiforme* in steady state dinitrogen-grown heterocyst-containing cultures, and at single time points during the differentiation of akinetes and hormogonia. J Bacteriol 189:5247–5256

Campbell EL, Wong FCY, Meeks JC (2003) DNA binding properties of the HrmR protein of *Nostoc punctiforme* responsible for transcriptional regulation of genes involved in hormogonium differentiation. Mol Microbiol 47:573–582

Canini A, Grilli Caiola M, Mascini M (1990) Ammonium content, nitrogenase activity and heterocyst frequency with the leaf cavities of *Azolla filiculoides* Lam. FEMS Microbiol Lett 71:205–210

Carrapico F, Costa MH, Costa ML, Teixeira G, Frazão AA, Santos MCR, Baioa MV (1996) The uncontrolled growth of *Azolla* in the Guadiana River. Aquaphyte 16:11

Castenholz RW (2001) Phylux BX. Cyanobacteria oxygenic photosynthetic bacteria. In: Boone DR, Castenholz RW (eds) Bergey's manual of systematic bacteriology, vol 1: the Archaea and the deeply branching and phototrophic bacteria, 2nd edn. Springer, Berlin Heidelberg New York, pp 473–599

Cohen MF, Meeks JC (1996) A hormogonium regulating locus, *hrmUA*, of the cyanobacterium *Nostoc punctiforme* strain ATCC 29133 and its response to an extract of a symbiotic plant partner *Anthoceros punctatus*. Mol Plant-Microbe Interact 10:280–289

Cohen MF, Wallis JG, Campbell EL, Meeks JC (1994) Transposon mutagenesis of *Nostoc* sp strain ATCC 29133, a filamentous cyanobacterium with multiple cellular differentiation alternatives. Microbiology 140:3233–3240

Cohen MF, Yamasaki H (2000) Flavonoid-induced expression of a symbiosis-related gene in the cyanobacterium *Nostoc punctiforme*. J Bacteriol 182:4644–4646

Costa J-L, Lindblad P (2002) Cyanobacteria in symbiosis with cycads. In: Rai A, Bergman B, Rasmussen U (eds) Cyanobacteria in symbiosis. Kluwer, Dordrecht, pp 195–205

Costa J-L, Paulsrud P, Rikkinen J, Lindblad P (2001) Genetic diversity of *Nostoc* symbionts endophytically associated with two bryophyte species. Appl Environ Microbiol 67:4393–4396

Ehira S, Ohmori M (2006) NrrA, a nitrogen-responsive response regulator facilitates heterocyst development in the cyanobacterium *Anabaena* sp. strain PCC 7120. Mol Microbiol 59:1692–1703

Enderlin CS, Meeks JC (1983) Pure culture and reconstitution of the *Anthoceros-Nostoc* symbiotic association. Planta 158:157–165

Fay P (1992) Oxygen relations of nitrogen fixation in cyanobacteria. Microbiol Rev 56:340–373

Flores E, Herrero A (1994) Assimilatory nitrogen metabolism. In: Bryant DA (ed) The molecular biology of cyanobacteria. Kluwer, Dordrecht, pp 487–517

Gallon JR (1992) Reconciling the incompatible: N_2 fixation and O_2. New Phytol 122:571–609

Gantar M, Kerby NW, Rowell P (1993) Colonization of wheat (*Triticum vulgare* L.) by N_2-fixing cyanobacteria; III. The role of a hormogonium-promoting factor. New Phytol 124:505–513

Golden JW, Yoon HS (2003) Heterocyst development in *Anabaena*. Curr Opin Microbiol 6:557–563

Herdman M, Rippka R (1988) Cellular differentiation: hormogonia and baeocytes. Methods Enzymol 167:232–242

Herrero A, Muro-Pastor AM, Flores E (2001) Nitrogen control in cyanobacteria. J Bacteriol 183:411–425

Herrero A, Muro-Pastor AM, Valladares A, Flores E (2004) Cellular differentiation and the NtcA trnscription factor in filamentous cyanobacteria. FEMS Microbiol Rev 28:469–487

Hill DJ (1975) The pattern of development of *Anabaena* in the *Azolla-Anabaena* symbiosis. Planta 122:179–184

Johansson C, Bergman B (1992) Early events during the establishment of the *Gunnera/Nostoc* symbiosis. Planta 188:403–413

Johansson C, Bergman B (1994) Reconstitution of the symbiosis of *Gunnera mannicata* Linden: cyanobacterial specificity. New Phytol 126:643–652

Jones KM, Buikema WJ, Haselkorn R (2003) Heterocyst-specific expression of *patB*, a gene required for nitrogen fixation in *Anabaena* sp. strain PCC 7120. J Bacteriol 185:2306–2314

Joseph CM, Meeks JC (1987) Regulation of expression of glutamine synthetase in a symbiotic *Nostoc* strain associated with *Anthoceros punctatus*. J Bacteriol 169:2471–2475

Kaplan D, Calvert HE, Peters GA (1986) The *Azolla-Anabaena* relationship. VII. Nitrogenase activity and phycobiliproteins of the endophyte as a function of leaf age and cell type. Plant Physiol 80:884–890

Kaplan D, Peters GA (1988) Interaction of carbon metabolism in the *Azolla-Anabaena* symbiosis. Symbiosis 6:53–68

Kimura J, Nakano T (1990) Reconstitution of a *Blasia-Nostoc* symbiotic association under axenic conditions. Nova Hedwigia 50:191–200

Knight CD, Adams DG (1996) A method for studying chemotaxis in nitrogen fixing cyanobacterium–plant symbioses. Physiol Mol Plant Path 49:73–77

Lazaroff N (1973) Photomorphogenesis and Nostocacean development. In: Carr NG, Whitton BA (eds) The biology of blue-green algae. Blackwell, Oxford, pp 279–329

Lee K-Y, Joseph CM, Meeks JC (1988) Glutamine synthetase specific activity and protein concentration in symbiotic *Anabaena* associated with *Azolla caroliniana*. Antonie van Leeuwenhoek 54:345–355

Lindblad P, Bergman B (1986) Glutamine synthetase: activity and localization in cyanobacteria of the cycads *Cycas revoluta* and *Zamia skinneri*. Plant 169:1–7

Lindblad P, Hällbom L, Bergman B (1985) The cyanobacterium-*Zamia* symbiosis: C_2H_2-reduction and heterocyst frequency. Symbiosis 1:19–28

Lindblad P, Rai AN, Bergman B (1987) The *Cycas revoluta-Nostoc* symbiosis: Enzyme activities of nitrogen and carbon metabolism in the cyanobiont. J Gen Microbiol 133:1695–1699

Meeks JC (1990) Cyanobacterial-bryophyte associations. In: Rai AN (ed) Handbook of symbiotic cyanobacteria. CRC, Boca Raton, FL, pp 43–63

Meeks JC (1998) Symbiosis between nitrogen-fixing cyanobacteria and plants. BioScience 48:266–276

Meeks JC (2003) Symbiotic interactions between *Nostoc punctiforme*, a multicellular cyanobacterium, and the hornwort *Anthoceros punctatus*. Symbiosis 34:55–71

Meeks JC (2005a) Molecular mechanisms in the nitrogen-fixing *Nostoc*-Bryophyte symbiosis. In: Overmann J (ed) Molecular basis of symbiosis. Springer, Berlin Heidelberg New York, pp 165–196

Meeks JC (2005b) The genome of the filamentous cyanobacterium *Nostoc punctiforme*, what can we learn from it about free-living and symbiotic nitrogen fixation? In: Palacios R, Newton WE (eds) Nitrogen fixation: 1888–2001, vol VI: Genomes and genomics of nitrogen-fixing organisms. Springer, Berlin Heidelberg New York, pp 27–70

Meeks JC, Campbell EL, Summers ML, Wong FC (2002) Cellular differentiation in the cyanobacterium *Nostoc punctiforme*. Arch Microbiol 178:395–403

Meeks JC, Elhai J (2002) Regulation of cellular differentiation in filamentous cyanobacteria in free-living and plant associated symbiotic growth states. Microbiol Mol Biol Rev 65:94–121

Meeks JC, Elhai J, Thiel T, Potts M, Larimer F, Lamerdin J, Predki P, Atlas R (2001) An overview of the genome of *Nostoc punctiforme*, a multicellular, symbiotic cyanobacterium. Photosyn Res 70:85–106

Meeks JC, Enderlin CS, Joseph CM, Chapman JS, Lollar MWL (1985). Fixation of [^{13}N]N$_2$ and transfer of fixed nitrogen in the *Anthoceros-Nostoc* association. Planta 164:406–414

Meeks JC, Joseph CM, Haselkorn R (1988) Organization of the *nif* genes in cyanobacteria in symbiotic association with *Azolla* and *Anthoceros*. Arch Microbiol 150:61–71

Meeks JC, Steinberg N, Enderlin CS, Joseph CN, Peters GA (1987) *Azolla-Anabaena* relationship. XIII. Fixation of [^{13}N]N$_2$. Plant Physiol 84:883–886

Meeks JC, Wycoff KL, Chapman JS, Enderlin CS (1982) Regulation of expression of nitrate and dinitrogen assimilation by *Anabaena* species. Appl Environ Microbiol 45:1351–1359

Mollenhauer D, Mollenhauer R, Kluge M (1996) Studies on initiation and development of the partner association in *Geosiphon pyriforme* (Kutz.) v. Wettstein, a unique encocytobiotic system of a fungus (Glomales) and the cyanobacterium *Nostoc punctiforme* (Kutz.). Hariot Protoplasma 193:3–9

Muro-Pastor MI, Reyes HC, Florencio FJ (2001) Cyanobacteria perceive nitrogen status by sensing intracellular 2-oxoglutarate levels. J Biol Chem 276:38320–38328

Nierzwicki-Bauer SA, Haselkorn R (1986) Differences in mRNA levels in *Anabaena* living freely or in symbiotic association with *Azolla*. EMBO J 5:29–35

Nilsson M, Rasmussen U, Bergman B (2006) Cyanobacterial chemotaxis to extracts of host and nonhost plants. FEMS Microbiol Ecol 55:382–390

Ow MC, Gantar M, Elhai J (1999) Reconstitution of a cycad-cyanobacterial association. Symbiosis 27:125–134

Pate JS, Lindblad P, Atkins CA (1988) Pathways of assimilation and transfer of fixed nitrogen in coralloid roots of cycad-*Nostoc* symbioses. Planta 176:461–471

Paulsrud P, Rikkinen J, Lindblad P (1998) Cyanobiont specificity in some *Nostoc*-containing lichens and in a *Peltigera aphthosa* photosymbiodeme. New Phytol 139:517–524

Peters GA, Meeks JC (1989) The *Azolla-Anabaena* symbiosis: basic biology. Annu Rev Plant Physiol Plant Mol Biol 40:193–210

Peters GA, Toia RE Jr, Evans WR, Crist DK, Mayne BC, Poole RE (1980) Characterization and comparisons of five N$_2$-fixing *Azolla-Anabaena* associations, I. Optimization of growth conditions for biomass increase and N content in a controlled environment. Plant Cell Environ 3:261–269

Rai AN, Bergman B, Rasmussen U (2002) Cyanobacteria in symbiosis. Kluwer, Dordrecht

Rai AN, Söderbäck E, Bergman B (2000) Cyanobacterium—plant symbioses. New Phytol 147:449–481

Rasmussen U, Johansson C, Bergman B (1994) Early communication in the *Gunnera-Nostoc* symbiosis: plant induced cell differentiation and protein synthesis in the cyanobacterium. Mol Plant-Microbe Interact 7:696–702

Rasmussen U, Nilsson M (2002) Cyanobacterial diversity and specificity in plant symbioses. In: Rai A, Bergman B, Rasmussen U (eds) Cyanobacteria in symbiosis. Kluwer, Dordrecht, pp 313–328

Rasmussen U, Svenning MM (2001) Characterization by genotypic methods of symbiotic *Nostoc* strains isolated from five species of *Gunnera*. Arch Microbiol 176:204–210

Raven JA (2002) Evolution of cyanobacterial symbioses. In: Rai A, Bergman B, Rasmussen U (eds) Cyanobacteria in symbiosis. Kluwer, Dordrecht, pp 329–346

Ray TB, Peters GA, Toia RE Jr, Mayne BC (1978) *Azolla-Anabaena* relationship. VII. Distribution of ammonia-assimilating enzymes, protein and chlorophyll between host and symbiont. Plant Physiol 62:463–467

Ray TB, Mayne BC, Toia RE Jr, Peters GA (1979) *Azolla-Anabaena* relationship. VIII. Photosynthetic characterization of the association and individual partners. Plant Physiol 64:791–795

Rippka R, Deruelles J, Waterbury JB, Herdman M, Stanier RY (1979) Generic assignments, strain histories and properties of pure cultures of cyanobacteria. J Gen Microbiol 111:1–61

Rippka R, Herdman M (1992) Pasteur culture collection of cyanobacteria in axenic culture. Institute Pasteur, Paris

Schopf JW (2000) The fossil record: tracing the roots of the cyanobacterial lineage. In: Whitton BA, Potts M (eds) The ecology of cyanobacteria, their diversity in time and space. Kluwer, Dordrecht, pp 13–35

Silvester WB, Parsons R, Watt PW (1996) Direct measurement of release and assimilation of ammonia in the *Gunnera-Nostoc* symbiosis. New Phytol 132:617–625

Söderbäck E, Bergman B (1992) The *Nostoc-Gunnera magellanica* symbiosis: Phycobiliproteins, carboxysomes and rubisco in the cyanobiont. Physiol Planta 84:425–432

Söderbäck E, Bergman B (1993) The *Nostoc-Gunnera* symbiosis: Carbon fixation and translocation. Physiol Planta 89:125–132

Söderbäck E, Lindblad P, Bergman B (1990) Developmental patterns related to nitrogen fixation in the *Nostoc-Gunnera magellanica* Lam. symbiosis. Planta 182:355–362

Steinberg NA, Meeks JC (1989) Photosynthetic CO_2 fixation and ribulose bisphosphate carboxylase/oxygenase activity of *Nostoc* sp. strain UCD 7801 in symbiotic association with *Anthoceros punctatus*. J Bacteriol 171:6227–6233

Steinberg NA, Meeks JC (1991) Physiological sources of reductant for nitrogen fixation activity in *Nostoc* sp. strain UCD 7801 in symbiotic association with *Anthoceros punctatus*. J Bacteriol 173:7324–7329

Stewart WDP, Rodgers GA (1977) The cyanophyte-hepatic symbiosis II. Nitrogen fixation and the interchange of nitrogen and carbon. New Phytol 78:459–471

Summers ML, Wallis JG, Campbell EL, Meeks JC (1995) Genetic evidence of a major role for glucose-6-phosphate dehydrogenase in nitrogen fixation and dark growth of the cyanobacterium/*Nostoc*/sp. strain ATCC 29133. J Bacteriol 177:6184–6194

Tandeau de Marsac N (1994) Differentiation of hormogonia and relationships with other biological processes. In: Bryant DA (ed) The molecular biology of cyanobacteria. Kluwer, Boston, pp 825–842

Tang LF, Watanabe I, Liu CC (1990) Limited multiplication of symbiotic cyanobacteria of *Azolla* spp. on artificial media. Appl Environ Microbiol 56:3623–3626

Tomitani A, Knoll AH, Cavanaugh CM, Ohno T (2006) The evolutionary diversification of cyanobacteria: molecular-phylogenetic and paleontological perspectives. Proc Natl Acad Sci USA 103:5442–5447

Vazquez-Bermudez MF, Herrero A, Flores E (2002) 2-Oxoglutarate increases the binding affinity of the NtcA (nitrogen control) transcription factor for the *Synechococcus glnA* promoter. FEBS Lett 512:71–74

Walsby AE (2007) Cyanobacterial heterocysts: terminal pores proposed as sites of gas exchange. Trends Microbiol 15:340–349

West N, Adams DG (1997) Phenotypic and genotypic comparisons of symbiotic and free-living cyanobacteria from a single field site. Appl Environ Microbiol 63:4479–4484

Wolk CP, Ernst A, Elhai J (1994) Heterocyst metabolism and development. In: Bryant DA (ed) The molecular biology of cyanobacteria. Kluwer, Dordrecht, pp 769–823

Wong FC, Meeks JC (2001) The *hetF* gene product is essential to heterocyst differentiation and affects HetR function in the cyanobacterium *Nostoc punctiforme*. J Bacteriol 183:2654–2661

Wong FC, Meeks JC (2002) Establishment of a functional symbiosis between the cyanobacterium *Nostoc punctiforme* and the bryophyte hornwort *Anthoceros punctatus* requires genes involved in nitrogen control and initiation of heterocyst differentiation. Microbiol 148:315–323

Yoon H-S, Golden JW (1998) Heterocyst pattern formation controlled by a diffusible peptide. Science 282:935–938

Yoon H-S, Golden JW (2001) PatS and products of nitrogen fixation control heterocyst pattern. J Bacteriol 183:2605–2613

Zhang C-C, Laurent S, Sakr S, Peng L, Bédu S (2006) Heterocyst differentiation and pattern formation in cyanobacteria: a chorus of signals. Mol Microbiol 59:367–375

Why Does *Gunnera* Do It and Other Angiosperms Don't? An Evolutionary Perspective on the *Gunnera–Nostoc* Symbiosis

Bruce Osborne[1] (✉) · Birgitta Bergman[2]

[1]UCD School of Biology and Environmental Science, University College Dublin, Belfield, Dublin 4, Ireland
Bruce.Osborne@ucd.ie

[2]Department of Botany, University of Stockholm, SE-10691 Stockholm, Sweden

1	Introduction	208
2	The Symbiotic Background	208
3	The Symbiotic Basics	210
4	A More Significant Past?	211
5	Evolutionary Drivers	212
6	Host Factors	215
7	Cyanobiont Factors	218
8	Conclusions	219
	References	221

Abstract The *Gunnera–Nostoc* symbiosis is an enigmatic plant–cyanobacterial symbiosis: the only known angiosperm–cyanobacterial symbiosis. We postulate that this symbiosis, together with perhaps all other plant–cyanobacterial symbioses, was more important in the geological past and was a response to a unique suite of environmental conditions that are uncommon today. Phylogenetic analyses indicate a distinct origin for the evolution of the *Gunnera–Nostoc* symbiosis within the angiosperms, although we suggest that this symbiosis may share more common features with both rhizobial and *Frankia* symbioses than might have been expected. Whilst we can only speculate on the evolutionary drivers that led to the establishment of the *Gunnera–Nostoc* symbiosis, there is plausible evidence that this could have been related to low oxygen-induced nitrogen deficiency. There is even evidence that low oxygen conditions can induce a number of factors that could be related to the establishment of plant–bacterial symbioses (rhizobia and *Frankia*). A particularly important goal for the future is the identification of the origin and development of specialised *Gunnera* glands, which are the conduit through which cyanobacteria enter cortical tissue. Currently we have little understanding of the functional or evolutionary significance of this structure. Far from being glands, in the strictest sense, there is evidence for an origin associated with adventitious root formation, a feature that also has parallels with nodule formation in legumes. This requires more detailed investigation.

1
Introduction

The *Gunnera–Nostoc* symbiosis is still the only known angiosperm–cyanobacterial nitrogen-fixing symbiosis. Whilst there has, in our view, been no systematic attempt at examining material of closely related genera and species as likely candidates for forming symbioses, and other angiosperm–cyanobacterial symbioses may await discovery, it is also possible that this will remain a unique angiosperm–cyanobacterial partnership. In particular, any systematic examination of possible contenders requires a better understanding of the phylogenetic relationships of *Gunnera* with other angiosperms. Phylogenetic analyses, based on molecular data, place *Gunnera* in its own family, the Gunneraceae, within the core eudicot clade, in the order Gunnerales (Soltis et al. 2005). This order, or subclade, includes only the Myrothamnaceae, a family of xerophytic shrubs with only two species, where there is no evidence of associations with cyanobacteria (Wilkinson 2000). Based on morphological/structural criteria (see Soltis et al. 2005), this relationship is at best equivocal; in terms of ecological relationships it is non-existent, with the dry arid environments occupied by *Myrothamnus* in marked contrast to the waterlogged or high rainfall habitats occupied by all members of the Gunneraceae (Osborne and Sprent 2002). This information tends to further emphasise the uniqueness of *Gunnera* and its cyanobacterial symbiosis, even within the same order (Gunnerales). Clearly, however, these relationships may change after further analyses. There is some evidence for an even closer relationship between the Gunneraceae and Myrthamnaceae that might not justify the existence of two separate families (see Soltis et al. 2005).

The question is why is this an apparently unique relationship? Whilst we cannot provide a direct answer to this question, we believe it is worthwhile to explore some of the factors that make *Gunnera* so unique and to examine possible reasons for its evolution, from both a host and symbiont (cyanobiont) perspective. This may provide clues as to its origin and form the basis of future investigations.

Initially, however, we examine the *Gunnera–Nostoc* symbiosis in relation to other bacterial or cyanobacterial N_2-fixing symbiosis in order to place *Gunnera* and its symbiosis into some context, both within the angiosperms and within the plant kingdom in general.

2
The Symbiotic Background

Nitrogen-fixing symbioses with angiosperms involving bacteria (*Frankia* strains or rhizobia) or cyanobacteria are rather uncommon in nature (see

Sprent 2001, 2006; Soltis et al. 2005), despite the widespread belief that they have the potential to confer significant nutritional advantages in many natural ecosystems. Whilst cyanobacterial symbioses with plants are widespread (Rai et al. 2000; Adams et al. 2006; Bergman et al. 2007) they are poorly represented within the major plant groups; only in the gymnosperms do they make a significant contribution (Vessey et al. 2004) and then only because of the large number of cycad species (Table 1).

Given the complex signalling systems and structural modifications that generally have to occur in both host and symbiont in order to form and maintain a functional N_2-fixing system (see Oldroyd et al. 2005; Barnett and Fisher 2006; Adams et al. 2006; Bergman et al. 2007) this may preclude the widespread development of symbiotic partnerships. Although some symbiosis may work without the "recognised" signalling system they may use other, as yet unidentified, mechanisms (Sprent 2008). We may also be rather naïve to assume that symbiosis involves mutually beneficial partnerships; there is now convincing evidence for a more complex story with exploitation by one partner, without appropriate reciprocation, being perhaps the common feature (Osborne 2007). This may require us to look at symbioses and the establishment of symbioses from a rather different perspective. In nature, the long-term exposure to a range of habitat factors, including low inoculum potential, less than optimal environmental conditions, such as water deficits (Sprent and Sprent 1990), or the absence of any advantage or even a penalty associated with partnership formation, could, in the long term, have resulted in the breakdown of any earlier associations. Were plant–cyanobacterial symbioses more common in the past and have they been lost during the course of evolution? Before turning to this question we provide some more specific details on the *Gunnera–Nostoc* symbiosis in order to provide the necessary grounding required for understanding many of the points discussed in sub-

Table 1 Estimated proportion of plant species relative to putative nitrogen-fixing cyanobacterial symbioses (based on information in Norstog and Nicholls 1997; Page 1997; Rai et al. 2002; Soltis et al. 2005; Jones 2002)

Plant group	Plant species : Putative N fixers	Comments
Bryophytes	~40 : 1	Probably many more epiphytic associations. Common in Hornworts (4 out of 6 genera)
Pteridophytes	>1500 : 1	*Azolla* the only representative?
Gymnosperms	>3 : 1	All belong to the cycads
Angiosperms	>4000 : 1	All belong to *Gunnera*[a]

[a] Proportion of angiosperms : nodulated legumes = 85 : 1

sequent sections. For more detailed reviews on the biology and ecology of the *Gunnera–Nostoc* symbiosis, see Bergman (2002) and Osborne and Sprent (2002). For recent information on plant–cyanobacterial symbioses in general, see Adams et al. (2006) and Bergman et al. (2007).

3
The Symbiotic Basics

Some of the basic features of the *Gunnera–Nostoc* symbiosis are detailed in Fig. 1. Establishment of the symbiosis is associated with shoot-associated "glands" (Figs. 2 and 3) formed at right angles to successive petioles and associated at later stages of development with adventitious roots (shoot-borne roots; see Barlow 1986 for a definition of similar root structures). These glands are the only known entry point for motile cyanobacteria (hormogonia) which are then able to move through the gland, finally penetrating cortical tissue (Towata 1985a; Osborne et al. 1991; Bergman et al. 1992; Bergman 2002). Morphological and biochemical changes in the cyanobacteria occur after entry and motility is lost, resulting in the formation of a very effective N-fixing partnership (Osborne et al. 1992) supported by carbon transfer from the host (Black et al. 2002). Rather surprisingly, however, most of the photosynthetic components are retained despite the dark heterotrophic conditions associated with the cyanobacterial colonies, and despite any evidence for photosynthetic carbon assimilation (Söderbäck and Bergman 1992; Black and Osborne 2004). It has been proposed that this is due, in part, to the modification of the D1 protein associated with PSII (Black and Osborne 2004). In mature plants the cyanobacterial colonies are associated with well-defined areas of host tissue, often with multiple populations, particularly in the larger species that have large rhizomes (see Osborne et al. 1991). The cyanobacterial colonies present in host tissue show various stages of development, ranging from young, metabolically active cells to older, often senescent populations (see Towata 1985b; Söderbäck et al. 1990; Black and Osborne 2004; Wang et al. 2004; Ekman et al. 2006), indicating that they were formed through successive and separate infection processes involving different glands and probably have a finite existence within host tissue. The ultimate fate of the cyanobacterial colonies in host tissue is, however, not known. Examination of mature plants of *G. tinctoria* growing in the field suggests that these are persistent within rhizomatous tissue and retained until these structures decompose (Osborne, unpublished results). There is also evidence of a close association between gland initiation and the ultimate location of the cyanobacterial colonies, suggesting that the presence of the gland in some way determines the location and extent of cyanobacterial infection in host tissue.

To date all examined *Gunnera* species have been shown to form cyanobacterial partnerships and, unlike rhizobia or *Frankia* symbioses, this appears to

> - Only known angiosperm cyanobacterial N_2-fixing symbiosis
> - Symbiotic cyanobacteria (cyanobionts) belong to genus *Nostoc*
> - Cyanobionts occur intracellularly within host tissue
> - Cyanobionts characterised intracellularly by morphological changes, including a greatly enhanced heterocyst production
> - Cyanobionts can supply all of the nitrogen requirements of the host plant
> - Establishment of the symbiosis associated with small stem glands and requires the formation of motile hormogonia
> - Considerable variation in host size, ranging from giant herbs to small stoloniferous plants; one annual species, *G. herteri*
> - All species associated with wet or high rainfall habitats

Fig. 1 Basic facts about the *Gunnera* symbiosis

be unaffected by combined nitrogen availability (Osborne et al. 1992). Ecologically an ability to assimilate N_2 may be an advantage in the largely wet or high rainfall habitats in which *Gunnera* grows, given the restricted decomposition of organic material, but this trait seems to be of minor importance given that *Gunnera* is not a common feature of wetland habitats worldwide (Osborne and Sprent 2002).

4
A More Significant Past?

All extant plant–cyanobacterial symbioses are associated with hosts with a long and largely continuous geological history, ranging from bryophytes/ferns ~400–440 MYA, through the cycads >250 MYA, to the most recent, *Gunnera*, ~90 MYA (see also Usher et al. 2007). In contrast, N_2-fixing associations involving bacteria are more recent at ~60 MYA (Schrire et al. 2005; Sprent 2006, 2008). Based on the globally dispersed occurrence of macrofossils of cycads, and the presence of *Gunnera* pollen in Cretaceous sediments in North America, where it accounted for 25–36% of the total pollen, these symbioses comprised a significant component of the flora where they occurred (see Osborne and Sprent 2002). Today, however, both cycads and *Gunnera* have a very restricted distribution (Osborne and Sprent 2002). Globally extensive and widespread *Azolla* deposits have also been described in the Cretaceous, around 100 MYA (Hall and Swanson 1968) and, more recently, at the Palaeocene/Eocene thermal maximum 50–55 MYA (Brinkhuis et al. 2006; Sluijs et al. 2006). Today there are around six species of *Azolla* (Lechno-Yossef and Nierzwicki-Bauer 2002; Evrard and van Hove 2002), whilst ~30 species have been described from Cretaceous deposits (Hall and Swanson 1968), representing a significant decrease in species diversity over the last ~100 MY. Clearly, there is plausible evidence that all of these symbioses constituted a significant component of the landscape during particular geological periods up to even the recent past, but are no longer so successful today. But why was *Gunnera*, in particular, more important in the past and why did the significance of this genus decline? To address this question we look at the possible environmental factors that may have been important in the evolution of the *Gunnera–Nostoc* symbiosis.

5
Evolutionary Drivers

Based on the occurrence of fossil pollen (Osborne and Sprent 2002) and phylogenetic analyses (Soltis et al. 2005), *Gunnera* is thought to have evolved in the Cretaceous, 90–100 MYA. This was an unusual period of high sea levels with globally extensive, shallow epi-continental (epeiric) seas or seaways (see Scotese 2001; Osborne and Sprent 2002; Miller et al. 2005; Peters 2007). As well as having a moderating impact on continental climates these interior seaways provided unique habitats for both plants and animals (Everhart 2005; Peters 2007). Given the particular conditions this would also have been true for microbes. Many animals that were unique to these habitats became extinct when the seaways disappeared (Everhart 2005; Peters 2007). Interestingly, the majority of fossil pollen records for *Gunnera* from Cretaceous deposits are correlated with these seas and seaways (Osborne and Sprent 2002). Perhaps the best documented seaway is the North American interior seaway, but similar seas or seaways were present throughout the world on what are now mostly continental land masses (Peters 2007).

Both biomarker and geological evidence suggests that these habitats were associated with sediment and photic zone anoxia/hypoxia, because of the particular features of these shallow basins. These include restricted circulation patterns and high temperatures (Allison and Wright 2005; Peters 2007), but perhaps less because of high algal productivity (Petsch 2005). They were also often sulphidic (see Simons et al. 2003) and provided environments that have little parallel today. The particular significance of anoxic/hypoxic conditions is that this limits the availability of inorganic (combined) forms of nitrogen, because of the restricted microbial oxidation of organic compounds. This provides a rationale for the evolution and increased abundance of nitrogen-fixing species. In fact wetland environments, in general, are often considered to be nitrogen limited (Mitsch and Gosselink 2000). There is also evidence that significant denitrification can occur during anoxia events, with the production of superabundant concentrations of N_2 gas (Jenkyns et al. 2001). Partnership formation under these conditions may have also favoured the symbiont (cyanobiont), as rhizosphere oxidation, a common feature of many wetland plants, would increase the mobilisation of Fe and Mo from precipitated sulphides. These are two important co-factors required for N_2 fixation, in particular. This works because SO_4^{3-} is oxidised before precipitation of ferric iron ($Fe^{2+} \rightarrow Fe^{3+}$; see Mitsch and Gosselink 2000).

Whilst it is unlikely that members of the *Gunnera* genus occupied deeper areas of the shallow seas, based on contemporary ecology (Osborne and Sprent 2002), they would have been associated with fringing marshes, swamps and streams that could have regularly received incursions of water containing low concentrations of oxygen. Interestingly, there is even evidence for shallow water anoxia along epeiric shorelines (Allison and Wright 2005) that could have provided suitable habitats in which N_2-fixing plant–cyanobacterial associations may have had an advantage. More information on the spatial distribution of pollen in these sites in relation to sediment chemistry is required, as is information on sediment microbiology. Clearly *Nostoc* can be a common feature of similar environments today (Dodds et al. 1995), and their predominance in these ancient sediments may have increased the likelihood of them being the "preferred" symbiont.

What the associations looked like in these early environments is still a matter of speculation because of the absence of macrofossil remains. Contemporary environments that are characterised by the presence of streams, swamps or marshes are more typical of the smaller members of the *Misandra*, *Malligania* or *Ostenigunnera* subgenera (see Osborne and Sprent 2002). Based on phylogenetic analysis these early species might have been similar to the extant annual *G. herteri* (*Ostenigunnera*), as this is considered to be sister to the remaining *Gunnera* species (see Wanntorp and Wanntorp 2003). Given the current preponderance ($\sim$50 species; see Osborne and Sprent 2002) of giant, large leafed, perennial species belonging to the subgenus *Panke*, evolution has favoured large size and a perennial life form.

The disappearance of shallow seas and subsequent increases in aridity, particularly in inner continental areas, was presumably only counteracted by high rainfall/high humidity conditions, with often a shift in distribution from low to high altitudes consistent with the restricted range of habitats that *Gunnera* occupies today (Osborne and Sprent 2002). This may have been exacerbated by an increase in atmospheric oxygen concentration towards the end of the Cretaceous period (see Petsch 2005; Beerling 2007). This could have enhanced the mineralisation of nitrogen and reduced any benefits associated with N_2 fixation in many wetland sites. Clearly, particular features of the physiology of the host, in particular, limit the utility of this symbiosis in a wider range of nutrient-poor environments. This appears to be largely because of the high water demand and poor control over water balance by the host (Osborne and Sprent 2002), but other factors may also be involved.

Although this argument for the evolution of the *Gunnera–Nostoc* symbiosis is still speculative, it does fit with some of the facts and provides an explanation for the decline in importance of this association. It also indicates why this symbiosis may confer little widespread benefit today. This suggestion, however, contrasts with ideas about the evolution of bacterial nitrogen-fixing symbioses with plants, which may have originated in response to a period of high atmospheric carbon dioxide concentration and, perhaps, reduced water availability (Schrire et al. 2005; Sprent 2006). These proposed differences in the evolution of bacterial and cyanobacterial-plant symbioses may be consistent with phylogenetic analyses that clearly indicate distinct and separate origins. However, there may be a wider argument for anoxia driven evolution, given that exposure of non-symbiotic plants to low oxygen conditions induces a number of changes that are related to those associated with the establishment of nitrogen-fixing symbioses. These include the synthesis of an ancestral form of symbiotic haemoglobin (called class 1 haemoglobin), alterations in transcription profiles and metabolism associated with inorganic nitrogen assimilation, carbon transport, ATP production and signalling pathways (Dordas et al. 2002; Igamberdiev et al. 2005). Cytoplasmic calcium oscillations, a particular feature of early signalling events in symbioses, may also be involved in the response to low oxygen conditions (Liu et al. 2005; Nie et al. 2006). This could indicate that processes associated with the response to hypoxia may be a common feature of all symbiotic N_2-fixing associations, including both bacterial and cyanobacterial symbioses. However, more specific information is required on the particular habitats in which both legumes and *Gunnera* evolved. An origin for legumes around the margins of the Tethys seaway in the Early Tertiary (see Schrire et al. 2005) could have been associated with environmental conditions not too dissimilar from those proposed for the origin of *Gunnera* in the vicinity of Cretaceous seaways (Osborne and Sprent 2002). It depends very much on what those local environmental conditions were. In fact there is evidence for considerable heterogeneity in the habitats present in the general vicinity of the Tethys seaway in Africa in the early Cenozoic, with a warm wet belt in the north contrasting markedly with interior aridity in more southerly areas (Bobe 2006).

Whilst the above arguments point to a possible explanation for the origin of the *Gunnera–Nostoc* symbiosis, they do not identify particular features of the host or cyanobiont that facilitated the establishment of this unusual partnership. Apparently, an ability to form these partnerships was either never a feature of angiosperms, or if it was, it has been lost during the course of evolution (see Usher et al. 2007). In the next section we explore possible features of the host and cyanobiont that may be related to the formation of the *Gunnera–Nostoc* symbiosis. In most cases we can only highlight features that may be important, given that there is little information on the role of the host or cyanobiont in the *Gunnera–Nostoc* symbiosis (Vessey et al. 2004; Adams et al. 2006; Bergman 2002).

6
Host Factors

In the *Gunnera–Nostoc* symbiosis there is good evidence for the host as the master player in the partnership—allowing the entry of cyanobacteria, but restricting their growth and extent of colonisation to a very small proportion of well-delineated areas of host tissue (see Osborne et al. 1991; Bergman et al. 1992, 2007; Bergman 2002; Rutishauser et al. 2004; Vessey et al. 2004). The mechanism by which the host controls this is, however, not known. Despite this restriction on the extent of colonisation of host tissue, the symbiont appears to be fully supported by host metabolic processes, enabling the cyanobacterium to grow, albeit slowly, differentiate and fix atmospheric nitrogen in quantities sufficient to meet all the demands of the host (Osborne et al. 1992).

Perhaps a common feature of symbioses is the ability of the symbiont to somehow overcome the host's defence responses. Although this is an attractive idea and could confer a high degree of specificity to the host colonised, there is little evidence that any plant defence reactions are stimulated in response to infection in the *Gunnera–Nostoc* symbiosis. Nor is there any information as yet on a possible role for programmed cell death as has been implicated in other plant symbioses (see Deshmukh et al. 2006). Also of interest is the relatively small alteration in the structure of host tissue in response to infection, with no external modifications that are comparable to either rhizobia or *Frankia* symbioses. However, a new internal "organ" is formed comprising both cyanobacterial and plant tissue with a considerably enhanced plasma membrane production (Bruce 1997). Part of the reason for this could be because of either a greater compatibility between host and symbiont or the absence of a requirement for the complex differentiation of tissues required to produce the structures and associated metabolic processes needed to control oxygen exchange. The cyanobiont (*Nostoc*) carries with it its own oxygen regulating system, the heterocyst, which is also the site of N_2 fixation. However, *Frankia* does have a structure, the vesicle, that may afford some oxygen protection, but perhaps this is not sufficient itself for complete control, necessitating some host involvement (Vessey et al. 2004).

Much attention has been directed at the unusual stem glands associated with all identified *Gunnera* species that are the initial entry points for subsequent colonisation of cortical tissue (Figs. 2 and 3). These are generally of a similar morphology, irrespective of plant size and life form, with only relatively small variations between individual species that have so far been examined (see Bonnett 1990; Bergman et al. 1992; Rutishauser et al. 2004; Bergman 2002). Typically, they comprise a variable number of raised mounds of tissue (papillae) in a circular arrangement around a central spike (Fig. 3c). The central spike may, however, be absent in the annual *G. herteri*. Also, the gland in *G. herteri* does not show the same differentiation into distinct outgrowths or mounds of tissue as reported for the larger species (Rutishauser

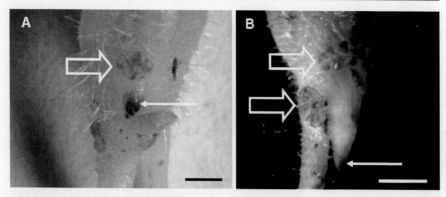

Fig. 2 Relationship between gland formation in *G. tinctoria* and adventitious root production (~10 weeks after germination). **A** Note close association between the gland (*open arrow*) in the centre of the picture and a newly emerging adventitious root (*thin arrow*) below, with a prominent *red* tip. **B** Somewhat later stage showing two mature glands (*above* and to the *left*) and an older adventitious root. *Scale bars* represent 1 cm

et al. 2004). These more "primitive" developmental features are probably consistent with phylogenetic analyses that place *G. herteri* as sister to all of the other species (Rutishauser et al. 2004). It is evident, however, that a more systematic examination of gland morphology and development within the genus is required in order to identify evolutionary relationships.

The gland (Fig. 3) originates internally from a lateral meristem and eventually ruptures the epidermis, a process that may be aided by the central spike (Fig. 3c), exposing the surfaces of cortical cells (Bergman et al. 1992). Within the gland there are several channels or crevices through which the cyanobacteria are able to invade host tissue. The channels of the gland are lined with cytoplasmic-rich cells (Johansson and Bergman 1992; Bergman et al. 1992), as are the cells at the interior terminal ends of the channels, and may represent a host response prior to establishment (Uheda and Silvester 2001; Rutishauser et al. 2004).

How this response is controlled is, however, not known but may be a way in which the host determines the particular cells via which the cyanobacterium will subsequently enter cortical cells (Bergman et al. 1992; Rutishauser et al. 2004). Clearly the glands play an essential role in the establishment of the association. But what is their origin and what exactly do they do?

Angiosperms are associated with a remarkable diversity of stem/tissue glands (Fahn 1979, 1988), all of which are potential entry points for cyanobacteria. Many, like *Gunnera*, have glands that secrete a copious quantity of mucilage, which in *Gunnera* can stimulate the production of motile hormogonia that are essential for the infection process (Bergman 2002; Bergman et al. 2007). However, as far as we are aware, there is little evidence that any of these other angiosperm glands serve to attract cyanobacteria or induce

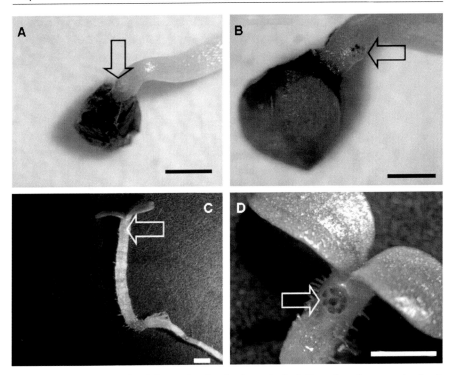

Fig. 3 Early stages in gland formation in *G. tinctoria* (~2–7 weeks after germination). **A** Note that early gland formation is associated with a dome-shaped structure (*open arrow*) close to the point of emergence, with localised *red* (anthocyanin) pigmentation that delineates the position of the mounds of tissue, called papillae, in the mature gland. **B** Intermediate stage of gland formation showing increased but localised *red* pigmentation (*open arrow*) in the now chlorophyll containing tissue. **C** Later stage of gland formation showing a *red* pigmented gland (*open arrow*) with papillae and a prominent central spike and two cotyledon leaves. **D** Seedling showing a more mature shoot-associated gland. *Scale bars* represent 1 cm

hormogonia formation. A search of the literature indicates that morphologically at least the *Gunnera* gland appears to bear little resemblance to other angiosperm glands. Interestingly, gland formation is closely associated with adventitious (shoot-associated) root formation (Fig. 2a and b; see Bergman 2002). During development, the main root(s) withers and dies (see Rutishauser et al. 2004; Osborne, personal observations) so the adult plant is dependent solely on adventitious roots for nutrient acquisition. Gland formation itself is initiated prior to shoot emergence and is evident early on as a dome-shaped structure with the red pigmentation that characterises the mature gland (Fig. 3a and b; Johansson and Bergman 1992).

Some of these features are clearly similar to adventitious root formation in other species, including the early initiation in pre-hypocotyl tissue prior to seedling emergence (Barker et al. 1986). In *Gunnera* the red anthocyanin pigment found in the glands is also mimicked in the tips of adventitious roots (Fig. 2a and b). The production of carbohydrate-rich mucilage is also a common root feature (e.g. Chaboud 1983). Gland formation is also thought to be controlled by plant growth regulators, such as auxin (Chui et al. 2005), clearly paralleling adventitious root formation. It has also been reported that adventitious roots may have a "glandular" function, with an ability to synthesise plant growth substances, such as cytokinins as well as amino acids (see Barlow 1986). Interestingly, cyanobacteria, including symbiotically competent species, are able to synthesise auxin (Sergeva et al. 2002), providing them with the potential at least to influence root/gland formation. If it turns out to be true that the stem glands are really modified adventitious roots, there is considerably more commonality among all plant nitrogen-fixing associations, all of which would then be root-associated and all of which may share many common features involved in infection and establishment.

7
Cyanobiont Factors

It is generally agreed that *Nostoc* is the only known genus of cyanobacteria to infect *Gunnera*. But why *Nostoc* and why a heterocystous nitrogen-fixer? This could be partly due to historical reasons associated with the availability of *Nostoc* in habitats occupied by *Gunnera* in the past. *Nostoc* would be a common cyanobacterium in near-shore freshwater-brackish aquatic systems, perhaps where ancestral forms of *Gunnera* were also present. If there was, and still is, a benefit associated with the formation of this partnership what was/is it? *Nostoc* clearly has a remarkable ability to colonise a wide range of habitats (Dodds et al. 1995), has the metabolic flexibility to survive under both heterotrophic and autotrophic conditions with the potential to survive under diverse environmental conditions and can assimilate both nitrogen gas and combined forms of nitrogen (Ran et al. 2006). What, therefore, does the host do to "convince" the cyanobacterium that the formation of the *Gunnera*–*Nostoc* symbiosis confers some benefit to the symbiont? In common with the majority of symbioses (Osborne 2007) one partner, in this case the host, clearly benefits through the supply of nitrogen, whilst reciprocal benefits are uncertain. These benefits are perhaps more significant with *Gunnera* because of a restricted ability to utilise inorganic nitrogen (Osborne et al. 1992), suggesting a more obligate dependency on the symbiotic cyanobacterium. Although organic carbon is supplied by the host in the symbiosis (Söderbäck and Bergman 1993), the cyanobiont has the capacity to photosynthesise in the free-living state and has no obligate dependency for host-derived carbon. The control of developmental patterns in the cyanobiont, and the controlled exploitation of the cyanobiont's metabolic capabilities by the host (Rasmussen et al. 1994; Wang et al. 2004; Ekman et al. 2006), suggest a subtle and quite sophisticated regulatory mechanism that restricts growth and colonisation but achieves an adequate supply of nitrogen that matches the need for growth (Osborne et al. 1992; Wang et al. 2004; Ekman et al. 2006). Rather surprisingly this occurs without any major changes in a number of photosynthetic components, despite the fact that the cyanobiont is unable to photosynthesise in situ (Black and Osborne 2004). An interesting possibility is that the host exerts its control through regulation of the availability of essential nutrients such as phosphate or iron, as both are required in large amounts by cyanobacteria. Iron is particularly attractive in this regard as it is an essential micronutrient in a range of metabolic processes, is generally in short supply and has been suggested to be involved in signalling processes in the rhizobium–legume symbiosis (Johnston et al. 2001). This requires further investigation.

8
Conclusions

It should be obvious from this review that we still have a very poor understanding of the factors that led to the establishment of the *Gunnera–Nostoc* symbiosis or why it was apparently more successful in an ecological context in the past than it is today. Nevertheless, this review does indicate areas that would address these two issues; one of the most important of these is a better understanding of the origin, development and role of the "gland" and its possible relationship with adventitious root formation. Interestingly, developmental responses associated with lateral root emergence may elicit responses in nearby cortical cells that are required for nodule formation in legumes (e.g. Mathesius et al. 2000). There is also increasing evidence that rhizobia are able to hijack a number of normal plant developmental processes during infection (Osborne 2007). One of the ways in which rhizobia are able to do this is through an effect on nodule formation through auxin production and the regulation of auxin transport by host tissue (see Pii et al. 2007). Given that cyanobacteria have also been shown to produce auxin (Sergeva et al. 2002), this phytohormone may also be involved in developmental changes associated with the formation of the *Gunnera–Nostoc* symbiosis. Whatever these putative auxin effects might be they would have to be quite subtle, as they are not related to any gross external developmental changes comparable to nodule formation, nor is cyanobacterially produced auxin required for gland formation (see Fig. 2). Endogenous production and transport of auxin by the host and its regulation by nitrogen and carbohydrate supply are, however, likely to be associated with gland development (Chui et al. 2005).

Although angiosperms are associated with a vast range of glands that secrete mucilage containing putative signalling compounds these are, as far as we are aware, never infected by cyanobacteria. Comparison of the secretions produced by these glands with that secreted by *Gunnera* (see Bergman 2002) may help to identify compounds that are important during the early phases of infection and establishment. Clearly glands are not required for the establishment of all plant–cyanobacterial symbioses, and the reason why this appears to be essential for the establishment of angiosperm–cyanobacterial symbioses deserves more attention. Despite differences between *Gunnera* and other plant–cyanobacterial symbioses, and between *Gunnera* and rhizobia or *Frankia* associations, common factors may be involved in the establishment and maintenance of all nitrogen-fixing symbioses. The separation of *Gunnera* from the main nitrogen-fixing clade, comprising rhizobia and *Frankia* associations, within the angiosperms may relate to very subtle differences. There are also plausible, although as yet unexamined, reasons why all plant symbioses might have originated in response to anoxia/low oxygen-induced nitrogen deficiency and this warrants further investigation. It is interesting that there may be common genetic determinants underpinning arbuscular mycorrhizal (AM) and rhizobial symbioses, and that the establishment of legume associations may have utilised pre-existing pathways that are related to the much older AM symbioses (see Szczyglowski and Amyot 2003). Whilst AM symbioses are ubiquitous (Smith and Read 1997), N_2-fixing symbioses involving either bacteria or cyanobacteria are uncommon (see Soltis et al. 2005), suggesting that these steps, although essential, do not confer an ability to form N_2-fixing associations with all plants.

In terms of the functional significance of the *Gunnera–Nostoc* symbiosis, further work on extant plants is required. Based on limited information of macrofossil remains of *Gunnera* (see Wilkinson 2000), the basic morphological features of the genus may have been retained during the course of evolution, making it more likely that at least some of the traits associated with its past success have been retained in extant material. However, we will still need, in addition, more information on the fossil record of this genus and how this is related to environmental factors. Further developments in geochemistry, such as the use of new biomarkers, may help us to obtain more specific information on the habitats and environmental conditions occupied by this genus in the past. What would also be helpful is evidence for the symbioses in fossil material. Confirmation that this genus did form symbioses with cyanobacteria in the past is still required.

And what of the evolutionary significance of the *Gunnera–Nostoc* symbiosis? It has been suggested that this could have been the result of a response to particular conditions—low oxygen, high water availability and low combined nitrogen–that were more widespread in Cretaceous environments (Osborne and Sprent 2002). Indeed, hypoxia has been a consistent feature of earth history (see Jacobs and Lindberg 1998; Peters 2007) and oxygen availability is increasingly being recognised as a major evolutionary driver (Peters 2007; Hallam and Wignall 1997). It is only in recent geological times that hypoxia as a globally significant environmental factor has been rare. Hypoxia could then be used to explain the evolution of other plant–cyanobacterial symbioses within major plant groups: the liverworts/hornworts representing the bryophyte response, the cycads representing the gymnosperm response, *Azolla* the fern response and *Gunnera* the angiosperm response. This would also account for the widespread but infrequent occurrence of cyanobacterial symbioses in different plant groups. Based on this information all plant–cyanobacterial symbioses may represent evolutionary "dead ends", reflecting a response to conditions that are no longer prevalent today (Osborne and Sprent 2002). Assessment of this proposal and the elucidation of factors that are crucial to the formation and establishment of the *Gunnera–Nostoc* symbiosis represent major research challenges for the future.

Acknowledgements One of us (BO) thanks Margherita Gioria and Tommy Gallagher for information. We would also like to thank Livia Wanntorp for discussions about gland origins and adventitious root formation. The Swedish Research Council is acknowledged for financial support (to BB).

References

Adams DG, Bergman B, Nierzwicki-Bauer SA, Rai AN, Schüssler A (2006) In: Dworkin M, Falkow S, Rosenberg E, Schleifer K-H, Stackebrandt E (eds) Cyanobacterial–plant symbiosis. The prokaryotes: a handbook on the biology of bacteria, 3rd edn, vol 1. Symbiotic associations, biotechnology, applied microbiology. Springer, New York, pp 331–363

Allison PA, Wright VP (2005) Switching off the carbonate factory: a-tidality, stratification and brackish wedges in epeiric seas. Sediment Geol 179:175–184

Barker WG, Hussey RB, Horton RF (1986) Involvement of the cotyledon in adventitious root initiation in Impatiens balsamina L. seedlings. Ann Bot 58:397–405

Barlow PW (1986) Adventitious roots of whole plants: their forms, functions and evolution. In: Jackson MB (ed) New root formation in plants and cuttings. Kluwer, Dordrecht, pp 67–110

Barnett MJ, Fisher RF (2006) Global gene expression in the rhizobial–legume symbiosis. Symbiosis 42:1–24

Beerling D (2007) The emerald planet. Oxford University Press, Oxford

Bergman B (2002) The Nostoc–Gunnera symbiosis. In: Rai AN, Bergman B, Rasmussen U (eds) Cyanobacteria in symbiosis. Kluwer, Dordrecht, pp 207–232

Bergman B, Johansson C, Söderbäck E (1992) The Nostoc–Gunnera symbiosis. New Phytol 122:379–400

Bergman B, Rasmussen U, Rai AN (2007) In: Elmerich C, Newton WE (eds) Cyanobacterial associations. Associative and endophytic nitrogen-fixing bacteria and cyanobacterial associations, vol 5. Kluwer, Dordrecht, pp 257–301

Berner RA (2004) The Phanerozoic carbon cycle. Oxford University Press, Oxford

Black K, Osborne BA (2004) An assessment of photosynthetic downregulation in cyanobacteria from the Gunnera–Nostoc symbiosis. New Phytol 162:125–132

Black K, Parsons R, Osborne BA (2002) Uptake and metabolism of glucose in the Gunnera–Nostoc symbiosis. New Phytol 153:297–305

Bonnett HT (1990) The Nostoc–Gunnera association. In: Rai AN (ed) Handbook of symbiotic cyanobacteria. CRC, Boca Raton, pp 161–171

Brinkhuis et al (2006) Episodic fresh surface waters in the Eocene Arctic Ocean. Nature 441:606–609
Bruce C (1997) An investigation into the process of gland development in the Gunnera tinctoria–Nostoc punctiforme symbiosis using light and electron microscopy. PhD thesis, University College Dublin
Chaboud A (1983) Isolation, purification and chemical composition of maize root cap slime. Plant Soil 73:395–402
Chui WL, Peters GA, Levieille G, Still PC, Cousins S, Osborne BA, Elhai J (2005) Nitrogen deprivation stimulates symbiotic gland development in Gunnera manicata. Plant Physiol 139:224–230
Deshmukh S, Huckelhoven R, Schafer P, Imani J, Sharma M, Weiss M, Waller F, Kogel K-H (2006) The root endophytic fungus Piriformospora indica requires host cell death for proliferation during mutualistic symbiosis with barley. Proc Natl Acad Sci USA 103:18450–18457
Dodds WK, Gudder DA, Mollenhauer D (1995) The ecology of Nostoc. J Phycol 31:2–18
Dordas C, Rivoal J, Hill RD (2003) Plant haemoglobins, nitric oxide and hypoxic stress. Ann Bot 91:173–178
Ekman M, Tollbäck P, Klint J, Bergman B (2006) Protein expression profiles in an endosymbiotic cyanobacterium revealed by a proteomic approach. Mol Plant Microbe Interact 19:1251–1261
Everhart MJ (2005) Oceans of Kansas, 2nd edn. Indiana University Press, Indianapolis
Evrard C, van Hove C (2004) Taxonomy of the American Azolla species (Azollaceae): a critical review. Syst Geogr Plants 74:301–318
Fahn A (1979) Secretory tissues in plants. Academic, London
Fahn A (1988) Secretory tissues in plants. New Phytol 108:229–257
Hall JW, Swanson NP (1968) Studies on fossil Azolla: Azolla montana, a Cretaceous megaspore with many small floats. Am J Bot 55:1055–1061
Hallam A, Wignall PB (1997) Mass extinctions and their aftermath. Oxford University Press, Oxford
Igamberdiev AU, Baron K, Manac'h-Little N, Stoimenova M, Hill RD (2005) The haemoglobin/nitric oxide cycle: involvement in flooding stress and effects on hormone signalling. Ann Bot 96:557–564
Jacobs DK, Lindberg DR (1998) Oxygen and evolutionary patterns in the sea; onshore/offshore trends and recent recruitment of deep sea fauna. Proc Natl Acad Sci USA 95:9396–9401
Jenkyns HC, Groke DR, Hesselbo SP (2001) Nitrogen isotope evidence for water mass denitrification during the early Toarcian (Jurassic) oceanic anoxic event. Paleoceanography 16:593–603
Johansson C, Bergman B (1992) Early events during the establishment of the Gunnera–Nostoc symbiosis. Planta 188:403–413
Johnston AWB, Yeoman KH, Wexler M (2001) Metals and the rhizobial–legume symbiosis—uptake, utilization and signalling. Adv Microb Physiol 45:13–156
Jones DL (2002) Cycads of the world. Smithsonian Institution Press, Washington
Lechno-Yossef S, Nierzwicki-Bauer SA (2002) In: Rai AN, Bergman B, Rasmussen U (eds) Azolla–Anabaena symbiosis. Cyanobacteria in symbiosis. Kluwer, Dordrecht, pp 153–178
Liu F, Vantoai T, Moy LP, Bock G, Linford LD, Quackenbush J (2005) Global transcriptional profiling reveals comprehensive insights into hypoxic response in Arabidopsis. Plant Physiol 137:1115–1129

Mathesius U, Weinman JJ, Rolfe BG, Djordjevic MA (2000) Rhizobia can induce nodules in white clover by highjacking mature cortical cells activated during lateral root development. Mol Plant Microbe Interact 13:170–182

Miller KG, Kominz MA, Browning JV, Wright JD, Mountain GS, Katz ME, Sugarman PJ, Cramer BS, Christie-Blick N, Pekar SF (2005) The Phanerozoic record of global sea level change. Science 25:1293–1298

Mitsch WJ, Gosselink JG (2000) Wetlands, 3rd edn. Van Nostrand Reinhold, New York

Nie X, Durnin DC, Igamberdiev AU, Hill RD (2006) Cytosolic calcium is involved in the regulation of barley haemoglobin gene expression. Planta 223:542–549

Norstog K, Nicholls TJ (1997) The biology of the cycads. Cornell University Press, Ithaca

Oldroyd GED, Harrison MJ, Udvardi M (2005) Peace talks and trade deals. Key to long-term harmony in legume–microbe symbioses. Plant Physiol 137:1205–1210

Osborne BA (2007) Holy alliances? New Phytol 175:602–605

Osborne BA, Sprent JI (2002) In: Rai AN, Bergman B, Rasmussen U (eds) Ecology of the Nostoc–Gunnera symbiosis. Cyanobacteria in symbiosis. Kluwer, Dordrecht, pp 233–251

Osborne BA, Doris F, Cullen A, McDonald R, Campbell G, Steer M (1991) Gunnera tinctoria: an unusual nitrogen-fixing invader. BioScience 41:224–234

Osborne BA, Cullen A, Jones PW, Campbell GJ (1992) Use of nitrogen by the Nostoc–Gunnera tinctoria (Molina) Mirbel symbiosis. New Phytol 120:481–487

Page CN (1997) The ferns of Britain and Ireland, 2nd edn. Cambridge University Press, Cambridge

Peters SE (2007) The problem with the Paleozoic. Paleobiology 33:165–181

Petsch ST (2005) In: Schlesinger WH (ed) The global oxygen cycle: biogeochemistry. Treatise on geochemistry, vol 8. Elsevier, Amsterdam, pp 515–555

Rai AN, Söderbäck E, Bergman B (2000) Cyanobacterium–plant symbiosis. New Phytol 147:449–481

Ran L, Huang F, Ekman M, Klint J, Bergman B (2007) Proteomic analyses of the photoauto and diazotrophically grown cyanobacterium Nostoc sp. PCC 73102. Microbiology 153:608–618

Rasmussen U, Johansson C, Bergman B (1994) Early communication in the Gunnera–Nostoc symbiosis: plant-induced cell differentiation and protein synthesis in the cyanobacterium. Mol Plant Microbe Interact 7:696–702

Rustishauser R, Wanntorp L, Pfeifer E (2004) Gunnera herteri—developmental morphology of a dwarf from Uruguay and S Brazil (Gunneraceae). Plant Syst Evol 248:219–241

Pii Y, Crimi M, Cremonese G, Spena A, Pandolfini T (2007) Auxin and nitric oxide control indeterminate nodule formation. BMC Plant Biol 7:21 doi: 10.1186/1471-2229-7-21

Schrire BD, Lavin M, Lewis GP (2005) Global distribution patterns of the Leguminosae: insights from recent phylogenies. Biol Skrifter 55:375–422

Scotese R (2001) Atlas of Earth history, vol 1. Palaeogeography. PALEOMAP project, University of Texas, Arlington

Segeeva E, Liaimer A, Bergman B (2002) Evidence for biosynthesis and release of the phytohormone indole-3-acetic acid by cyanobacteria. Planta 215:229–238

Simons D-JH, Kenig F, Schroder-Adams CJ (2003) An organic geochemical study of Cenomanian–Turonian sediments from the Western Interior Seaway, Canada. Org Geochem 34:1177–1198

Sluijs et al. (2006) Subtropical Arctic Ocean temperatures during the Palaeocene/Eocene thermal maximum. Nature 441:610–613

Smith SE, Read DJ (1997) Mycorrhizal symbiosis, 2nd edn. Academic, London
Söderbäck E, Bergman B (1992) The Nostoc-Gunnera magellanica symbiosis: phycobiliproteins, carboxysomes and Rubisco in the microsymbiont. Physiol Plant 84:425–432
Söderbäck E, Bergman B (1993) The Nostoc-Gunnera symbiosis: carbon fixation and translocation. Physiol Plant 89:125–132
Söderbäck E, Lindblad P, Bergman B (1990) Developmental patterns related to nitrogen fixation in the Nostoc-Gunnera magellanica Lam. symbiosis. Planta 182:355–362
Soltis DE, Soltis PS, Endress PK, Chase MW (2005) Phylogeny and evolution of angiosperms. Sinauer, Sunderland
Sprent JI (2001) Nodulation in legumes. Royal Botanic Garden, Kew
Sprent JI (2006) Evolving ideas of legume evolution and diversity: a taxonomic perspective on the occurrence of nodulation. New Phytol 174:11–25
Sprent JI (2008) 60 million years of legume nodulation? What's new? What's changing? J Exp Bot (in press)
Sprent JI, Sprent P (1990) Nitrogen fixing organisms: pure and applied aspects. Chapman and Hall, London
Szczyglowski K, Amyot L (2003) Symbiosis, inventiveness by recruitment. Plant Physiol 131:935–940
Towata EM (1985a) Mucilage glands and cyanobacterial colonisation in Gunnera kaalensis (Haloragaceae). Bot Gaz 146:56–62
Towata EM (1985b) Morphometric and cytochemical ultrastructural analyses of the *Gunnera kaalensis/Nostoc* symbiosis. Bot Gaz 146:293–301
Uheda E, Silvester WB (2001) The role of papillae during the infection process in the Gunnera-Nostoc symbiosis. Plant Cell Physiol 42:780–783
Usher K, Bergman B, Raven J (2007) Exploring cyanobacterial mutualisms. Annu Rev Ecol Syst 38:255–273
Vessy JK, Pawlowski K, Bergman B (2004) Root-based N_2-fixing symbioses; legumes, actinorhizal plants, Parasponia sp. and cycads. Plant Soil 266:205–230
Usher K, Bergman B, Raven J (2007) Exploring cyanobacterial mutualisms. Annu Rev Ecol Evol Syst 38:255–273
Wang C-M, Ekman M, Bergman B (2004) Expression of ntcA, glnB, hetR and nifH and corresponding protein profiles along a developmental sequence in the Nostoc-Gunnera symbiosis. Mol Plant Microbe Interact 4:436–443
Wanntorp L, Wanntorp HE (2003) The biogeography of Gunnera L: vicariance and dispersal. J Biogeogr 30:979–987
Wilkinson HP (2000) A revision of the anatomy of Gunneraceae. Bot J Linn Soc 134:233–266

Cyanobacteria in Symbiosis with Cycads

Peter Lindblad

Department of Photochemistry and Molecular Science, The Angstrom Laboratories, Uppsala University, 751 20 Uppsala, Sweden
peter.lindblad@fotomol.uu.se

1	Introduction	226
2	Coralloid Roots	227
3	Establishment of the Symbiosis	227
4	Specificity and Diversity of the Symbiotic Cyanobacteria	229
5	Characteristics of the Symbiotic Cyanobacteria	230
5.1	The Symbiotic Filament	230
5.2	Nitrogen Fixation and Nitrogen Metabolism	230
5.3	Transfer of Fixed Nitrogen	231
5.4	Carbon Metabolism	232
References		232

Abstract Cycads establish symbioses with filamentous cyanobacteria in highly specialised lateral roots termed "coralloid roots". The coralloid roots, recorded in all cycad genera, show a marked negative geotropism, and grow laterally and upward toward the surface of the soil. The cyanobacteria are present in a specific cortical layer inside the root, the so-called cyanobacterial zone. The filamentous heterocystous cyanobacteria inside the root induces irreversible modifications to its growth and development due to a differentiation of elongated cycad cells, which have suggested to be responsible for the transfer of metabolites between the partners. The process of infection is still unclear. Invasion of filamentous cyanobacteria may occur at any stage of development of the root, but the precise time and location of the invasion is unpredictable. Using detailed molecular technques no genetic variation of the symbiotic cyanobacterial cells was observed within a single coralloid root. This is consistent with infection by a single cyanobiont. However, different coralloid roots from a single cycad specimen may harbour different cyanobacteria, and the same cyanobiont may be present in two different cycad specimens as well as in different cycad species. The filamentous heterocystous cyanobacteria within the cyanobacterial zone are located extracellularly between the elongated cycad cells and embedded in mucilage. All the molecular work is consistent with different *Nostoc* strains, the cyanobiont in cycad symbioses. The cyanobionts differentiate into vegetative cells and heterocysts, but rarely akinetes. In general, the symbiotic *Nostoc* filaments show only few modifications in comparison to its free-living counterparts: (1) an increased heterocyst frequency, (2) an increased level of nitrogen fixation, and (3) a transfer of fixed nitrogen from the cyanobacterial to the cycad cells. However, there is a distinct developmental gradient within the coralloid roots with "free-living like" filaments in the growing tip, which rapidly develop into symbiotic cells followed by older and metabolically less active cells in the older parts. Due to their location in coralloid roots (complete darkness), the cyanobionts are expected to have a heterotrophic mode of carbon nutrition. However, almost nothing is known about the heterotrophic metabolism in the symbiotic cells. The present review focuses on recent advances in the understanding of symbiotic cyanobacteria in cycads.

1
Introduction

Cycads are an ancient group of seed plants that first appeared in the Pennsylvanian period and have therefore existed for approximately 300 million years. Nearly 100 million years ago, cycads had a wide geographic distribution, extending from Alaska and Siberia to the Antarctic. Today, they occur naturally on every continent except Europe and Antarctica. However, they are usually restricted to small populations in the tropics and subtropics of both hemispheres. In addition, many are facing possible extinction in nature. The approximately 160 species of cycads are distributed in three families: Cycadaceae with a single genus, *Cycas*, Stangeriaceae with two genera, *Stangeria* and *Bowenia*, and Zamiaceae with eight genera, *Ceratozamia*, *Chigua*, *Dioon*, *Encephalartos*, *Lepidozamia*, *Macrozamia*, *Microcycas* and *Zamia*. Absolute majority of cycads are terrestrial and arborescent, except for a few examples of truly epiphytic species (e.g. *Zamia pseudoparasitica* Yates). Some species may be found as components of forests (both rainforest and seasonally dry forest) and others grow in loose strands in grasslands. Structurally, cycads have an aerial stem, which is normally columnar and woody with the exception of e.g. *Zamia pygmaea* Sims., which has a subterranean stem. The aerial stem may be branched and is covered by persistent leaf bases. In general, the vegetative shoots of cycads produce two types of leaves, scale leaves or cataphylls and foliage leaves. The leaves are born terminally and are mainly pinnate with the exception of the genus *Bowenia* that has bipinnate leaves. Reproductively, cycads are dioecious, producing either pollen-bearing or ovule-bearing modified leaves called sporophylls. The ovules, like those of true gymnosperms, are naked.

Cycads produce three types of roots: (i) a tap-root that is equivalent to the primary root system found in most types of plants (a special type of root restricted to the genus *Cycas*, which is adventitious, arises from the lower side of trunk offsets and grows downwards in close proximity to the trunk), (ii) lateral roots, and (iii) a highly specialised type of lateral roots usually termed "coralloid roots" and in which symbiotic filamentous cyanobacteria may be found (Norstog and Nicholls 1997). The present review focuses on recent advances in the understanding of symbiotic cyanobacteria in cycads. The reader is referred to earlier reviews (Lindblad 1990; Lindblad and Bergman

1990; Adams 2000; Rai et al. 2000; Costa and Lindblad 2002) for detailed discussions regarding overall symbiotic structures and a characterisation of the symbiotic cyanobacteria.

2
Coralloid Roots

Coralloid roots have been recorded in all genera, and in all cycad species examined. The ability to form this type of root is encoded by genes in the cycad, and the overall structure of the roots are formed before being invaded by symbiotic cyanobacteria. These roots show a marked negative geotropism, and grow laterally and upward toward the surface of the soil. In seedlings of *Macrozamia*, the process of coralloid root development begins with the initiation of papillose roots called "pre-coralloids" or non-infected coralloid roots (Ahern and Staff 1994). The first "pre-coralloid" roots are adventitious and emerge from the hypocotyl immediately below the cotyledonary petioles. Their subsequent development involves phases of maturation, cyanobacterial invasion, coralloid formation, senescence, and regeneration. At least in *Macrozamia*, coralloid roots with and without symbiotic cyanobacteria are structurally different (Fig. 1). The cyanobacteria are present in a specific cortical layer inside the root, the so-called cyanobacterial zone. The filamentous heterocystous cyanobacteria inside the root induces irreversible modifications to the growth and development of the root. The growth in length of these roots is much retarded in comparison to the normal roots of same age. In addition, some cycad cells in the cyanobacterial zone undergo marked differentiation, elongating radially to interconnect the two adjacent cortical layers (Lindblad et al. 1985a). It has been suggested that these elongated cells are specialised cells responsible for the transfer of metabolites between the partners (Lindblad et al. 1985a).

3
Establishment of the Symbiosis

The process of infection is still unclear. Invasion of filamentous cyanobacteria may occur at any stage of development of the root, but the precise time and location of the invasion is unpredictable. Several suggestions are present in the literature: (a) through injured parts of the root; (b) through lenticels; (c) through the papillose, and (d) through breaks in the dermal layer (Nathanielsz and Staff 1975a). More detailed studies are needed to establish if there is a single process of infection in cycads or if different cycad genera/families have developed different strate-

Fig. 1 *Macrozamia riedlei* (Fischer ex Gaudichaud-Beaupré) C.A. Gardner in its natural habitat in Western Australia. **A** Male specimen with three emerging cones. **B** Uninfected coralloid root. **C** Coralloid root with established symbiotic cyanobacteria. From Costa and Lindblad (2002) (with permission)

gies. Interestingly, the cyanobiont of *Zamia furfuracea* (*Nostoc* sp. strain FUR 94201) was isolated and grown in an axenic culture before being successfully reunited with a sterile seedling of the same cycad species (Ow et al. 1999).

4
Specificity and Diversity of the Symbiotic Cyanobacteria

Until recently very little was known about the biodiversity and specificity of symbiotic cyanobacteria in cycads. Heterologous Southern hybridisations using cloned genes from the free-living cyanobacterium *Anabaena* sp. strain PCC 7120 demonstrated a diversity when analysing cyanobacterial DNA prepared from a large number of pooled coralloid roots collected from cycads in their natural habitat in Mexico (Lindblad et al. 1989). In a more detailed study the diversity and host specificity of the cyanobionts of several cycad species (*Cycas circinalis* L., *C. rumphii* Miq., *Encephalartos lebomboensis* I. Verd., *E. villosus* Lem., and *Zamia pumila* L.) collected in Fairchild Tropical Garden (Florida, USA) were examined using the tRNA^{Leu}(UAA) intron sequence as a genetic marker (Costa et al. 1999). Nested PCR was used to specifically amplify the intron directly from the freshly isolated symbiotic cyanobionts. The intron sequences obtained from the cycad cyanobionts showed high similarities to the corresponding sequences in the free-living strains *Nostoc* sp. PCC 73102 and *N. muscorum* as well as in several lichen and bryophyte cyanobionts, indicating their *Nostoc* identity (Costa et al. 2002). Although different intron sequences were found, no variation was observed within a single coralloid root. This is consistent with infection by a single cyanobiont. However, different coralloid roots from a single *E. villosus* specimen may harbour different cyanobacteria. In addition, cyanobionts in coralloid roots of two different *Encephalartos* species were found to possess the same intron sequence, indicating that the same cyanobiont may be present in two different cycad species (Costa et al. 1999). The genetic diversity of cyanobacteria associated with cycads was examined using the tRNALeu (UAA) intron as a genetic marker. The same techniques were also used to analyse the cyanobionts of both natural populations of the cycad *Macrozamia riedlei* (Fischer ex Gaudichaud-Beaupré) C.A. Gardner growing in Perth, Australia and cycads growing in greenhouses, also in Perth. Several *Nostoc* strains were found to be involved in this symbiosis, both in natural populations and greenhouse-originated cycads (Costa et al. 2004). However, only one strain was present in individual coralloid roots and in individual plants, even when analyzing different coralloid roots from the same plant. Moreover, when examining plants growing close to each other (female plants and their respective offspring) the same cyanobacterium was consistently present in the different coralloid roots. Whether this reflects a selective mechanism or merely the availability of *Nostoc* strains remains to be ascertained.

5
Characteristics of the Symbiotic Cyanobacteria

Morphological, physiological and biochemical characterisations of the symbiotic cyanobacteria in cycads have a long tradition. In general, the symbiotic *Nostoc* filaments show only few modifications in comparision to its free-living counterparts; (1) an increased heterocyst frequency, (2) an increased level of nitrogen fixation, and (3) a transfer of fixed nitrogen from the cyanobacterial to the cycad cells. It is important to note that the developmental gradient within the coralloid roots with "free-living like" filaments in the growing tip rapidly develops into symbiotic cells followed by older and metabolically less active cells in the older parts.

5.1
The Symbiotic Filament

The filamentous heterocystous cyanobacteria within the cyanobacterial zone are located extracellularly between the elongated cycad cells and embedded in mucilage (Lindblad et al. 1985a). However, there are reports on an intracellular location in *Cycas revoluta* Thunb. and *Macrozamia communis* L. (Nathanielsz and Staff 1975b; Obukowicz et al. 1981). Although the cyanobiont has been variously reported as *Nostoc*, *Anabaena* or sometimes *Calotrix*, the classifications were exclusively based on morphological features. However, all the molecular work detailed above is consistent with different *Nostoc* strains being the cyanobiont in cycad symbioses. The cyanobiont differentiates into vegetative cells and heterocysts, but rarely akinetes (Lindblad et al. 1985a,b). Structurally, the vegetative cells of the cycad cyanobionts show only few modifications compared to free-living isolates. However, a high frequency of heterocysts has been described in the cyanobiont of several cycad species. Examination of successive sections of *Zamia skinneri* coralloid roots, collected from its natural habitat in Costa Rica, revealed an increased heterocyst frequency ranging from 16.7% at the growing tip to 46% in the older basal parts (Lindblad et al. 1985b). It was also apparent, from the relative frequency of mainly single heterocysts at the growing tip (86%), that more and more heterocysts occur in multiples (double to quadruple) as the root becomes older.

5.2
Nitrogen Fixation and Nitrogen Metabolism

Symbiotic cyanobacteria from coralloid roots have the ability to fix N_2. Analysis of nitrogenase activity, using the acetylene reduction assay, in sequential sections of coralloid roots clearly demonstrated a gradual decline in older parts of the coralloid root. The highest activity occurs in the growing tip of the coralloid root. Nitrogenase is localised in heterocysts (Bergman et al. 1986) and appears to be confined primarily to sections of the coralloid root where single heterocysts predominate (Lindblad et al. 1985b). Freshly isolated cyanobionts from coralloid roots of *Macrozamia riedlei*, from natural populations in Western Australia, retain nitrogenase activity even when separated from their host as long as the cells are not exposed to O_2 levels above 1% (Lindblad et al. 1991). Exposure to higher levels of O_2 resulted in a dose-dependent inhibition of the nitrogenase activity. Moreover, nitrogenase activity increases considerably when the cells are exposed to light, which has been suggested to be due to an increased availability of ATP produced by PS-I mediated cyclic photophosphorylation. It is reasonable to assume that the low levels of nitrogenase activity, observed when assaying freshly isolated cyanobionts in darkness, reflect loss or damage of heterotrophic mechanisms that provided ATP to the cyanobiont in the intact coralloid root. Separation of the cyanobacteria from the host tissue disrupts the intercellular microenvironment and any metabolic and biochemical interactions that the cyanobiont experienced in situ in the coralloid roots.

Free-living cyanobacteria assimilate the ammonia produced by nitrogen fixation primarily via glutamine synthetase (GS)–glutamate synthase (GOGAT) enzyme system. In contrast to other cyanobacterial symbioses, the symbiotic cyanobacteria of *Cycas revoluta*, *Ceratozamia mexicana*, and *Zamia skinneri* all show in vitro GS activities and relative GS protein contents, similar to those found in different free-living cyanobacteria, including strains originally isolated from cycads (Lindblad and Bergman 1986). In addition, the in vitro activity of GOGAT in the cyanobiont of *C. revoluta* is also similar to that of free-living cyanobacteria (Lindblad et al. 1987).

5.3
Transfer of Fixed Nitrogen

Using $^{15}N_2$, it has been clearly demonstrated that the fixed nitrogen is transferred from the cyanobiont to the cycad. By analysing the xylem sap from freshly detached coralloid roots, two strategies for further processing the fixed nitrogen have been described (Pate et al. 1988). Citrulline and glutamine are the principal translocated N-solutes in *Macrozamia*, *Lepidozamia* and *Encephalartos*, all belonging to Zamiaceae. Glutamine and, to a smaller extent, glutamic acid (but not citrulline) are present in xylem sap of *Bowenia* (Boweniaceae) and *Cycas* (Cycadaceae). In contrast to other cyanobacterial symbioses, it seems that in cycads fixed nitrogen is translocated to the host either as a combination of citrulline and glutamine, or as glutamine alone.

The fixation of $^{14}CO_2$ and formation of [^{14}C]-citrulline and [^{14}C]-arginine by freshly isolated cyanobionts of *Macrozamia riedlei* (a cycad exporting the fixed nitrogen as a combination of glutamine and citrulline) and *Cycas revoluta* (a cycad exporting the fixed nitrogen as glutamine only) have been examined. In the *Macrozamia* cyanobiont [^{14}C]-citrulline was readily synthesised, whereas in the *Cycas* symbiont [^{14}C]-citrulline was formed only when the cells were incubated together with exogenous ornithine. Similar results were obtained when determining the formation of [^{14}C]-arginine by freshly isolated cells from both *Macrozamia* and *Cycas*. It is interesting to note the general differences between cells incubated in light and in darkness and between freshly isolated cyanobionts and corresponding free-living strains (Lindblad et al. 1991). Further studies are needed to address the role of cycads in regulating the export of fixed nitrogen by cyanobionts as glutamine in some (even though the capacity to synthesise citrulline is present) and as a combination of glutamine and citrulline in others.

5.4
Carbon Metabolism

Due to their location in coralloid roots (complete darkness), the cyanobionts are expected to have a heterotrophic mode of carbon nutrition. The fixed-carbon may be provided either by the photosynthetic host (the cycad) and/or by the cyanobacteria themselves via dark CO_2 fixation (see above). Interestingly, the symbiotic cyanobacteria retain in vivo functional RuBisCO (Lindblad et al. 1987), and they do fix CO_2 (Lindblad et al. 1991). Further investigations are needed to explore the fixation and further metabolism of CO_2.

Acknowledgements The Swedish Natural Science Research Council/The Swedish Research Council has financially supported our research on symbiotic cyanobacteria.

References

Adams DG (2000) Symbiotic interactions. In: Whitton BA, Potts M (eds) The Ecology of Cyanobacteria. Kluwer Academic Publishers, Dordrecht, pp 523–561

Ahern CP, Staff IA (1994) Symbiosis in cycads: The origin and development of coralloid roots in *Macrozamia communis* (Cycadaceae). Am J Bot 81:1559–1570

Bergman B, Lindblad P, Rai AN (1986) Nitrogenase in free-living and symbiotic cyanobacteria: immunoelectron microscopic localization. FEMS Microbiol Lett 35:75–78

Costa J-L, Lindblad P (2002) Cyanobacteria in symbiosis with cycads. In: Rai AN, Bergman B, Rasmussen U (eds) Cyanobacteria in Symbiosis. Kluwer Academic Publishers, Dordrecht, pp 195–205

Costa J-L, Paulsrud P, Lindblad P (1999) Cyanobiont diversity within coralloid roots of selected cycad species. FEMS Microbiol Ecol 28:85–91

Costa J-L, Paulsrud P, Lindblad P (2002) The cyanobacterial tRNA^{Leu}(UAA) intron: Evolutionary patterns in a genetic marker. Mol Biol Evol 19:850–857

Costa J-L, Romero EM, Lindblad P (2004) Sequence based data supports a single *Nostoc* strain in individual coralloid roots of cycads. FEMS Microbiol Ecol 49:481–487

Lindblad P (1990) Nitrogen and carbon metabolism in coralloid roots of cycads. Advances in Cycad Research I. Memoirs New York Bot Garden 57:104–113

Lindblad P, Bergman B (1986) Glutamine synthetase: activity and localization in cyanobacteria of the cycads *Cycas revoluta* and *Zamia skinneri*. Planta 169:1-7

Lindblad P, Bergman B (1990) The cycad-cyanobacterial symbiosis. In: Rai AS (ed) Handbook of Symbiotic Cyanobacteria. CRC Press, Boca Raton, FL, pp 137-159

Lindblad P, Bergman B, Hofsten AV, Hällbom L, Nylund JE (l985a) The cyanobacterium-*Zamia* symbiosis: an ultrastructural study. New Phytol 101:707-716

Lindblad P, Hällbom L, Bergman B (1985b) The cyanobacterium-*Zamia* symbiosis: C_2H_2-reduction and heterocyst frequency. Symbiosis 1:19-28

Lindblad P, Rai AN, Bergman B (1987) The *Cycas revoluta-Nostoc* symbiosis: enzyme activities of nitrogen and carbon metabolism in the cyanobiont. J Gen Microbiol 133:1695-1699

Lindblad P, Haselkorn R, Bergman B, Nierzwicki-Bauer SA (1989) Comparison of DNA restriction fragment length polymorphisms of *Nostoc* strains in and from cycads. Arch Microbiol 152:20-24

Lindblad P, Atkins CA, Pate JS (1991) N_2-fixation by freshly isolated *Nostoc* from coralloid roots of the cycad *Macrozamia riedlei* (Fischer ex Gaudichaud-Beaupré) C.A. Gardner. Plant Physiol 95:753-759

Nathanielsz CP, Staff IA (1975a) A mode of entry of blue-green algae into the apogeotropic roots or *Macrozamia communis* (L. Johnson). Am J Bot 62:232-235

Nathanielsz CP, Staff IA (1975b) On the occurence of intracellular blue-green algae in cortical cells of the apogeotropic roots of *Macrozama communis* (L. Johnson). Ann Bot 39:363-368

Norstog KJ, Nicholls TJ (1997) The Biology of the Cycads. Cornell Univ Press, Ithaca, NY

Obukowicz M, Schaller M, Kennedy GS (1981) Ultrastructure and phenolic histochemistry of the *Cycas revoluta-Anabaena* symbiosis. New Phytol 87:751-760

Ow MC, Gantar M, Elhai J (1999) Reconstruction of a cycad-cyanobacterial association. Symbiosis 27:125-134

Pate JS, Lindblad P, Atkins CA (1988) Pathways of assimilation and transfer of the fixed nitrogen in coralloid roots of cycad-*Nostoc* symbioses. Planta 176:461-471

Rai AN, Söderbäck E, Bergman B (2000) Cyanobacterium-plant symbioses. New Phytol 147:449-481

Structural Characteristics of the Cyanobacterium–*Azolla* Symbioses

Weiwen Zheng[1] (✉) · Liang Rang[2] · Birgitta Bergman[2]

[1] Biotechnology Institute, Fujian Academy of Agricultural Sciences,
350003 Fuzhou, Fujian, China
Zheng@botan.su.se, bcfaas01@hotmail.com

[2] Department of Botany, Stockholm University, 10691 Stockholm, Sweden

1	Introduction	236
2	Structural Characteristics of the Cyanobacterial–*Azolla* Symbiosis	237
2.1	Morphology and Structure of the Leaf Cavity	237
2.1.1	Morphogenesis of the Leaf Cavity	237
2.1.2	Pockets and Envelopes	238
2.1.3	Physical and Chemical Conditions	239
2.1.4	The Pore and Teat Cells	239
2.2	Structure and Function of the Epidermal Plant Trichomes	240
2.2.1	The Epidermal Trichomes/Hairs at the Apical Meristem Region	240
2.2.2	The Trichomes of the Leaf Cavities	241
2.2.3	The Trichomes around Young Sporocarps	242
2.2.4	The Trichomes of the Embryonic Sporophyte	243
2.2.5	Potential Roles of the Trichomes	244
2.3	Structural Aspects of the Cyanobiont	246
2.3.1	Hormogonia at the Apical Region of *Azolla*	247
2.3.2	The Heterocystous Filaments	247
2.3.3	Akinetes within Leaf Cavities and Sporocarps	249
2.4	The *Azolla*–Cyanobacterium Interactions	251
2.4.1	Horizontal Transfer of the Cyanobiont During Asexual Reproduction of *Azolla*	251
2.4.2	Vertical Transfer of the Cyanobiont during Sexual Reproduction	253
2.4.3	Re-establishment of *Azolla*–Cyanobacterial Symbiosis	255
2.5	The Third Partner in the *Azolla* Symbiosis	255
2.5.1	Occurrence and Identification of the Eubacteria Within *Azolla*	255
2.5.2	Morphology and Ultrastructure of the Bactobionts	256
2.5.3	Potential Role of the Eubacteria	258
3	Conclusions and Outlook	259
	References	260

Abstract Structure is a fundamental base for all life forms, whether plants or microbes. The development of special structures results in unique functions. Though structures of the mutualistic *Azolla*–cyanobacterial symbiosis are still largely un-explored, they have attracted attention of researchers in the past two decades. The occurrence of the leaf cavity and trichomes within the water-fern *Azolla* are hallmarks for the two cyanobacterial-plant symbiotic systems. The trichomes and the multicellular filaments are suggested to be involved in metabolic exchange between the cyanobionts and host plants due to their cell wall ingrowths, i.e., transfer cell characteristics. At the apical region of the *Azolla* plant, the primary branched trichomes touch each other, thereby forming linked bridge-like structures that lead to partitioning of the cyanobacteria into the young leaf cavities, thus promoting horizontal transfer of the cyanobiont during vegetative growth and asexual reproduction via sporophyte fragmentation. The trichomes developing during the sexual reproduction stages of *Azolla* facilitate the partitioning of the motile cyanobiont into the sporocarps, thus promoting a vertical transfer of cyanobacteria between *Azolla* generations, a capacity unique among cyanobacterial–plant symbioses. The cyanobionts in *Azolla*, *Blasia* and *Anthoceros* undergo pronounced morphological, physiological and molecular modifications to keep a synchronized development with the plant partner and to meet needs for maintaining a mutualistic symbiosis.

1
Introduction

Cyanobacteria possess an unprecedented capacity to enter into symbiotic relationships with members of the plant kingdom. These include plants from the divisions: Bryophyta (mosses, liverworts and hornworts), Pteridohyta (aquatic ferns of the genus *Azolla*), gymnosperms of the family Cycadaceae and angiosperms of the family Gunneraceae (Bergman et al. 1996, 2007). Thus far, the nitrogen-fixing *Azolla*–cyanobacterial symbiosis is only of economical importance to farming systems. The genus *Azolla*, identified by Lamarck in 1873, has been grouped into two sections based on the structure of the megasporocarp and is assumed to consist of seven species (Lumpkin and Plucknett 1980): section *Euazolla* includes *Azolla caroliniana* Willdenow, *A. filiculoides* Lamarck, and *A. mexicana* Presl. *A. microphylla* Kaulfuss, *A. rubra* Brown, and the section *Rhizosperma* include *A. nilotica* Decaisne and *A. pinnata* Brown.

Morphological and structural characteristics of the *Azolla*-cyanobacterial symbiosis were first given by Strasburger in 1783 and were subsequently described by several investigators (Sadebeck 1902; Smith 1938; Schaede 1947; Shen 1961; Moore 1969). Based on light microscopic observations, a detailed anatomical description of the symbiotic association was made by Konar and Kapoor (1972), followed by a series of scanning and transmission electron microscopy observations of the leaf cavity, the inoculation chamber of the sporocarps and the sporling, as well as the trichomes of different *Azolla* species and/or their cyanobacterium-free fronds (Peters et al. 1978; Duckett et al. 1975; Calvert et al. 1981, 1984, 1985; Zheng et al. 1986, 1987, 1988, 1990; Zheng and Huang 1994; Becking 1987; Nierzwicki-Bauer et al. 1989; Plazinski 1990). More recently, the pore and the associated teat cells of the leaf cavity were examined morphologically and cytochemically (Veys et al. 1999, 2000, 2002). The *Azolla*-cyanobacterium symbiosis has been proposed to be a natural microcosm in which, besides cyanobacteria, diverse eubacteria may function as a third symbiotic partner. The ultrastructure of these bacteria has by now been documented using light and electron microscopy in several studies since the 1970's (see Lechno-Yossef and Nierzwicki-Bauer 2002).

Fundamental morphological, physiological and molecular aspects of the *Azolla* hosts and their cyanobionts have been well covered in several reviews and book chapters (Shi and Hall 1988; Peters and Meeks 1989; Braun-Howland and Nierzwicki-Bauer 1990; Adams 2000; Lechno-Yossef and Nierzwicki-Bauer 2002). Here we attempt to summarize current understanding of these cyanobacterial symbioses with focus on recent developments regarding the morphogenesis, ultrastructure and functional analysis of the *Azolla*–cyanobacterium–bacteria association.

2
Structural Characteristics of the Cyanobacterial–*Azolla* Symbiosis

The most noticeable feature of cyanobacterial–*Azolla* symbiosis is the formation of the leaf cavity and its appendages, including the trichomes, the teat cells and the envelope. These specialized organs/structures are unique not only in cyanobacteria–plant symbiosis but also in plant–microbe symbiosis. It is suggested that they play important roles in the establishment, maintenance and continuity of the cyanobacterial–*Azolla* symbiosis.

2.1
Morphology and Structure of the Leaf Cavity

The leaf cavity harboring the cyanobacteria and bacteria is the essential structure in the *Azolla*–cyanobacterial symbiosis. It is also unique among nitrogen-fixing microbe–plant symbioses being the only site of symbiosis located in photosynthetic leaves. Most other microbes colonize structures located in non-photosynthetic plant organs, primarily in root nodules or their equivalents (Vessey et al. 2005).

2.1.1
Morphogenesis of the Leaf Cavity

The formation of the leaf cavity does not require the presence of the cyanobiont, since the cavities also develop in cyanobacterial-free plants (Fig. 1a; Walmsley et al. 1973; Peters et al. 1978). A slight depression at the ventral side of the developing dorsal leaf lobe appears to be the first recognizable sign of leaf cavity formation. In the younger leaves (leaves approximately 2nd–5th from the shoot apex), the cavities are more developed and start to open towards the outside (Fig. 1b). The opening of the leaf cavity closes as the leaf matures, usually occurring at leaf 6 in *A. microphylla* (Fig. 1c). A ma-

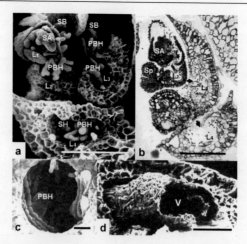

Fig. 1 Scanning and light micrographs illustrating the ontogeny of the *Azolla* leaf cavity. **a** The development of a leaf cavity in a cyanobacterium-free *A. filiculoides*. Note the occurrence of the trichomes (hairs) at the apical region and in the leaf cavities, SA: shoot apex; BA: branched shoot apex; L2–L4: leaf No. 2 to 4; PBH: primary branched hair. *Bar*: 100 µm. **b** Cross-section of the apex of a cyanobacterial colonized *A. microphylla* showing the development of the four first leaf cavities. S: stem, *star*: a trichome developing at the base of the sporophyte, *arrow*: the development of the opening/pore. *Bar*: 100 µm. **c** Close-up of the pore of a leaf cavity colonized by cyanobacterial filaments. *Bar*: 10 µm. **d** Cross sectioned leaf cavity with cyanobacteria occupying the periphery of the cavity leaving a central gas and mucilage-filled "void", CB: the inter-mingled cyanobacterial filaments, V: the void. *Bar*: 100 µm

ture leaf cavity is an extracellular leaf compartment, ellipsoidal in shape and approximately 0.3 mm in length (Adams 2000). The symbionts, embedded within a mucilaginous material, mostly line the periphery of the cavity wall in mature leaves (Fig. 1d).

2.1.2
Pockets and Envelopes

In *Azolla*, leaf "pockets" packed with cyanobionts and bacteria (bactobionts) are observed within each cavity in each individual leaf located along the main stem axis and its branches. Following an enzymatic digestion procedure, the pockets, which are bound by a limiting envelope, may be individually isolated as shown for *A. caroliniana* (Peters 1976). The envelope is composed of an inner and outer membrane that shows characteristics of membranes rather than solely an accumulation of mucilaginous material, although it lacks the typical tripartite membrane ultrastructure. The presence of the inner envelope in cyanobacterium-free *Azolla* eliminates a cyanobacterial origin (Nierzwicki-Bauer et al. 1989). It is assumed that the envelopes have a role in the interchange of metabolites between the host and the symbionts, since the foliar trace is separated from the cavity by one more layer of highly vacuolated plant cells (Duckett 1975; Peters et al. 1978). The permeability of the envelope may be of importance for the metabolic exchange between the host and symbionts (Nierzwicki-Bauer et al. 1989).

2.1.3
Physical and Chemical Conditions

The use of a cryochamber attached to a scanning-electron microscope, which allows observation of frozen hydrated specimens, demonstrated that the center of the leaf cavity is gaseous (Peters and Meeks 1989). Employing a microprobe, the oxygen concentration inside the leaf cavity of *A. filiculoides* was shown to be below the ambient oxygen concentration (Grilli Caiola et al. 1989). Lower concentrations of oxygen were also found by Uheda et al. (1995) who collected and analyzed gas bubbles in the leaf cavity of *A. pinnata*.

Ammonium concentration was comparatively low in leaves 1–4 (0.8 to 2 mM) but increased up to 6 mM in leaves 13 to 16 of *A. filiculoides*. This suggests that nitrogen fixation, taking place in heterocysts, is not inhibited by certain levels of ammonium as activity increased from leaf 1 to 12, and it was proposed that there is a mechanism for the removal of ammonium by the epidermal trichomes and potentially also by associated bacteria (Canini et al. 1990, 1993). High concentrations of cations (calcium, potassium and sodium) were found in older leaves, while the concentration of chloride ions, and pH were lower in older than in younger leaves (Canini et al. 1992, 1993). The dramatic change in calcium concentration was suggested to be due to inefficient utilization of the ion by disintegrating hair cells and to be a possible reason for the observed decreased activity of nitrogenase in old leaves (Canini et al. 1992).

2.1.4
The Pore and Teat Cells

The opening (circular pore, see Figs. 1c, 2a,b) in the cavity towards the surrounding environment is located in the adaxial epidermis of the *Azolla* leaf. Studies of the development of the cavity opening showed that four cell layers surround the pore in *A. filiculoides* (Fig. 2a). The layer inside the pore is made of teat-shaped cells extending from the initial epidermis layer (Veys et al. 1999, 2000, 2002). In *A. pinnata*, however, the pore was surrounded by four cell layers, which corroborate our own observations of the same species (Fig. 2b).

Three to four tiers of teat cells form a cone-like pore with an average diameter of 80 μm at the base. No mucilage was observed in the pore. Once fully elongated, the teat cells show typical ultrastructural features of secretory cells and produce cell wall projections relevant to their secretory activity.

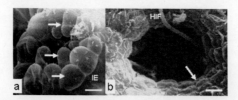

Fig. 2 Scanning electron micrographs (SEM) of the pore and teat cells in *Azolla*. **a** Two teat cell layers surround the pore of a leaf cavity in *A. filiculoides*. The teat cells (*white arrows*) develop from the inner epidermis (IE) of the leaf cavity. *Bar*: 12.5 μm. **b** Four cell layers surrounding the opening of leaf cavity of *A. pinnata imbricata*. Note the cyanobacterial filament entering the cavity (HLF). *Bar*: 10 μm

Recently, it was demonstrated that teat cell projections excrete compounds by an exocytotic mechanism. Based on the structure of the teat cells and the pore, it was suggested that they may serve the following functions: (i) act as a physical mechanism for gas exchange and water repulsion; (ii) play a defense role by preventing particles and other microorganisms from invading the cavity; (iii) serve as a barrier for maintaining the symbiosis by preventing the cyanobacterial, and possibly bacterial, symbionts from exiting the cavity (Veys et al. 2000, 2002).

2.2
Structure and Function of the Epidermal Plant Trichomes

It is apparent from studies of all *Azolla* species that the numerous trichomes extending from the plant cell walls surrounding the cavity are fundamental structures for the functioning of this cyanobacterial–plant symbiotic system and that the development of trichomes is initiated by the plant.

2.2.1
The Epidermal Trichomes/Hairs at the Apical Meristem Region

In cyanobacterial-free *A. pinnata* (Fig. 3a) and *A. microphylla* (Fig. 3b), the first epidermal trichome or hair appears at the base of the dome derived from the apical cell of the meristem. The second trichome arises once the first has formed 2–3 terminal cells, and is located at the adaxial end of the dorsal lobe primordium. In *A. mexicana*, however, the second trichome develops when the first epidermal trichome is fully engulfed in the developing leaf cavity (Peters and Meeks 1989). Mature trichomes have one basal cell that supports 5–8 terminal cells (Figs. 1a,c and 3b) and is termed a "primary branched hair" (PBH; Peters et al. 1978). As the *Azolla* sporophyte develops, the terminal cells of PBHs (at the main shoot apex (SA), branched apex (BA) and young leaves) undergo enlargement, elongation and eventually contact with each other, resulting in the formation of a bridge-like structure (BlS), linking the apical

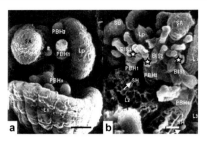

Fig. 3 SEM micrographs illustrating the development of epidermal trichomes in cyanobacterial-free *Azolla*. **a** An initial of the first epidermal trichome (It) arises near the main shoot apex (SA) in *A. pinnata*, and two trichomes (PBH1 and PBH2) are seen located at the leaf primordium. *Bar*: 10 µm. Lp: leaf primordium. **b** Apical region of *A. microphylla* showing that the PBHs developing at SA, branch apex (BA) and in young leaves appear close together. Second generation simple hairs (*arrows*) arising in young leaf cavities are also apparent. *Star*: the bridge-like structure. *Bar* 10 µm

meristem with young leaf cavities (Fig. 3b). When two BlSs at the shoot apex (or within a young cavity) are linked to a young cavity (or link the shoot apex with a branch apex), this is termed "alternate bridging".

Mature PBH contains abundant dictyosomes, endoplasmatic reticula, mitochondria and modest numbers of chloroplasts with poorly developed thylakoids. This suggests that the hair cells have a high metabolism but low photosynthetic capacity. A most typical feature of PBH is the numerous wall ingrowths that exhibit transfer cell ultrastructure (TCU), implying a vital role in nutrient/chemical signal exchange between the host and its symbionts.

2.2.2
The Trichomes of the Leaf Cavities

As *Azolla* leaf development and maturation proceed, two PBHs are "engulfed" by the developing leaf cavity in which other trichomes also start to differentiate via division of the epidermal cells surrounding the leaf cavity (Figs. 3a,b). The latter are composed of two types of cells: one terminal and one basal cell (Fig. 4a and b). These are termed "simple hair cells" (SH; Calvert et al. 1985). About 25 SHs are randomly distributed around the walls of mature leaf cavities, while the two PBHs are always located on the path of the foliar trace. During the PBH ontogeny, both cells undergo transfer cell differentiation followed by senescence, which begins in their terminal cell (Peters 1976; Calvert and Peters 1981; Calvert et al. 1985). The resulting terminal cell and basal cells have large nuclei located between pairs of large vacuoles. The terminal cell quickly differentiates TCU. The vacuoles decrease as the cytoplasm becomes more organelle-rich (Fig. 4c).

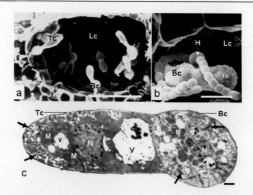

Fig. 4 Micrographs showing simple hairs/trichomes in *Azolla* leaf cavities. **a** Simple hairs extending from the cavity cell wall in cyanobacterial-free *A. filiculoides*. *Bar*: 100 μm. **b** A simple hair surrounded by cyanobiont filaments in colonized *A. pinnata*. H: heterocyst, Tc: terminal cell, Bc: basal cell. *Bar*: 10 μm. **c** A longitudinal TEM section of simple hair exhibiting ingrowths of cell wall (*black arrows*), indicating transfer cell ultrastructure, and abundant organelles. N: nucleus, M: mitochondria, V: vacuole. *Bar*: 2 μm

2.2.3
The Trichomes around Young Sporocarps

During the sexual reproduction of the *Azolla* sporophyte, a second round of trichome initials develop at the ventral surface, near the sporangial initials, at the *Azolla* apical meristem. The sporangia always develop in pairs. In *A. microphylla*, the first sporangial trichome flanks the proximal sporangium and is adaxial to the older leaf of the main stem axis. The second sporangial trichome is positioned just behind the developing sporocarp pair. Both trichomes have a basal cell, two to three terminal cells and exhibit TCU similar to that of PBHs at the apex and SHs in the leaf cavity (Fig. 5a). In *A. mexi-*

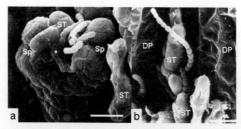

Fig. 5 SEM micrographs depicting primary sporangial trichomes in relation to a pair of sporangium initials in *A. microphylla*. **a** One of the two sporangial trichomes referred to as ST, which emerge from one side of the base of the sporangial initial and the other from the space between the sporangial pair (Sp). Note the cyanobacterial hormogonia adhere to the ST. *Bar*: 10 μm. **b** Sporangial trichomes arising at the base of a pair of developing sporocarps. Note that the STs are entangled by hormogonia. *Bar*: 10 μm

cana four trichomes, composed of a stalk cell, two body cells and 2–3 terminal cells were observed by Perkins and Peters (1993) during early developmental stages of the sporocarps.

As the sporocarp pair develops, more than 20 trichomes, with only one basal cell and one terminal cell arise around the base of the sporocarp pairs. These sporangial trichomes are morphologically analogous to the simple hairs in the leaf cavities (Fig. 5b). The ontogeny of the sporangial trichomes appears to be even more rapid than that of SHs in leaf cavities (Calvert et al. 1985). Structurally both cells of each of the sporangial trichome exhibit TCU (see Fig. 4c).

2.2.4
The Trichomes of the Embryonic Sporophyte

Azolla plants are capable of both sexual and asexual reproduction. During sexual reproduction each megasporocarp (female) is colonized by a small number of cyanobacterial akinetes (or spores; see below). These are located in the space between the megasporangium indusium and its apical membrane, which is a one-cell thick, distal remnant of the megasporangial wall.

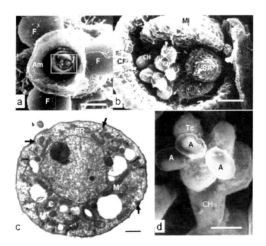

Fig. 6 Electron micrographs of the cotyledonary sporophyte of *A. microphylla*. **a** Top view of the embryonic sporophyte, showing the cotyledonary leaf (Cl), the shoot apex (SA), a modified true leaf (Ml) and cotyledonary hairs (CH), which break through the apical membrane (Am) of the embryonic sporophyte. F: float, b: bacterium. *Bar*: 100 μm. **b** Enlargement of the box in Fig. 6a. Note the six cotyledonary hairs emerging from the apical region of the embryonic sporophyte. *Bar*: 25 μm. **c** A cross section of simple trichome showing ingrowths in the cell wall (*arrow*) and abundant organelles: endoplasmic reticulum (ER), mitochondrion (M), rudimentary chloroplast (C), nucleus (N). *Bar*: 1 μm. **d** A multibranched cotyledonary hair within the modified true leaf No. 4. Note the cyanobiont akinetes (A) located close to the CH. Tc: terminal cell of CH. *Bar*: 10 μm

As the embryonic sporophyte breaks through the thin membrane of the germinating megasporocarp, four (in *A. mexicana*, Peters and Perkins 1993) or six (in *A. microphylla*, see Fig. 6a and b; Zheng et al. 1990) seriated epidermal trichomes, which develop along the axis of the cotyledonary leaf, are apparent. These trichomes are referred to as "cotyledonary hairs" (Peters and Perkins 1993). Based on analyses of serial sections of developing sporophytes, other cotyledonary hairs (CH) also occur within the modified true leaves. The first modified true leaf of the developing sporophyte contains two branched hairs, each of which is comprised of six to seven cells. The second, third and fourth modified true leaves each contain a single multibranched hair comprised of six to eight cells (Peters and Perkins 1993).

The cotyledonary hairs, which exhibit TCU (Fig. 6c), become directly associated with the germinating cyanobacterial akinetes (Fig. 6d). It is postulated that the cyanobacteria, which at this stage appear as akinetes (spores), are associated with the cotyledonary hairs in the germinating megasporocarp indusium chambers and might receive a chemical signal from this hair to trigger their germination. The hairs in the modified true leaf cavities might provide the nutrients required of the actively dividing cyanobionts before they differentiate nitrogen-fixing heterocysts (Zheng et al. 1990). Alternatively, they may play a role in exchange of metabolites between the partners (Peters and Perkins 1993).

2.2.5
Potential Roles of the Trichomes

The trichomes are major "organs" in the *Azolla* symbiosis and apparently important as they occur at most developmental stages and as they are always in close contact with the cyano- and bactobionts throughout the life cycle of the plant. Their multitude of cell types, including PBHs, SHs, STs and CHs, and distinct transfer cell ultrastructure have led to the following general suggestions in regards to their roles (Table 1).

(i) The PBHs participate in the exchange of the nitrogen fixed by the cyanobiont and the host, whereas SHs are specifically involved in the exchange of the carbon fixed by the plant in the opposite direction (see Calvert et al. 1985).

(ii) The PBHs and STs that occur at the apical region of the *Azolla* plants may attract, guide and facilitate the entrance of the cyanobacterial hormogonia into young non-colonized leaf cavities and developing sporocarps, respectively (see below).

(iii) All/some trichomes may be involved in attraction and/or recognition between the plant (hair cells) and the cyanobionts via secretion of unknown chemical signals. This is also supported by the presence of glycoprotein and ATPase activity in the hairs (Fig. 7).

Table 1 Types of trichomes in the *Azolla*–cyanobacterial symbiosis and their potential functions

Types of epidermal hairs/trichomes	Occurrence and location	Morphology and Structure	Suggested function(s)
Primary branched hair (PBH)	Initiated at the shoot apex of *Azolla* and located at the developing leaf cavity	One body cell supporting 5–8 terminal cells, exhibiting transfer cell ultra-structure (TCU)	Attracts and facilitates the hormogonia entering into the leaf cavity and provide them with nitrogenous compounds
Simple hair (SH)	Develop within the leaf cavity as cavity formation progresses	One body cell supporting one terminal cell exhibiting TCU	Promotes differentiation of the vegetative cells to heterocysts, and are involved in carbon exchange
Sporangial trichome (ST)	Initiated near the sporangial initials and also emerges at the base of the pair of developing sporocarps	One body cell and one or several terminal cells exhibiting TCU	Attracts and facilitates the entrance of hormogonia into the developing sporocarps
Cotyledonary hair (CH)	Initiated at the axis of the cotyledonary leaf and subsequently at the modified true leaf of the embryonic sporophyte	One body cell and several terminal cells exhibiting TCU	Facilitates germination of akinetes by providing unknown chemical signals

As seen in Fig. 7a, the glycoprotein layer positioned at the ingrowths of the cell wall and plasma membrane suggests that the trichomes may be involved in recognition of the endosymbionts via signaling. ATPases play an important role in the transportation and absorption of substances. Figure 7b and c shows that lead phosphate precipitates, resulting from the reaction of ATP hydrolytic enzymes, are distributed in the cell wall and cytoplasm of the trichome, implying that trichomes are involved in the metabolic exchange between the host and the cyanobionts.

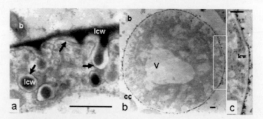

Fig. 7 Cytochemical localization of glycoproteins and ATPase in the trichomes. **a** Portion of the leaf cavity trichome treated with 500 ppm of Ruthenium red (RR), which stains glycoproteins or polysaccharides. Glycoproteins occur at the ingrowths of the cell wall (Icw), which exhibit transfer cell ultrastructure (TCU). *Arrow* denotes cell membrane. *Bar*: 1 μm. **b** A cross section of one leaf cavity trichome showing lead phosphate (staining ATPase) deposited at the surface of the cell wall and in the cytoplasm between the ingrowths of the cell wall (fresh *Azolla* leaves were fixed in 4% and 2.5% glutaradehyde in 50 mM cacodylate (pH 7.2), then washed with TRIS-malate buffer and subsequently incubated in Wachstein–Meisel reaction solution (3 mM Pb(NO$_3$)$_2$, 5 mM MgSO$_4$, 2 mM ATP, pH 7.2) at 22 °C for 3 h). cc: cyanobacterial cells, V: vacuole. *Bar*: 1 μm. **c** Magnification of the box in (**b**). *Bar*: 1 μm

2.3
Structural Aspects of the Cyanobiont

Like free-living cyanobacteria those symbiotically competent also may form four types of cells: hormogonium-like filaments (HLFs, motile infective units), vegetative cells (sites for the oxygenic photosynthesis), heterocysts (sites for nitrogen fixation), and akinetes (resting spores). As seen in Fig. 8,

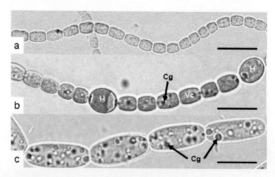

Fig. 8 Light micrographs of the various developmental stages shown by cyanobacterial filaments associated with *A. microphylla*. **a** A hormogonium-like filament at the apical region of *Azolla*. Note the actively dividing smaller cells and the lack of heterocysts. Cyanophycin granules (Cg; N-storages) are small and rare. *Bar*: 10 μm. **b** A mature heterocystous filament within a leaf cavity. Note the larger size of vegetative cells (Vc) and the presence of heterocysts (H). Cyanophycin granules are common. *Bar*: 10 μm. **c** Akinetes within a developing sporocarp. The considerably large cell size and the numerous cyanophycin granules are apparent at this stage. *Bar*: 10 μm

the cyanobacterial cells in Azolla cavities vary in size, shape and cell contents, when viewed under the light microscope. In general, cells of the HLFs of the cyanobiont are 2–3 μm in width and 3–4 μm in length. The cells within the heterocystous filament are larger, with the vegetative cells being 3–4 μm in width and 4–5 μm and heterocysts 4–6 × 5–7 μm^2. The size of the akinetes, only present in old leaves and in megasporocarps, range between 4–6 × 8–15 μm^2 in size. The akinetes typically possess many large cyanophycin granules compared to HLFs and vegetative cells.

2.3.1
Hormogonia at the Apical Region of *Azolla*

With plant growth, hormogonia (Fig. 8a) are always carried along the shoot apex, where they are associated with the PBH. The hormogonia are identified as actively dividing and undifferentiated filaments, lacking heterocysts but possessing motility. They are uniform in size but the cells are smaller (3–4 μm in length and 3 μm in width) than cells in vegetative filaments within mature leaf cavities (Fig. 8b). The ultrastructural characteristics of cells in the hormogonia-like filaments are similar to that of vegetative cells in leaf cavities, which exhibit typical characteristics of Gram negative bacteria with a peptidoglycan cell wall layer between an inner and an outer membrane. They usually contain thylakoid membrane (photosynthetic lamellae), ribosomes, numerous carboxysomes (polyhedral bodies with ribulose-1,5-bisphosphate carboxylase/oxygenase), a nucleoplasmic area and polymorphic electron dense material (Lang 1965; Neumüller and Bergman 1981). The photosynthetic thylakoid membranes often run in parallel to one another and to the cell wall, and are typically distributed in the peripheral portion of the cell. Glycogen granules appear between the parallel thylakoid membranes, and the larger cyanophycin granules are often located at the cross walls of the cell (Zheng et al. 1987; Braun-Howland and Nierzwicki-Bauer 1990). The presence of gas vesicles have been reported, although rarely. Being motile, hormogonia function as the plant infection units, being packed into the young leaf cavities/young sporocarps as host development proceeds (Meeks 1998).

2.3.2
The Heterocystous Filaments

The cyanobacterial filaments in young to intermediate leaf cavities are composed of vegetative cell and heterocysts. Within mature and older cavities, the relative shape and size of the vegetative cells are larger and contain fewer carboxysomes but more cyanophycin granules than do cells located in younger cavities. In contrast to the apically located vegetative cells, with thylakoids running parallel to the outer cell membrane, the vegetative cells found within mature to old cavities have thylakoids distributed throughout the cell cytoplasm. These thylakoids have frequently been observed to form whorls or lattice structures, especially near the formation of new cross walls (Hill 1977; Braun-Howland and Nierzwicki-Bauer 1990).

Nitrogen fixation is confined to the heterocysts that differentiate from vegetative cells within the leaf cavities, and at higher frequencies than in free-living cyanobacteria (Fig. 8b). The structural changes accompanying differentiation of vegetative cells into heterocysts (Fig. 9d–f) can be classified into three categories (Braun-Howland et al. 1988): at early stages, the most noticeable morphological and structural changes are elongation of the cells, appearance of a loosely organized layer of envelope material outside of the cell wall, narrowing of the area of contact with adjacent veg-

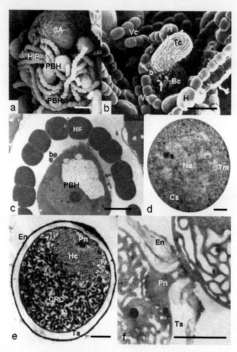

Fig. 9 EM micrographs of cyanobacteria in *Azolla microphylla*. **a** Hormogonium-like filaments (HLFs) are typically tightly packed and adhering to PBH at the *Azolla* apical region. *Bar*: 10 μm. **b** Heterocystous filaments in a mature leaf cavity. H: heterocyst, Vc: vegetative cell. *Bar*: 10 μm. **c** TEM micrograph of a young PBH surrounded by an actively dividing HLF at the apical region. *Bar*: 10 μm. **d** A cross sectioned vegetative cell. Tm: thylakoid membranes, Cs: carboxysomes (medium electron transparent polyhedral inclusions with RuBisC/O), Na: nuclear area. *Bar*: 10 μm. **e** A cross sectioned heterocyst, lacking the vegetative cellular inclusions. CRC: contorted reticulate lamellae, Hc: honeycomb configuration, Pn: polar nodule, En: envelope, Ta: electron transparent area. *Bar*: 1 μm. **f** Magnification of the connection between a heterocyst (*left*) and an adjacent vegetative cell (*right*). *Bar*: 1 μm

etative cells, appearance of polar nodules (equivalent to cyanophycin granules) at both ends of the cell (Fig. 9f), fragmentation of thylakoid membranes and decrease in the number of glycogen granules and carboxysomes. In the intermediate stage, the heterocyst cell width increases to approximately 5–6 µm, a thick envelope layer surrounding the cell wall but separated by an electron transparent area develops and polar nodules and honey-comb structures form (Fig. 9e). Later, glycogen granules disappear, carboxysomes become fewer and the contorted membranes appear throughout the cytoplasm region. Matured heterocysts are characterized by a well-developed honey-comb configuration, large polar nodules and tightly packed contorted reticulate lamellae (Fig. 9e and f). Morphological and structural changes are elongation of cells, appearance of a loosely organized layer of envelope material outside of the cell wall, formation of the necks via the narrowing of areas in contact with adjacent vegetative cells, appearance of the polar nodules at both ends of the cell, fragmentation of thylakoid membranes and decrease in cyanophycin, glycogen granules numbers and carboxysomes.

The number of heterocysts formed varies to some extent between species, but frequencies may reach as high as 30% in mature leaf cavities (Hill 1975), compared to 5–10% in free-living cyanobacteria grown in the absence of a combined nitrogen source (Braun-Howland and Nierzwicki-Bauer 1990).

2.3.3
Akinetes within Leaf Cavities and Sporocarps

Vegetative cells of cyanobacteria in mature-aged *Azolla* leaf cavities may differentiate into akinetes (Fig. 10a). Whereas akinetes comprise a significant proportion (about 17%) of the cyanobacterial population in *A. caroliniana*, they have not been observed in *A. filiculoides* (Hill 1977) or *A. pinnata* (Neumüller and Bergman 1981). Akinetes are surrounded by a thick external envelope, and their size is considerably larger than that of the vegetative cells. The function of akinetes in *Azolla* leaf cavities is unknown.

During sexual reproduction of *A. filiculoides*, *A. mexicana*, *A. microphylla* and *A. pinnata pinnata*, the hormogonium-like filaments from the apical colony become entrapped into the developing sporocarps. Once inside, the packed cyanobacterial cells rapidly differentiate into akinetes (Fig. 10b) (Zheng et al. 1990; Zheng and Huang 1994, 1998; Grilli Caiola et al. 1993; Peters and Perkins 1993). According to our observations of the akinete differentiation in sporocarps of *A. microphylla* using transmission electron microscopy, the transition of the hormogonium-like filaments into akinetes may be divided into the following phases: (1) enlargement, elongation and cytoplasm disorganization of the hormogonium cells to give rise to proakinetes (Fig. 10c, left); (2) secretion and deposition of membrane vesicles from the cytoplasm through the cell wall, forming a con-

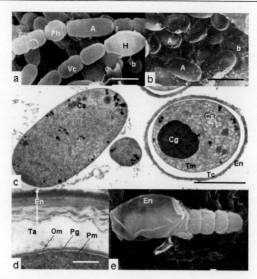

Fig. 10 EM micrographs of the akinete development in *A. microphylla*. **a** A cyanobactrial akinete (A), recognized by its size, shape and location, formed within a cyanobacterial filament and adjacent to a pro-heterocyst (Ph). Mature heterocyst (H) and vegetative cells (Vc). *Bar*: 5 μm. **b** Akinetes within a mature megasporocarp. *Bar*: 10 μm. **c** A pro-akinete (*left*), recognized by its oblong cell shape and lack of a distinct cell envelope, and more mature akinete (*right*) within a developing megasporocarp. *Bar*: 5 μm. **d** The multilayered external envelop of a mature akinete. Pg: peptidoglycan, Pm: plasma membrane, Om: outer membrane. *Bar*: 10 nm. **e** A germinating akinete showing the cyanobiont germlings emerging through the ruptured envelope of their parent akinete. *Bar*: 1 μm

tinuous thick fibrillar envelope around the cell; (3) rearrangement of the thylakoid-, plasma-, and outer membranes and cell inclusion reorganizations; (4) enlargement of the cyanophycin granules (Cg) and a widening of the transparent area (Ta) between the outer membrane and the envelope, now extending to its maximum (Fig. 10c, right); and (5) appearance of a multi-layer envelope with different electron densities outside the outer membrane (Fig. 10d), which indicates the emergence of a fully mature akinete.

Germination patterns (Fig. 10e) of akinetes within the embryonic sporocarp of *Azolla* were shown to be similar to that in akinetes of free-living cyanobacteria (Wildman et al. 1975; Sutherland et al. 1985). After undergoing a process of cytoplasmic reorganization and membrane rearrangement, the cyanobacterial germlings emerge by breaking through the envelope of the akinete (Peters and Perkins 1993). The development of akinetes and their preservation in the host megasporocarp are likely to play critical roles for the cyanobacterial survival during their vertical transfer between the *Azolla* generations.

2.4
The *Azolla*-Cyanobacterium Interactions

Interactions between *Azolla* and cyanobacteria take place throughout the life cycle of *Azolla* and are always accompanied by the occurrence of trichomes/hair cells. Owing to their transfer cell ultrastructure, the trichomes emerging at different developmental stages of the *Azolla* symbiosis are probably involved in signaling, recognition and metabolic exchange between *Azolla* and the cyanobionts (Table 1). It is proposed that *Azolla* attracts, guides and governs development and cyanobacterial cell differentiation, transfer of nutrients between organs and the behavior of the cyanobiont through the trichomes, which are the main host organs/structures in contact with the cyanobiont.

2.4.1
Horizontal Transfer of the Cyanobiont during Asexual Reproduction of *Azolla*

Previous observations have shown that the hormogonium-like cyanobacterial colony is always associated with the PBHs at the plant apical region (Peters et al. 1978; Zheng et al. 1986, 1987). Once the initial PBH arise from the apex, the hormogonium-like filaments (HLF) adhere selectively to the PBH (Fig. 9a). As the growth and development of the PBHs at the shoot apex continues, the association and enlargement of the HLF in the apical colony increases in parallel.

When a PBH of a young leaf reaches the PBHs at the apical meristem, a bridge-like structure is formed (BLS, see Fig. 3b), which facilitates transfer of some of the HLFs from the apical colony to an adaxial depression of a young leaf or a young leaf cavity. This process is repeated for each developing leaf. The BLS appears to be involved in the sequential partitioning of HLFs, and the repeated bridging results in partitioning of HLFs from the apex into the developing components of the apical segments (leaf depression, leaf cavities and branch apex) (Fig. 11a and b). As growth progresses the "older" leaves are displaced from the meristem, and their PBHs dissociate from the PBHs at the shoot apex, or those within younger leaves. This leads to an entrapment of the HLFs, previously at the apical region, into the leaf cavities.

Following the entrance of HLFs into the developing leaf cavity, they appear to migrate to the periphery of the cavity, and, as the cavity enlarges, they adhere to the simple hairs (SH) (Fig. 11c). Commensurate with the development of a leaf and maturation of the cavity, is differentiation of some vegetative cells of the HLFs into heterocyst. In mature cavities, the now multiheterocystous filaments remain in association with the hair cells and form a layer of entangled filaments around the periphery of the cavity (Nierzwicki-Bauer et al. 1989; Fig. 1d).

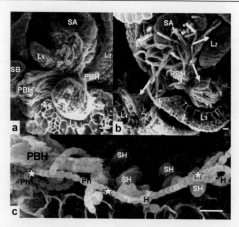

Fig. 11 Partitioning of the cyanobacteria into the *Azolla* leaf cavity via horizontal transfer. **a** Transmission of HLFs to the shoot apex, branched shoot apex and a leaf cavity in *A. filiculoides*. *Bar*: 5 µm. **b** Multiple partitioning (*white arrows*) of HLFs from the shoot apex into leaf cavities of *A. microphylla*. *Bar*: 10 µm. **c** A cyanobacterial filament (*stars*) extends from PBH to SHs via division and growth. Note that some of the vegetative cells have already differentiated into pro-heterocysts (Ph) or into large celled heterocysts (H). *Bar*: 10 µm

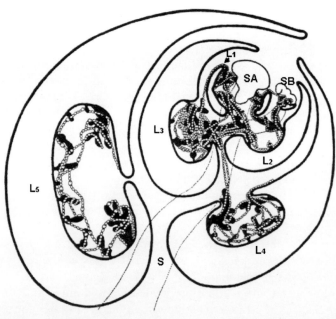

Fig. 12 Schematic illustration of the ontogeny of the trichomes (*black*) and the incorporation of the cyanobacteria into each developing leaf cavity. S: stem, SA: shoot apex; SB: branched apex; L1–L5: leaf 1 to 5

A diagrammatic model, illustrating the incorporation of the cyanobacterium into each leaf cavity, and their close association to the numerous hair cells is presented in Fig. 12. This model is based on our own (Zheng 1987) as well as previous observations (Calvert and Peters 1981; Peters and Calvert 1983; Nierzwicki-Bauer et al. 1989). The model builds on the observations that: (i) the inoculation of a leaf cavity with HLFs from the *Azolla* shoot apex is guided and facilitated by BLS; (ii) the alternate bridging results in multiple partitioning of the cyanobacteria and (iii) a synchronous development occurs between the leaf cavity and its cyanobiont. It is notable that the multiple partitioning of cyanobiont promotes a horizontal transfer of the cyanobiont from the main shoot apex to the branched apex of the sporophyte during the asexual propagation of the *Azolla* plants.

2.4.2
Vertical Transfer of the Cyanobiont during Sexual Reproduction

The development of the megasporocarp of *Azolla* and the cyanobacterial infection process has previously been examined using light and electron microscopy (Calvert et al. 1985; Zheng et al. 1988; Perkins and Peters 1983). These works showed that the partitioning of the cyanobionts into developing sporocarps was facilitated by sporangial trichomes (ST) (Fig. 13b). A schematic drawing based on these observations is presented in Fig. 14.

The colonization of the *Azolla* sporocarps by cyanobacteria start at a very early stage of the sporocarps development, i.e., already as the ventral leaf lobe primordium at the growing plant apex undergoes divisions to give rise to a pair of sporangial initials. In *A. microphylla*, two trichome initials (a basal trichome and an inter-sporangial trichome) are visible as soon as the sporangial initials emerge (Zheng et al. 1988; Zheng and Huang 1994) (Fig. 13a;

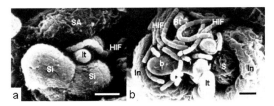

Fig. 13 SEM micrographs of the development of sporocarps in *A. microphylla*. *Bar*: 10 µm. **a** The ventral lobe primordium undergoes division to give rise to two sporangial initials (Si) which are ellipsoid in shape. Some hormogonium-like filaments (HLF) from the apical colony of the plant are already associated with the trichome initials (It). **b** A top view of a pair of young sporocarps showing the megasporangium (S) now surrounded by the developing indusia (In), and with additional inter-sporangial trichomes (Its) and basal trichomes (Bts) simultaneously emerging. Numerous HLFs are now associated with the trichomes and appear to proceed towards the megasporangium

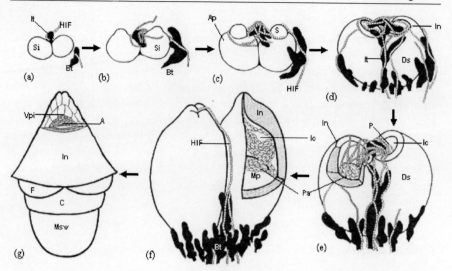

Fig. 14 Schematic diagram illustrating the infection of the *Azolla* sporocarps with the cyanobacterial hormogonium-like filaments. Ap: annular protuberance, S: sporangium, In: indusium; Ic: indusium chamber; Ds: developing sporocarps; P: pore of the indusium chamber; P: proakinetes; Mp: megaspore; Vpi: vertical profile of the indusium; A: akinete; F: float; C: collar; Msw: megasporangium wall

Fig. 14a), while in *A. mexicana*, four epidermal trichome initials develop (Perkins and Peters 1993).

Hormogonium-like filaments originating from the *Azolla* plant apical colony associate with the two sporangial initials once they emerge (Fig. 13a and b). Subsequently, annular protuberances develop around the base of the sporangial apical cells of the paired sporangial initials (Fig. 14c), which then develop into a two-layered pro-indusium, starting to surround, and eventually enclose, the developing megasporangium (Figs. 13b, 14d). Simultaneously, the indusium chambers are closing up but leaving an open pore at the tip (Fig. 14e). The HLFs eventually gain entry, facilitated by the trichomes, into the sporocarps through this pore (Fig. 14e). As the development proceeds, the partitioning of the long HLFs continue but now without the involvement of the sporangial trichomes, and, eventually, the pores close completely (Fig. 14f).

In *A. microphylla*, the vegetative cells of the hormogonia-like filaments differentiate into pro-akinetes soon after being entrapped in the indusium chamber of the developing sporocarps (Zheng et al. 1988). In *A. mexicana*, however, differentiation of pro-akinetes occasionally commence already outside the developing sporocarps (Perkins and Peters 1993). Concomitant with the maturation of the megasporocarp, the akinetes become located in the top part of the indusium chamber (Fig. 14g).

2.4.3
Re-establishment of *Azolla*-Cyanobacterial Symbiosis

The events leading to the re-establishment of *Azolla*–cyanobacteria symbiosis during sexual reproduction were described in the early 90s (Zheng et al. 1990; Peters and Perkins 1993). Following fertilization and embryogenesis, the cotyledonary leaf along with four (in *A. mexicana*, Peters and Perkins 1993) or six (in *A. microphylla*, Zheng et al. 1990) unbranched epidermal trichomes develop. Simultaneously, the cyanobacterial akinetes embedded in the mucilaginous matrix at the top of the indusium chamber are released and some start to germinate (Fig. 10e). As the cotyledonary leaf and the trichomes grow upward and extend into the indusium chamber, the now often germinating akinetes become intimately associated with the growing trichomes (Peters and Perkins 1993) (also see Fig. 8d). Some of the akinetes associated with the trichomes also germinate at this stage in *A. microphylla* (Zheng et al. 1990). Following the partitioning pattern described in 2.4.1 in this chapter, the germinating akinetes associated with the PBHs at the apical meristem are next packed into the leaf cavities, along with the PBHs displaced from apical region.

2.5
The Third Partner in the *Azolla* Symbiosis

Besides cyanobacteria, a second prokaryotic group is always present in the leaf cavities in all *Azolla* species, these bacteria are termed bactobionts (Lindblad et al. 1991). This is in contrast to the situation in all other cyanobacterial-plant symbioses in which cyanobacteria are the exclusive micro-symbionts.

2.5.1
Occurrence and Identification of the Eubacteria within *Azolla*

Since the 1960s, the occurrence of eubacteria has been recognized within *A. caroliniana*, *A. filiculoides*, *A. mexicana*, *A. microphylla* and *A. pinnata pinnata* (Grilli Caiola 1964; Gates et al. 1980; Wallance and Gates 1986; Forni et al. 1989, 1990; Plazinski et al. 1990; Nierzwicki-Bauer et al. 1990, 1991; Carrapico 1991; Zheng 1991; Serrano et al. 1999) and has by now been well-documented (Braun-Howland and Nierzwicki-Bauer 1990; Lechno-Yossef and Nierzwiski-Bauer 2002). The occurrence of the associated bacteria is illustrated in Figs. 6c, 7a,b, 9b, 10a,b and 13b.

In light of the above and our own observations (unpublished data), it is conclusively demonstrated: (1) that eubacteria are present in all *Azolla* species; (2) that the bacteria are present not only in leaf cavities but also at the shoot apex, inside the indusium chambers of both the mega- and microsporocarps and germlings, and even inside *Azolla* root and stem tissues; (3) that bacteria are more abundant and versatile in section *Azolla* than in section *Rhizosperma*; (4) that the bacteria inside section *Azolla* exhibit high genetic diversity: not only Gram negative but also Gram positive representatives, (5) that the bacteria may either be diazotrophic or non-diazotrophic; (6) that some bacteria are closely associated to the surface of the cyanobacteria and the *Azolla* trichomes, or are found embedded in the mucilaginous matrix; (7) that the bacterial community, as opposed to the cyanobacterial, are represented by different morphotypes and exhibit different ultrastructural characteristics in one *Azolla* plant, particularly within the section *Azolla* (Fig. 12); and (8) that the bacterial community is retained even in cyanobacteria-free *A. filiculoides*, *A. mexicana* and *A. microphylla* (author's unpublished observations), but rarely in *Azolla* species of the section *Rhizosperma* (Fig. 3a; Zheng et al. 1986).

The identity of the bacteria present in *Azolla* is not fully established and the range of bacteria identified is substantial (Table 2). The use of different bacterial isolation techniques, bacterial growth media and growth conditions have resulted in a diverse array of bacteria being identified. In addition, *Azolla* species from different geographical origins and the use of different identification methods for the host plant may have contributed to the disparate identities of the bacteria published. There is also the risk that some bacteria may represent epiphytes. Table 2 summarizes the various bacteria identified in two *Azolla* species (*A. filiculoides* and *A. caroliniana*). Nierzwicki-Bauer and colleagues, for instance, identified the isolate from *A. caroliniana* CA3001 as *Arthrobacter* sp. with conventional microbiology and biochemical methods. Later, when using 16S rRNA-PCR technique the bacteria in the same *Azolla* strain were identified as not less than four distinct bacterial genera *Variovorax* sp., *Agrobacterium* sp., *Rhizobium* sp. and *Frateuria* sp. It therefore appears that a great variation in genera may exist with no distinct specificity patterns between the host and the bacteria being apparent (Table 2).

It is estimated that only 1% of all bacteria are cultivable (Amman et al. 1995). The isolates identified so far may therefore represent only part of the bacterial community potentially capable of entering the *Azolla*–cyanobacterium associations. With the development of molecular techniques it should now be possible to reveal the genetic composition of the bacterial community without the need to isolate and culture the bacteria.

2.5.2
Morphology and Ultrastructure of the Bactobionts

Different morphological features of the bacteria in the *Azolla* leaf cavities can be seen in Figs. 6, 9, 13 and 15. The ultrastructure of bacteria in leaf cavities of different ages have been extensively examined in *A. caroliniana* and *A. mexicana*. The bacterial cells exhibit different cell shapes (rod, coccus and

Table 2 Identification of bacteria isolated from *A. caroliniana* and *A. filiculoides*

Investigators	Newton et al. 1979	Getes et al. 1980	Forni et al. 1989	Nierzwicki-Bauer et al. 1991	Shannon et al. 1993	Nierzwicki-Bauer et al. 2002
Azolla species or strains	*A. caroliniana* (from G.A. Peters, no accession number)	*A. caroliniana* (from G.A. Peters, no accession number)	*A. caroliniana*	*A. caroliniana* CA3001	*A. caroliniana* (from G.A. Peters, no accession number)	*A. caroliniana* CA3001
Bacteria identified	*Alcaligenes feacalis*, *Caulobacter fusiformis*	*Arthrobacter globiformis*, *Corynebacterium*	*Arthrobacter globiformis*, *A. nicotianae*, *A. autescens*, *A. cristallopoietes*, *A. pasces*	*Arthrobacter* sp.	*Arthrobacter* sp.	*Variovorax* sp., *Agrobacterium* sp., *Rhizobium* sp., *Frateuria* sp.

Investigators	Forni et al. 1989	Forni et al. 1990	Plazinski et al. 1990	Zheng et al. 1991	Shannon et al. 1993	Serrano et al. 1999
Azolla species or strains	*A. filiculoides*	*A. filiculoides* (5 strains)	*A. filiculoides* Wild, Australia	*A. microphylla*	*A. filiculoides*	*A. filiculoides* Wild, Portugal
Bacteria identified	*Arthrobacter* sp., *A. globiformis*, *A. nicotianae*	*Arthrobacter* sp., *A. globiformis*, *A. nicotianae*, *A. autescens*, *A. cristallopoietes*, *A. pasces*	*Agrobacterium* sp.	*Alcaligenes feacalis*	*Arthrobacter* sp.	*Agrobacterium radiobacter*, *Staphylococcus* sp., *Rhodococcus* sp., *R. equi*, *Corynebacterium jeikeium*

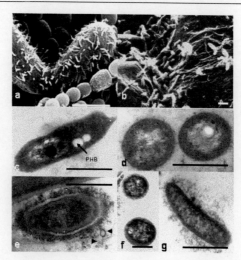

Fig. 15 EM micrographs of bacteria within *Azolla*. **a** Bacterial colony in a leaf cavity of *A. microphylla*. Note the numerous spiral shaped bacteria that adhere to the surface of single hair (SH) within the leaf cavity. Note also the size difference to the cyanobacterial cells (*right hand corner*). Bacteria of both the short rod and cocci types are apparent. *Bar*: 10 μm. **b** Bacteria adhering to the inner surface of the leaf cavity of cyanobacterial-free *A. filiculoides*. *Bar*: 0.1 μm. (**c–g**). Bacteria residing in the *Azolla* megasporocarp showing different ultrastructural characteristics, such as **c** electron transparent poly-β-hydroxybutyrate granules (PHB) and **d** multiple cell membranes. **e** A bacterium surrounded by a thick fibrillar capsule and membrane vesicles (*arrows*). **f** A typical gram-negative bacterium, and **g** a gram-positive bacterium with a thick fibrous layer. *Bar*: 500 nm

spiral), cell sizes (0.25–0.5 μm in width, 0.5–4 μm in length), different cell wall structures (with and without peptidoglycan layer) and thickness, as well as a variation in cytoplasmic organization and accumulation of storage material (e.g., glycogen, polyphosphate and poly-β-hydroxybutyrate granules). Based on these observations, five ultrastructurally distinct types of bacteria were recognized by Nierzwicki-Bauer and Aulfinger (1990, 1991).

Using TEM and SEM, ultrastructural characterization of bacteria residing within developing sporocarps of the two species *A. microphylla* and *A. pinnata pinnata* was recently investigated (authors' lab). Besides the presence of the distinct ultrastructural features of bacterial cells, some of the bacteria within the developing sporocarps were often seen surrounded by thick envelopes (Fig. 15c–g).

2.5.3
Potential Role of the Eubacteria

The fact that bacteria are located within different organs and tissues and present throughout the life cycle of *Azolla*, suggest that they may play an important role in *Azolla*–cyanobacteria symbioses. Using an antibody against both the Fe and MoFe proteins of the nitrogenase complex and immunogold labeling, the nitrogenase enzyme was detected in a subset of the bacteria within the leaf cavities of *A. filiculoides* and *A. caroliniana* (Lindblad et al. 1991) suggesting that some of the bacteria may be fixing nitrogen. *Arthrobacter* spp. isolated from *Azolla* were shown to have the ability to release the phytohormon auxin in culture medium (Forni et al. 1992a) and possibly therefore also in *Azolla* (Lechno-Yossef and Nierzwicki-Bauer 2002). The polysaccharides, which are a major component of the mucilaginous matrices in the leaf cavities, were proposed to be contributed to by the bactobionts and to be composed of glucose, galactose and fucose (Forni et al. 1992b, 1998). Hence, the bacteria may contribute to the formation of the mucilaginous matrices in the leaf cavities, and perhaps also in the indusium chambers of the sporocarps and in the sporling. It is also proposed that the bactobionts may contribute to the low oxygen micro-environment via respiration (Grilli Caiola et al. 1989), thereby stimulating nitrogen fixation of the cyano- and, potentially, the bactobionts.

Together these findings suggest that the *Azolla*–cyanobacterium–bacteria system may represent a natural microcosm in which all partners develop and interact co-coordinately to maintain and continue the beneficial mutual symbiosis through *Azolla* generations.

3
Conclusions and Outlook

The *Azolla*–cyanobacteria symbioses are unique not only among plant–cyanobacterial association but also among plant–microbe symbioses in general. *Azolla* has evolved several mechanisms to house their symbiotic partners. The most remarkable feature of those is the leaf cavities of *Azolla*. As in many other cyanobacterial–plant symbioses the leaf cavities contain mucilaginous slime and provide a relatively micro-aerobic environment suitable for nitrogen fixation. A second specialized structure for the symbiotic interactions in *Azolla* is the branched and non-branched trichomes, both of which exhibit, to some extent, transfer cell ultrastructure (TCU) which suggests the involvement in exchange of metabolites between the partners, such as carbon and nitrogen. In addition, the trichomes occur at specific developmental stages of *Azolla* and apparently play crucial roles in promoting both horizontal and vertical transfer of the cyanobiont during asexual and sexual reproduction of *Azolla*.

The cyanobionts associated with *Azolla* seem to be passive partners. Their growth, differentiation, and even death are probably governed by the host plant. The *Azolla* plant can be grown on media with and without their symbiotic partners, provided combined nitrogen is available. Appearance of the specialized symbiotic structures is independent of the presence of the cyanobionts. Once the symbiosis is established, the cyanobiont adapt their size, shape, differentiation frequency and patterns as well as behavior to stay in synchronous development with the plant partner and to maintain the mutual symbiosis. It is possible that the host control the cyanobiont via the trichomes or multicellular hairs by releasing nutrients or chemical signals throughout the life span of the symbiosis to achieve a stable symbiosis. For some reason the cyanobionts cannot survive in culture when separated from the *Azolla* plant. In addition, there are few reports regarding the reestablishment of the *Azolla* symbiosis in reconstitution experiments.

Compared to rhizobia-legume symbioses, studies on structure and function of cyanobacteria-*Azolla* symbioses are still in their infancy. We are just beginning to understand how host plants manage to house their prokaryotic partners, and the induction of their specialized symbiotic structures. It is now a challenge to identify genes encoding the morphogenesis and functional characteristics of these specialized plant structures and to disclose the regulatory network governing the transcription of the corresponding genes. In the *Azolla* symbioses, the cyanobacterial community (and the bacterial population) is embedded in the mucilage matrices of the leaf cavity. To monitor and understand the formation of these "biofilms", as well as the presence of any quorum sensing and cell death events during establishment and development of the symbioses is also needed. Recent cellular and molecular techniques have opened possibilities to also deeply exploit the dynamic changes in cytoskeletal elements and in periplasmic systems during *Azolla*-cyanobiont and cyanobiont-bactobiont interactions. If we understand the molecular and cellular mechanisms by which the plant hosts recognize, attract and control the cyanobacterial partners through specialized structures, and how the cyanobiont and bactobiont communities in turn adapt and coordinate their development and collective behavior, we may be able to extend these abilities to economically important crops such rice and wheat via the establishment of artificial nitrogen-fixing symbioses.

References

Adams DG (2000) Symbiotic interaction. In: Whitton B, Potts M (eds) Ecology of Cyanobacteria: Their Diversity in Time and Space. Kluwer Academic Publishers, Dordrecht, pp 523–561

Amman RI, Ludwig W, Schleifer K-H (1995) Phylogenetic identification and in situ detection of individual microbial cells without cultivation. Microbiol Rev 59:143–169

Becking JH (1987) Endophyte transmission and activity in the *Anabaena-Azolla* association. Plant Soil 100:183–212

Bergman B, Matveyev A, Rasmussen U (1996) Chemical signaling in cyanobacterial–plant symbioses. Trends Plant Sci 1:191–197

Bergman B, Rai AN, Rasmussen U (2007) Cyanobacteria associations. In: Elmerich C, Newton WE (eds) Associative and Endophytic Nitrogen-fixing Bacteria and Cyanobacterial Associations. Springer, Dordrecht, pp 257–302

Braun-Howland EB, Lindblad P, Nierzwicki-Bauer SA (1988) Dinitrogenase reductase (Fe protein) of nitrogenase in the cyanobacterial symbints of three *Azolla* species: localization and sequence of appearance during heterocyst differentiation. Planta 176:319–332

Braun-Howland EB, Nierzwicki-Bauer SA (1990) *Azolla–Anabaena* symbiosis: biochemistry, physiology, ultrastructure and molecular biology. In: Rai AN (ed) CRC Handbook of Symbiotic Cyanobacteria. CRC Press, Boca Raton, FL, pp 65–117

Calvert HE, Perkins SK, Peters GA (1984) Involvement of epidermal trichomes in the continuity of the *Azolla–Anabaena* symbiosis through the *Azolla* life cycle. Am J Bot 72:808 (Abstr)

Calvert HE, Pence MK, Peters GA (1985) Ultrastructural ontogeny of leaf cavity trichomes in *Azolla* implies a functional role in metabolite exchange. Protoplasma 129:10–27

Calvert HE, Peters GA (1981) The *Azolla–Anabaena azollae* relationship. IX. Morphological analysis of leaf cavity hair populations. New Phytol 89:327–335

Canini A, Grilli Caiola MG, Mascini M (1990) Ammonium content, nitrogenase activity and heterocyst frequency within the leaf cavies of *Azolla filiculoides* Lam. FEMS Microbiol Lett 71:205–210

Canini A, Grilli Caiola MG, Bertocchi P, Lavagnini MG, Mascini M (1992) Ion determinations within *Azolla* leaf cavities by microelectrodes. Sens Actuator B 7:431–435

Canini A, Albertano P, Grilli Caiola MG (1993) Subcellular localization of calcium in *Azolla–Anabaena* symbiosis by chlorotetracycline, ESI and EELS. Bot Acta 106:146–153

Carrapico F (1991) Are bacteria the 3rd partner of the *Azolla–Anabaena* symbiosis? Plant Soil 137:157–160

Duckett JG, Toth R, Soni SL (1975) An ultrastructural study of the *Azolla–Anabaena* relationship. New Phytol 75:111–118

Forni C, Grilli Caiola MG, Gentili S (1989) Bacteria in the *Azolla–Anabaena* symbiosis. In: Skinner AF, Boddey RM, Fendrik I (eds) Nitrogen Fixation with Non-legume. Kluwer Academic Publishers, Dordrecht, pp 83–88

Forni C, Gentili S, van Hove C, Grilli Caiola MG (1990) Isolation and characterization of the bacteria living in the sporocarps of *Azolla filiculoides* Lam. Ann Microbiol 40:235–243

Forni C, Riov J, Grilli Caiola MG, Tel-or E (1992a) Indole-3 acetic acid (IAA) production by *Arthrobacter* species isolated from *Azolla*. J Gen Microbiol 138:377–381

Forni C, Haegi A, Delogallo M, Grilli Caiola MG (1992b) Production of polysaccharides by *Arthrobacter globiformis* associated with *Anabaena azollae* in *Azolla* leaf cavity. FEMS Microbiol Lett 93:269–274

Forni C, Gaegi A, Del Gallo M (1998) Polysaccharide composition of the mucilage of *Azolla* algal packet. Symbiosis 24:303–313

Gates JE, Fisher RW, Candle RA (1980) The occurrence of corynoforme bacteria in the leaf cavity of *Azolla*. Arch Microbiol 127:163–165

Grilli Caiola MG (1964) Infrastructure di *Anabaena azollae* vivente nelle foglioline di *Azolla caroliniana*. Ann Microbiol 14:69–90

Grilli Caiola MG, Canini A, Moscone D (1989) Oxygen concentration, nitrogenase activity and heterocyst frequency in the leaf cavities of *Azolla filiculoides* Lam. FEMS Microbiol Lett 59:283–288

Grilli Caiola MG, Forni C, Castagnola M (1993) *Anabaena azollae* akinetes in the sporocarps of *Azolla filiculoides* Lam. Symbiosis 14:247-264

Hill DJ (1975) The pattern of development of *Anabaena* in the *Azolla-Anabaena* symbiosis. Planta 122:179-184

Hill DJ (1977) The role of *Anabaena* in the *Azolla-Anabaena* symbiosis. New Phytol 78:611-616

Konar RN, Kapoor RK (1972) An anatomic study of *Azolla pinnata*. Phytomorphology 22:211-223

Lang NJ (1965) Electron microscopic study of heterocyst development in *Anabaena azollae* Strasburger. J Phycol 1:127-134

Lechno-Yossef S, Nierzwicki-Bauer SA (2002) *Azolla-Anabaena* symbiosis. In: Rai AN, Bergman B, Rasmussen U (eds) Cyanobacteria in Symbiosis. Kluwer Academic Publishers, Dordrecht, pp 153-178

Lindblad P, Bergman B, Nierzwicki-Bauer SA (1991) Immunocytochemical localization of nitrogenase in bacteria symbiotically associated with *Azolla* spp. Appl Environ Microbiol 57:3637-3640

Lumpkin TA, Plucknett DL (1980) *Azolla*: botany, physiology, and use as a green manure. Econ Bot 34:111-153

Meeks JC (1998) Symbiosis between nitrogen-fixing cyanobacteria and plants. Bioscience 48:266-276

Moore AW (1969) *Azolla*: biology and agronomic significance. Bot Rev 35:17-34

Neumüller M, Bergman B (1981) The ultrastructure of *Anabaena azollae* in *Azolla pinnata*. Physiol Plant 51:69-76

Nierzwicki-Bauer SA, Aulfinger H (1990) Ultrastructural characterization of eubacteria residing within leaf cavities of symbiotic and cyanobiont-free *Azolla mexicana*. Curr Microbiol 21:123-129

Nierzwicki-Bauer SA, Aulfinger H (1991) Occurrence and ultrastructural characterization of bacteria in association with *Azolla*. Appl Environ Microbiol 57:3629-3636

Nierzwicki-Bauer SA, Aulfinger H, Braun-Howland EB (1989) Ultrastructural characterization of an inner envelope that confines *Azolla* endosymbionts to the leaf cavity periphery. Can J Bot 67:2711-2719

Perkins SK, Peters GA (1993) The *Azolla-Anabaena* symbiosis: endophyte continuity in the *Azolla* life-cycle is facilitated by epidermal trichomes. I. Partitioning of the endophytic *Anabaena* into developing sporocarps. New Phytol 123:53-64

Peters GA (1976) Studies on the *Azolla-Anabaena azollae* symbiosis. In: Newton WE, Nyman CJ (eds) Proc 1st International Symposium on Nitrogen Fixation. Washington University Press, Washington DC, pp 592-610

Peters GA, Calvert HE (1983) The *Azolla-Anabaena azollae* symbiosis. In: Goff LJ (ed) Algal Symbiosis. Combridge University Press, New York, pp 109-145

Peters GA, Toia RE Jr, Raveed D, Levine NJ (1978) The *Azolla-Anabaena azollae* relationship. VI. Morphological aspects of the association. New Phytol 80:583-593

Peters GA, Meeks JC (1989) The *Azolla-Anabaena* symbiosis: basic biology. Annu Rev Plant Physiol Plant Mol Biol 40:193-210

Peters GA, Perkins SK (1993) The *Azolla-Anabaena* symbiosis: endophyte continuity in the *Azolla* life-cycle is facilitated by epidermal trichomes. II. Re-establishment of the symbiosis following gametogenesis and embryogenesis. New Phytol 123:65-75

Plazinski J (1990) The *Azolla-Anabaena* symbiosis. In: Gresshoff PM (eds) The Molecular Biology of Symbiotic Nitrogen Fixation. CRC Press, Boca Raton, pp 51-75

Plazinski J, Taylor R, Shaw W, Croft L, Rolfe BG, Gunning BES (1990) Isolation of *Agrobacterium* sp. strain from the *Azolla* leaf cavity. FEMS Microbiol Lett 70:55-59

Sadebeck R (1902) Salviniaceae. In: Engler A, Prantl K (eds) Natürliche Pflanzenfamilien, I. Teil, 4. Abt. Wilhelm Engelmann, Leipzig, pp 383–402

Schaede R (1947) Untersuchungen über *Azolla* und ihre Symbiose mit Blaualgen. Planta 35:319–330

Serrano R, Carrapico F, Vidal R (1999) The presence of lectins in bacteria associated with *Azolla-Anabaena* symbiosis. Symbiosis 15:169–178

Shen YET (1961) Concerning *Azolla imbricate*. Am Fern J 51:151–155

Shi DJ, Hall DO (1988) The *Azolla-Anabaena* association: historical perspective, symbiosis and energy metabolism. Bot Rev 54:353–386

Smith GM (1938) Salviniaceae. In: Cryptogamic Botany, Vol 2. McGraw-Hill Inc., New York, pp 353–362

Sutherland JM, Stewart WDP, Herdman M (1985) Akinetes of the cyanobacterium *Nostoc* PCC 7524: morphological changes during synchronous germination. Arch Microbiol 122:269–274

Uheda E, Kitoh S, Dohmanu T, Shiomi N (1995) Isolation and analysis of gas bubbles in the cavities of *Azolla* leaves. Physiol Plant 93:1–4

Vessey JK, Pawlowski K, Bergman B (2005) Root-based N_2-fixing symbioses: legumes, actinorhizal plants, *Parasponia* sp. and cycads. Plant Soil 274:51–78

Veys P, Waterkeyn L, Lejeune A, van Hove C (1999) The pore of the leaf cavity of *Azolla*: morphology, cytochemistry and possible functions. Symbiosis 27:33–57

Veys P, Lejeune A, van Hove C (2000) The pore of the leaf cavity of *Azolla*: Interspecific morphological differences and continuity between the cavity envelopes. Symbiosis 29:33–47

Veys P, Lejeune A, van Hove C (2002) The of leaf cavity of *Azolla*: Teat-cell differentiation and cell wall projections. Protoplasma 219:31–42

Walmsley RD, Breen CM, Kyle E (1973) Aspects of the fern–algal relationship in *Azolla filiculoides*. News Lett Limnol Soc S Afr 20:13–16

Wallace WH, Gates JE (1986) Identification of eubacteria isolated from leaf cavities of 4 species the N_2-fixing *Azolla* fern as *Arthrobacter* Conn and Dimmick. Appl Environ Microbiol 55:425–429

Wildman RB, Loescher JH, Winger CL (1975) Development and germination of akinetes of *Aphanizomenon flos-aquae*. J Phycol 11:96–104

Zheng WW (1991) Occurrence of bacteria in *Azolla-Anabaena* association and their interaction. J Electron Microsc China 15:54–56

Zheng WW, Huang JH (1994) New data on the infection of the developing sporocarps with the symbionts of *Azolla* sporulation. J Fujian Acad Agric Sci 9:49–54

Zheng WW, Huang JH, Lu PJ, Liu CC (1986) The ultrastructure of *Azolla* and *Anabaena*-free *Azolla*. J Fujian Agric Coll 15:211–219

Zheng WW, Lin YH, Lu PJ, Liu CC (1987) Scanning electron microscopic observation of symbiotic relationship of *Azolla-Anabaena azollae* during the vegetative growth. Acta Bot Sinica 29:588–593

Zheng WW, Lin YH, Lu PJ, Liu CC (1988) Electron microscopic observation of symbiotic relationship between *Azolla* and *Anabaena azollae* during the formation of the sporocarp of *Azolla*. Acta Bot Sin 30:664–666

Zheng WW, Lin YH, Lin YH, Lu PJ, Liu CC (1990) Electron microscopic observation of symbiotic relationship between *Azolla* and *Anabaena* during the megaspore germination and sporeling development of *Azolla*. Acta Bot Sin 32:514–520

Zheng WW, Liu LH, Xiu WQ (1998) The fate and ultrastrucural changes of *Anabaena azollae* within the developing sporocarps of *Azolla*. Fujian J Agric 13:1–8

Relations Between Cyanobacterial Symbionts in Lichens and Plants

Jouko Rikkinen

Department of Biological and Environmental Sciences, University of Helsinki, 65, 00014 Helsinki, Finland
jouko.rikkinen@helsinki.fi

1	Introduction	265
2	Lichen-forming and Plant-associated *Nostoc* Strains	267
3	Concluding Remarks	268
	References	269

Abstract Cyanobacteria participate in many types of symbioses, either serving as a source of fixed carbon and nitrogen, as in many cyanolichens, or solely as a source of nitrogen, as in plant symbioses. Symbiotic strains of *Nostoc* are by far the most common cyanobionts in lichens and related strains are also found in thalloid bryophytes, cycads, and in the angiosperm *Gunnera*. This article provides a short summary of *Nostoc* diversity patterns in lichens and argues that symbiont-switches between lichen-forming fungi and plants may have played a role in the evolution of some extant cyanobacterial symbioses.

1
Introduction

Lichen-forming species represent a major ecological group among the Ascomycota, with over 13 500 lichenized species. While a great majority of these fungi associate with green algae, over 1500 species of lichen-forming fungi have cyanobacteria as primary or accessory photobionts. The general term "cyanolichens" refers to all types of lichens with cyanobacterial symbionts, either as the sole photosymbiotic component, or as an accessory photobiont in addition to green algae.

Cyanolichens are often divided into two artificial groups: the bipartite and tripartite species. Bipartite cyanolichens are stable symbioses between a single species of lichen-forming fungus and one cyanobacterial photobiont. In most of these lichens the cyanobiont forms a more or less continuous layer immediately below the upper cortex of the lichen thallus (Fig. 1). Tripartite cyanolichens, on the other hand, house both green algal and cyanobacterial symbionts. In these lichens the cyanobionts, which usually represent a minor proportion of photobiont biomass, are restricted to special structures called

Fig. 1 Main types of cyanolichens. **a** In the bipartite cyanolichen *Peltigera scabrosa* the *Nostoc* cyanobiont forms a continuous photobiont layer near the upper surface of the lichen thallus; **b** In the tripartite cyanolichen *Peltigera aphthosa* the *Nostoc* cyanobiont is restricted to wart-like cephalodia, and the green phycobiont *Coccomyxa* forms the photobiont layer; **c** A photosymbiodeme of the tripartite lichen *Peltigera aphthosa* has cyanobacteria (*Nostoc*) and green algae (*Coccomyxa*) as primary photobionts in different parts of the same thallus

cephalodia (Fig. 1). Over 500 lichen species are known to produce internal or external cephalodia. In addition, many green algal lichens associate with "free-living" cyanobacteria, presumably in order to access an extra supply of fixed nitrogen. Sometimes the free-living cyanobacteria are totally covered by fungal hyphae, and the resulting compound structures are called paracephalodia (Poelt and Mayhofer 1988).

The mycobionts of certain cephalodiate lichens may sometimes produce different thallus morphotypes in symbiosis with compatible green algae and cyanobacteria. Chimeroid lichens with green algae and cyanobacteria as primary photobionts in different parts of the same thallus are called photosymbiodemes (Fig. 1). The contrasting morphotypes may either combine into a compound thallus or, in some cases, live separate lives. Photosymbiodemes offer unique opportunities to compare the physiological performances of green algal and cyanobacterial lichen photobionts under almost identical conditions of habitat, growth history, and fungal association (Demmig-Adams et al. 1990; Green et al. 1993, 2002). On the whole, however, photosymbiodemes are rare and only occur in some species of *Lobaria*, *Nephroma*, *Peltigera*, *Pseudocyphellaria*, and *Sticta*.

Lichen-symbiotic cyanobacteria have the potential to give photosynthate and/or fixed nitrogen to their fungal partners (Palmqvist 2002; Rai 2002). The relative importance of these activities is known to vary between bi- and tripartite lichens. The cyanobionts of bipartite lichens usually show lower heterocyst frequencies and lower rates of nitrogen fixation than those of tripartite species. In tripartite cyanolichens, the cyanobiont often shows a high rate of nitrogen fixation, while the green algal photobiont typically produces most of the photosynthate. Both bipartite and tripartite lichen symbioses have clearly evolved repeatedly in different lineages of fungi and convergent evolution has often led to similar thallus structures in distantly related cyanolichens (Rikkinen 2002).

2
Lichen-forming and Plant-associated *Nostoc* Strains

Serious attempts to determine the strain identity of the cyanobacterial photobiont have only been made for a small fraction of all cyanolichen species. However, different strains of *Nostoc* are known to occur in many cyanolichens, especially in association with lichen-forming species of the Lecanorales (Ascomycota). Typical strains of *Nostoc* produce isopolar trichomes with no evidence of branching or meristematic zones, and their cells are cylindrical or spherical. Also the characteristic life-cycle, with motile hormogonia and vegetative filaments is shown by most strains in culture. However, some symbiotic strains do not produce hormogonia under common growth conditions.

Recent studies in Europe (Paulsrud and Lindblad 1998; Paulsrud et al. 1998, 2001; Oksanen et al. 2002, 2004; Lohtander et al. 2003), North America (Paulsrud et al. 2000; O'Brien et al. 2005), East Asia (Rikkinen et al. 2002), New Zealand (Summerfield et al. 2002), South-America (Stenroos et al. 2006), and Antarctica (Wirtz et al. 2003) have shown that cyanolichens house a wide variety of symbiotic *Nostoc* genotypes. Further diversity has been found from hornwort, liverwort and cycad symbioses (Costa et al. 1999, 2001), and from macroscopic colonies of non-symbiotic *Nostoc* (Wright et al. 2001).

All evidence indicates that the *Nostoc* symbionts of cyanolichens are closely related to plant-symbiotic and free-living strains. Together these organisms form a genetically diverse, but rather well-delimited, monophyletic group among the Nostocalean cyanobacteria. Most strains of *Nostoc* s.str. can be conveniently assigned to two broad groups. The first group is genetically heterogeneous and includes the cyanobionts of many predominately terricolous cyanolichens, including those of all tripartite *Nephroma* and *Peltigera* species, but also the *Nostoc* symbionts of thalloid bryophytes and cycads, many free-living strains, etc. The second group is genetically less diverse and seems to only contain cyanobionts of bipartite, predominately epiphytic or lithophytic cyanolichens—their fungal hosts, however, are variable and represent many different groups among the Ascomycota.

Studies have shown that lichen-forming fungi are highly selective with respect to their cyanobionts. Only a few closely related *Nostoc* strains typically serve as the appropriate symbiotic partners for each fungal species. On the other hand, many different fungi, often from different genera or even families, can exploit and potentially share specific cyanobacterial strains. Thus, some cyanolichens may form photobiont-mediated guilds, i.e. co-occurring populations of lichen-forming fungi that utilize a common symbiont. Some lichen-forming fungi may even depend on other guild members for the effective dispersal of their cyanobionts (Rikkinen et al. 2002). While the ecological boundaries between existing guilds may appear steep, on an evolutionary timescale they are easily crossed. For a lichen-forming fungus, each successful shift into a new guild has provided new opportunities for specialization and subsequent radiation. Furthermore, each shifting fungus has had the potential to change pre-existing ecological balances between previous guild members (Rikkinen 2003).

3
Concluding Remarks

In the present context it is particularly significant that during evolution many lichen-forming fungi may have formed novel associations with plant-symbiotic *Nostoc* strains and/or vice versa. One could even expect that "cyanobiont-switches" may have occurred regularly between terricolous lichens, thalloid bryophytes, and/or cycads. The cyanobionts of most terricolous lichens belong to a specific group of symbiotic *Nostoc* strains. Related cyanobionts are also found in the coralloid roots of cycads, in the symbiotic slime cavities of hornworts, and in the symbiotic auricles of thalloid liverworts. The two thalloid liverworts with *Nostoc* symbionts, *Blasia pusilla* and *Cavicularia densa*, occupy a very basal position among the complex thalloid liverworts. These plants, the hornworts and the cycads together represent a conspicuous proportion of all extant plant lineages that can be reliably traced back to pre-Permian times. Concurrently, among the Fungi, some terricolous lichens are believed to be of ancient origin. Considering all this antiquity, it may not be a coincidence that all the hosts rely on related cyanobacterial symbionts.

References

Costa JL, Paulsrud P, Lindblad P (1999) Cyanobiont diversity within coralloid roots of selected cycad species. FEBS Microbiol Ecol 28:85–91

Costa JL, Paulsrud P, Rikkinen J, Lindblad P (2001) Genetic diversity of *Nostoc* endophytically associated with two bryophyte species. Appl Environ Microbiol 67:4393–4396

Demmig-Adams B, Adams WW III, Green TGA, Czygan FC, Lange OL (1990) Differences in the susceptibility to light stress in two lichens forming a phycosymbiodeme, one partner possessing and one lacking the xanthophyll cycle. Oecologia 84:451–456

Green TGA, Büdel B, Heber U, Meyer A, Zellner H, Lange OL (1993) Differences in photosynthetic performance between cyanobacterial and green algal components of lichen photosymbiodemes measured in the field. New Phytol 125:723–731

Green TGA, Schlensog M, Sancho LG, Winkler JB, Broom FD, Schroeter B (2002) The photobiont determines the pattern of photosynthetic activity within a single lichen thallus containing cyanobacterial and green algal sectors (photosymbiodeme). Oecologia 130:191–198

Lohtander K, Oksanen I, Rikkinen J (2003) Genetic diversity of green algal and cyanobacterial photobionts in *Nephroma* (Peltigerales). Lichenologist 35:325–329

O'Brien H, Miadlikowska J, Lutzoni F (2005) Assessing host specialization in symbiotic cyanobacteria associated with four closely related species of the lichen fungus *Peltigera*. Eur J Phycol 40:363–378

Oksanen I, Lohtander K, Paulsrud P, Rikkinen J (2002) A molecular approach to cyanobacterial diversity in a rock pool community involving gelatinous lichens and free-living *Nostoc* colonies. Ann Bot Fenn 39:93–99

Oksanen I, Lohtander K, Sivonen K, Rikkinen J (2004) Repeat type distribution in trnL intron does not correspond with species phylogeny; comparison of the genetic markers 16S rRNA and trnL intron in heterocystous cyanobacteria. IJSEM 54:765–772

Palmqvist K (2002) Cyanolichens: Carbon metabolism. In: Rai A, Bergman B, Rasmussen U (eds) Cyanobacteria in symbiosis. Kluwer Academic Publishers, Dordrecht, pp 73–96

Paulsrud P, Lindblad P (1998) Sequence variation of the tRNA^{Leu} intron as a marker for genetic diversity and specificity of symbiotic cyanobacteria in some lichens. Appl Environ Microbiol 64:310–315

Paulsrud P, Rikkinen J, Lindblad P (1998) Cyanobiont specificity in some *Nostoc*-containing lichens and in a *Peltigera aphthosa* photosymbiodeme. New Phytol 139:517–524

Paulsrud P, Rikkinen J, Lindblad P (2000) Spatial patterns of photobiont diversity in some *Nostoc*-containing lichens. New Phytol 146:291–299

Paulsrud P, Rikkinen J, Lindblad P (2001) Field experiments on cyanobacterial specificity in *Peltigera aphthosa*. New Phytol 152:117–123

Poelt J, Mayhofer H (1987) Über Cyanotrophie bei Flechten. Plant Syst Evol 158:265–281

Rai A (2002) Cyanolichens: Nitrogen metabolism. In: Rai A, Bergman B, Rasmussen U (2002) Cyanobacteria in symbiosis. Kluwer Academic Publishers, Dordrecht, pp 97–115

Rikkinen J (2002) Cyanolichens: An evolutionary overview. In: Rai A, Bergman B, Rasmussen U (2002) Cyanobacteria in symbiosis. Kluwer Academic Publishers, Dordrecht, pp 31–72

Rikkinen J (2003) Ecological and evolutionary role of photobiont-mediated guilds in lichens. Symbiosis 34:99–110

Rikkinen J (2004) Ordination analysis of tRNA^{Leu} (UAA) intron sequences in lichen-forming *Nostoc* strains and other cyanobacteria. Symb Bot Upsalienses 34:377–391

Rikkinen J, Oksanen I, Lohtander K (2002) Lichen guilds share related cyanobacterial symbionts. Science 297:357

Stenroos S, Högnabba F, Myllys L, Hyvänen J, Thell A (2006) high selectivity in symbiotic associations of lichenized ascomycetes and cyanobacteria. Cladistics 22:230–238

Summerfield TC, Galloway DJ, Eaton-Rye JJ (2002) Species of cyanolichens from *Pseudocyphellaria* with indistinguishable ITS sequences have different photobionts. New Phytol 155:121–129

Wirtz N, Lumbsch T, Schroeter B, Türk R, Sancho L (2003) Lichen fungi have low cyanobiont selectivity in maritime Antarctica. New Phytol 160:177–183

Wright D, Prickett T, Helm RF, Potts M (2001) Form species *Nostoc commune* (Cyanobacteria). IJSEM 51:1839–1852

Part IV
Diazotrophic Endophytes

Diazotrophic Bacterial Endophytes in *Gramineae* and Other Plants

Michael Rothballer · Michael Schmid · Anton Hartmann (✉)

Department Microbe-Plant Interactions,
GSF – National Research Center for Environment and Health,
Ingolstädter Landstrasse 1, 85764 Neuherberg/Munich, Germany
anton.hartmann@gsf.de

1	Endophytic Bacteria – General Definitions and Methods of Isolation, Characterization, and Localization	274
2	Endophytic Bacteria as Biological Control and Plant Growth Promotion Agents	275
3	Diazotrophic Bacterial Endophytes	276
3.1	Introduction and General Considerations	277
3.2	Alpha-Proteobacteria	278
3.2.1	*Rhizobium* spp. and *Ochrobactrum* spp.	278
3.2.2	*Azospirillum* spp.	280
3.2.3	*Gluconacetobacter* spp.	282
3.3	Beta-Proteobacteria	284
3.3.1	*Herbaspirillum* spp.	284
3.3.2	*Azoarcus* spp.	286
3.3.3	*Burkholderia* spp.	288
3.4	Gamma-Proteobacteria	289
3.4.1	*Klebsiella pneumoniae*	290
3.4.2	*Pseudomonas stutzeri*	290
4	Conclusions and Outlook	291
	References	292

Abstract Almost every plant has endophytic bacteria, which do not cause any harm to the plant, but may exert supportive effects on plant development and health. Diazotrophic bacteria were found to reside in roots, stems, and leaves of *Gramineae* (e.g., sugarcane, rice, maize, wheat, *Miscanthus*, and *Pennisetum*) and other plants (e.g., coffee, sweet potato, pineapple, banana). *Gluconacetobacter diazotrophicus* is a typical endophytic diazotroph in sugarcane roots, stems, and leaves, where its plant growth promotion is supposedly based on a combination of nitrogen fixation and phytohormonal effect. Several Rhizobia were shown to colonize gramineous plants (rice, wheat, maize) endophytically which are grown in rotation or as mixed cultivation with legumes. They exert plant growth promoting effects in their non-leguminous hosts. Some *Azospirillum* strains are able to colonize the root cortex of *Gramineae* and other plants and act as plant growth promoting agents mostly via phytohormonal stimulation of root development and activity. *Herbaspirillum seropedicae* is residing inside the roots and stems of sugarcane, rice, sweet potato, and other plants. It has been shown to exert plant growth promotion and nitrogen fixation in planta. *Azoarcus* sp. BH72 and other species are endophytes of Kallar grass (*Leptochloa fusca*), a halophyte in Pakistan, and also of different rice species. They colonize the aerenchyma of roots, are able to fix nitrogen in planta, and may turn to a non-culturable state inside the host plant. Several diazotrophic *Burkholderia* species, *B. tropica*, *B. unamae*, and *B. brasilensis* were described as endophytes of sugarcane, rice, maize, and teosinte plants. Most interestingly, three quite closely related *Burkholderia* species (*B. phymatum*, *B. tuberum* and *B. mimosum*) are able to form nodules in legumes (e.g., *Mimosa* spp.) and fix nitrogen in a symbiotic state like Rhizobia. Among the gamma-proteobacteria, *Klebsiella pneumoniae* 342 and *Pseudomonas stutzeri* A1501 were identified as effective and systemic endophytes of maize and rice, respectively.

1
Endophytic Bacteria – General Definitions and Methods of Isolation, Characterization and Localization

The term "endophytes" is used quite broadly, from interactions of mycorrhizal or mycorrhizal-like beneficial fungi, to saprophytic or pathogenic fungi as well as bacteria. For this chapter we apply the following definition: Endophytic bacteria cause "... infections of plant parts which are inconspicuous, the infected host is at least transiently symptom less, and the microbial colonization can be demonstrated internally ..." (Stone et al. 2000). Despite the fact that this citation was used to characterize fungal endophytic interactions, it is equally valid for endophytic bacteria–plant interactions. This definition covers also avirulent microorganisms and latent pathogens and stresses the characterization that the endophytic bacteria do not cause apparent harm and therefore are characterized by an absence of macroscopically visible pathologic symptoms. In contrast, substantial improvement in plant growth and performance may occur in absolutely healthy plants.

The first indication for colonization of the interior of roots by non-symbiotic bacteria goes back to the scientific work of Lorenz Hiltner, when he defined the term "rhizosphere" in 1904. In this key publication, he also reported about microscopic observations of bacterial cells within the root cortex and in analogy to the term "mycorrhiza" he coined the term "bacteriorhiza". Hiltner stated that the occurrence of this "bacteriorhiza" is connected with healthy roots, which is in agreement with the endophyte definition. Independently, this observation was confirmed by R.L. Starkey in 1926 (Starkey 1958). The diazotrophic lifestyle of some endophytes was finally verified in the 1970s and 1980s using specific identification tools (see below).

Bacterial endophytes have been isolated from the roots of almost all plants as has been recently reviewed by Rosenblueth and Martinez-Romero (2006) and Hallmann and Berg (2006). The endophytic diversity found in different host plants varies. Some endophytic populations may occur ubiquitously or they have a certain host specificity. However, the term "specificity" should be reserved for endophytes, which colonize and grow only in one host (Schulz and Boyle 2006). If the interaction is less specific a term like "host preference" could be used. In any case, an adaptation of host and endophyte to one another has occurred. This adaptation may not only be directed towards a particular host plant, but even to the colonization of only one plant organ (roots vs. shoots or seeds).

The characterization of an endophytic agent appears to be an easy task, but in practice this presents severe problems to be solved. Some endophytes were characterized by entering a non-culturable state and therefore cannot be identified by culture techniques. Therefore, three approaches are generally applied to detect and identify endophytic bacteria in plant tissue: (i) surface sterilization of the host tissue and isolation of the microbes on appropriate growth media, (ii) PCR-amplification of 16S-rDNA or other bacterial or fungal marker genes from DNA or RNA extracted from colonized tissues, (iii) in situ detection with specific probes, like immunological methods, fluorescence tags (e.g., green fluorescence label, gfp) (Elbeltagy et al. 2001; Tombolini et al. 1999) or phylogenetic oligonucleotide probes and fluorescence in situ hybridization (Amann et al. 1995; Wagner et al. 2003).

The most commonly applied isolation procedures for endophytic microorganisms use a combination of surface sterilization of the plant tissue followed by either grinding of the plant tissue and subsequent plating on nutrient agar, or by placing small surface sterilized segments onto nutrient agar. Methods that avoid surface sterilization like vacuum or pressure extraction were also applied (Hallmann et al. 2006). The choice of the isolation and extraction procedure certainly affects the microbial spectrum retrieved, which can be characterized by cultivation and/or molecular detection methods for non-culturable endophytic microbes (van Overbeek et al. 2006). The direct in situ localization is regarded as an independent and necessary proof of an endophytic life style, because it cannot be excluded that after surface sterilization some organisms may have escaped the treatment in protected niches on the rhizoplane and are still able to grow out on culture media. On the other hand, an overly harsh treatment may affect the endophytic community in the cortex tissue. Confocal laser scanning microscopy and/or transmission electron microscopy are potent tools for clear-cut localization and interaction studies (Bloemberg and Camacho Carvajal 2006; Schmid et al. 2004).

2
Endophytic Bacteria as Biological Control and Plant Growth Promotion Agents

The application of some plant-associated bacteria, including endophytic bacteria, results in a reduction of incidence or severity of diseases. This phenomenon is termed "biological control" and the beneficial bacteria involved are usually understood as antagonists leading to an efficient reduction of the pathogen population or its disease-producing potential (Weller 1988). An alternative mechanism for biological control arises from bacterial metabolites affecting the plant by stimulating its resistance to pathogens, a process termed induced systemic resistance (ISR). Elicitors for such a response are components of the bacterial surface, like lipo- and exopolysaccharides and flagellin. More recently, bacterial signaling compounds such as the autoinducer N-acylhomoserine lactone of certain Gram-negative bacteria were identified to induce systemic pathogen resistance response in a tomato (Schuhegger et al. 2006). Resistance can also be elicited in plants after an attack by necrosis-producing pathogens, which is termed systemic acquired resistance (SAR). As reviewed by Kloepper and Ryu (2006) examples of endophytic bacteria, such as *Bacillus amyloliquefaciens* IN937a, *Bacillus pumilis* SE34, *Serratia marcescens* 90-166, *Pseudomonas fluorescens* 89B-61 (G8-4) and *P. fluorescens* CHAO to cause systemic resistance are to be found only in dicotyledonous plants such as tobacco, cucumber, tomato, cotton, or *Arabidopsis* and will not be dealt with in detail in this chapter. In monocots, the phenomenon of ISR is much less studied and poorly described to date, although diazotrophic endophytes may indeed harbor biological control properties of some kind.

Interestingly, many root-associated bacteria – whether isolated from the root surface or the root interior – have the potential to stimulate or modulate root growth and the architecture of the root system. These bacteria can be diazotrophs or able to support the plant in the suppression of pathogens. Growth promotion can be mediated by the production and excretion of plant hormones such as indole acetic acid and other auxins, or gibberellins by the bacteria (for review see Costacurta and Vanderleyden, 1995). Some root-associated bacteria exert a stimulation of root development by degrading the ethylene precursor ACC which relieves the inhibition of root development in many stress situations (for a review see (Glick et al. 1998). Further interesting features of endophytic microbes were collected in reviews and a book by Sturz et al. (2000), Rosenblueth and Martinez-Romero (2006) and Schulz et al. (2006).

3
Diazotrophic Bacterial Endophytes

The best-studied interactions of diazotrophic bacteria with plants leading to efficient biological nitrogen fixation (BNF) are the root nodule forming Rhizobia-legume and actinorhizal symbioses (Chapters 1–7). However, most important agricultural crops like wheat, rice, maize, and sugarcane, which belong to the monocotyledonous family *Gramineae (Poaceae)*, are not able to form these highly specialized symbiotic structures. Therefore, investigations about efficient nitrogen fixation in non-legumes have been carried out for many years leading to the most interesting discoveries of diazotrophic bacterial endophytes in these plants.

3.1
Introduction and General Considerations

In some genera within the *Gramineae*, as in some sugarcane varieties, quite substantial BNF does occur without any inoculation (Boddey 1995). In addition, there is also strong evidence from other gramineous plants like *Paspalum notatum* cv. batatais, the smooth cord grass (*Spartina alterniflora*), Kallar grass (*Leptochloa fusca*) in Pakistan, and dune grass (*Ammophila arenaria* and *Elymus mollis*) from Oregon (Dalton et al. 2004) growing in nitrogen-limited soils, that endophytic BNF can contribute substantially to the nitrogen need of *Gramineae*. In the case of *Paspalum notatum*, the ^{15}N isotope-dilution method showed that 10% of the total N accumulated in the grass originated from BNF (Boddey et al. 1983). A high diversity of diazotrophic bacteria have been isolated over the years from rhizosphere soil and the root surface (rhizoplane) of these plants, but none of these diazotrophic bacteria could be found to contribute considerable amounts of fixed nitrogen to its host plant, thus fulfilling Koch's postulates. Convincing evidence was collected by several research teams, that the endorhizosphere (i.e., the root interior with the central cylinder and the cortex with the root epidermis) could provide the habitat of effective diazotrophic populations. In the endorhizosphere, the bacteria have much better access to the carbon and energy flow from the plant host with much less competition as compared to the rhizoplane or rhizosphere soil. In addition, the products of BNF can find the way to the plant hosts much more easily.

Many decades after the original description of bacterial root endophytes by Hiltner in 1904 and Starkey in 1926, the evidence for diazotrophic bacterial endophytes gradually augmented in the 1970s and 1980s. Due to the fact that some *Azospirillum* isolates have been obtained from surface sterilized root samples, it was assumed that they should be present and highly active in root tissues (Döbereiner and Day 1976, Patriquin et al. 1983). Efforts to stain the bacteria with the non-specific activity dye tetrazolium red successfully showed endophytic bacterial colonization, but the results did not give information about the types of bacteria (Patriquin et al. 1983). There was a first immunofluorescence report of occurrence of *Azospirillum brasilense* in the root cortex of field-grown plants by Schank et al. (1979) and electron microscopic studies by Umali-Garcia et al. (1980). However, only with the application of more specific tools and techniques could the endophytic colonization of several diazotrophs be unequivocally shown for e.g., *Azospirillum brasilense*, e.g., Bashan and Levanony (1988) and *Azoarcus* sp., e.g., Reinhold and Hurek (1988). The discovery of *Herbaspirillum seropedicae* in maize roots by Baldani et al. (1986) and of *Acetobacter diazotrophicus* (now

Gluconacetobacter diazotrophicus) from within sugarcane roots (Döbereiner 1992) confirmed the success of the endophyte hypothesis for certain diazotrophs in non-leguminous plants. Since then, much of the ongoing research on nitrogen fixation with non-legumes is focused on endophytic diazotrophs. Endophytic *nifH* gene diversity was recently observed in rice roots using molecular cultivation independent techniques (Tan et al. 2003) and African sweet potato (Reiter et al. 2003). In addition to Gram-negative bacteria endophytic nitrogen fixing Gram-positive *Clostridia* could be shown for the first time to colonize endophytically the grass *Miscanthus sinensis* using t-RFLP analysis (Miyamoto et al. 2004).

Endophytic diazotrophic bacteria were reported to occur in high numbers (up to 10^8 bacteria per g root dry weight) when cultured from internal root tissues of plants that did not result from breeding for intensive agricultural practice (McClung et al. 1983, Reinhold et al. 1986). Diazotrophs were found to be highly enriched in the root interior as compared to the root surface and the soil. Therefore, this plant habitat should be highly advantageous for N_2-fixing bacteria because of the ready supply of substrate (McCully 2001). However, it can be questioned for several reasons, that all bacteria isolated from surface-sterilized roots of *Gramineae* can be considered true endophytes. Firstly, they may be very firmly attached to the root surface and embedded deeply in the mucilageous surface layer such that root washings and surface desinfection cannot efficiently inactivate them. Secondly, in the roots of gramineous plants, the cells of the cortex cell layer outside the vascular tissue and the pericycle degrade rather quickly. Therefore, only bacteria inside the central cylinder would be truly endophytic (McCully 2001). Considering this, the classification of an endophyte definitely requires detailed and specific localization studies, at best during the whole life span of a plant.

3.2
Alpha-Proteobacteria

Many nitrogen-fixing root endophytes belong to the alpha-proteobacteria, like the Rhizobiaceae, forming nitrogen-fixing symbioses with legumes. Rhizobia and other alpha-proteobacteria such as *Gluconacetobacter* spp. and *Azospirillum* spp. were also found as efficient endophytic root colonizers in *Gramineae* with different mode of colonization and functions.

3.2.1
***Rhizobium* spp. and *Ochrobactrum* spp.**

Among the diverse endophytic bacteria which are able to fix nitrogen (diazotrophs) and which can be isolated from healthy plant tissues of cereals are Rhizobia. Usually, these form specific symbiotic interactions with leguminous host plants (Chapters 1–4). It has been first shown independently by two laboratories, that *Rhizobium leguminosarum* bv. *trifolii* efficiently colonized roots of different monocots, such as wheat, maize, and rice (Schloter et al. 1997; Yanni et al. 1997). In the Egyptian Nile delta many hundred years' recorded history of rice/forage legume (Egyptian berseem clover, *Trifolium alexandrinum* L.) rotation existed on 60–70% of the 500 000 ha of land used for rice production. It has been estimated that this clover rotation with rice can replace 25–33% of the recommended amount of fertilizer-N needed for optimal rice production. Detailed studies have revealed that the clover rootnodule occupant *Rhizobium leguminosarum* bv. *trifolii* is forming endophytic associations with rice plants. Certain strains of endophytic Rhizobia as inoculants for a rice crop could significantly improve the vegetative growth, grain productivity, and agronomic fertilizer use efficiency (measured as kg grain yield per kg of fertilizer-N applied) (Dazzo and Yanni 2006). Detailed colonization studies using GFP-labeled *Rhizobium leguminosarum* have shown that the bacteria enter through cracks in the root cortex appearing at the root emergence site (Chi et al. 2005). Using a variety of endophytic GFP-labeled Rhizobia including *Sinorhizobium meliloti* USDA 1021, *Rhizobium leguminosarum* bv. *trifolii* USDA 2370, *Mesorhizobium huakui* 93, and *Azorhizobium caulinodans* ORS 571, an ascending migration from the roots to the leaves in rice plants was documented (Chi et al. 2005). In situ quantitation of bacteria using the CMEIAS image analysis system indicated local endophytic population densities reaching as high as 9×10^{10} Rhizobia per cm^3 in infected host tissues (Chi et al. 2005). The inoculated plants produced significantly higher root and shoot biomass, and increased their photosynthetic activity, stomatal conductance, water use efficiency and flag leaf area. The inoculated plants accumulated higher levels of growth-regulating phytohormones, such as indole acetic acid and gibberellin. Rhizobium-cereal associations are quite dynamic, including dissemination in tissues below and above ground, and enhanced the plant's root architecture as well as the overall growth physiology. Therefore, they have a high potential value as a biofertilizer for sustainable agricultural practice. A similar association was described between maize and *Rhizobium etli* (Gutierrez-Zamora and Martinez-Romero 2001). In the observed intercellular colonization of wheat roots by *Azorhizobium caulinodans* the plant flavonoid naringenin was shown to act as a stimulant (Webster et al. 1998). Recently, evidence for the presence of Rhizobia (16S-rDNA and *nifH* DNA) in surface sterilized roots of the desert grass *Lasiurus scindicus* Henrard in Rajastan, India were obtained, which is able to grow under very challenging and nutrient deficient conditions (Chowdhury et al. 2007).

Ochrobactrum oryzae sp. nov. belonging to the *Rhizobiaceae* family was recently characterized as an endophytic colonizer of deep water rice (Tripathi et al. 2006; Verma et al. 2004b). A nitrogen-fixing member of the genus *Ochrobactrum* could be isolated from nodules of *Acacia mangium* which forms efficient nodules in the roots of its host plant (Ngom et al. 2004). Recently, two diazotrophic nodule forming *Ochrobactrum* spp.,

O. lupini (Trujillo et al. 2005) and *O. cytisi* (Zurdo-Pineiro et al. 2007) were newly described as nodule-forming diazotrophic bacteria. Therefore, certain *Ochrobactrum* spp. can be considered also as endophytic root colonizers in *Gramineae* and also diazotrophic symbionts in certain legumes.

3.2.2
Azospirillum spp.

The original characterization of a spirillum-like bacterium dates back to Beijerinck in 1923 (Beijerinck 1925). Becking rediscovered a *Spirillum lipoferum*-like bacterium in 1963 (Becking 1963) and described its nitrogen fixation ability in pure culture (Becking 1982). Independent from these earlier findings, flagellated vibroid or spirillum-like bacteria were isolated again frequently in the 1970s from roots of grasses and cereal crops at the EMBRAPA-center in Seropedica, Rio de Janeiro, Brazil, by the group of Johanna Döbereiner, using malate-containing semisolid nitrogen free medium (NFB) (Döbereiner and Day 1976). The genus *Azospirillum* was originally described with the two species *A. lipoferum* and *A. brasilense* (Tarrand et al. 1978). In the following years, *A. amazonense* (Magalhaes et al. 1983), *A. irakense* (Khammas et al. 1989), *A. halopraeferens* (Reinhold et al. 1987), and *A. largimobile* (Sly and Stackebrandt 1999) followed. More recently, several new *Azospirillum* spp. have been characterized, but their root colonization behavior has not been studied in detail yet. These genera include *A. doebereinerae*, isolated from the roots of the giant C4-grass *Miscanthus sinensis* (Eckert et al. 2001), which is nowadays preferably cultivated as a so-called "energy plant" and renewable resource.

The taxonomy, physiology, and ecology of the genus *Azospirillum* have been intensively reviewed by Steenhoudt and Vanderleyden (2000) and Hartmann and Baldani (2006). *Azospirillum brasilense* and *A. lipoferum* are the most studied species of this genus up to date. Usually, these bacteria efficiently colonize the rhizoplane. For root surface colonizing strains such as *A. brasilense* Sp7, it could be demonstrated that they are connected with each other and the root surface with fibrillar material. They frequently form large aggregates of pleiomorphic cell forms at the emergence point of the side roots (Assmus et al. 1995), where obviously chemotactic behavior had guided them (Zhulin and Armitage 1993; Zhulin et al. 1996). *Azospirillum* spp. were sometimes found in superficial layers of the root cortex and isolations from surface-disinfected root samples were reported. Using highly specific detection and localization methods such as strain-specific monoclonal antibodies and immunogold staining as well as GFP-labeled strains or the FISH-method, it could be definitively proven that the *Azospirillum brasilense* strain Sp245 is able to colonize intercellular spaces inside the root cortex of wheat and even whole cortex cells (Assmus et al. 1995; Rothballer et al. 2003; Schloter and Hartmann 1998) (see also Fig. 1a and b). In contrast, the *A. brasilense* type strain Sp7, isolated from rhizosphere soil, was shown to colonize the rhizo-

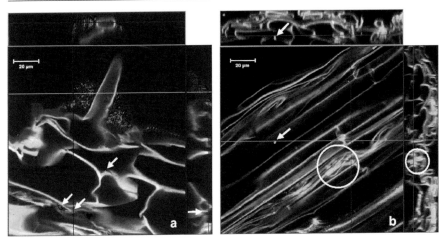

Fig. 1 The pictures show orthogonal views of a three-dimensional confocal laser scanning image created from a z-stack of xy-scans. The top view, framed in blue, depicts one picture from the middle of this z-stack. The red and green lines represent vertical optical cuts through the stack, which result in the side view images framed in red and green, respectively. In these side views the blue line marks the vertical position, where the top view image is located within the z-stack. **a** In situ detection by FISH analysis in a longitudinal root sliced from the root hair zone of a monoxenically grown, 4 week old Naxos wheat inoculated with *A. brasilense* strain Sp245. Arrows indicate endophytic colonization. Hybridization was performed with EUB-338-I, II, III-Cy3 (Amann et al. 1990; Daims et al. 1999) specific for the domain *Bacteria* and Abras-1420-Cy5 (Stoffels et al. 2001) specific for *A. brasilense*. The green framed side view shows an optical cut through the massively surface colonized root hair. **b** Root hair zone of a Naxos root, inoculated with *gfp* tagged *A. brasilense* strain Sp245. The *white arrow* marks an *A. brasilense* endophyte both in the top and the side view. The *white circle* encloses some cells, which could be easily mistaken for endophytic colonizers, as the top view shows a picture of the z-stack, which is from its position already quite deep in the root interior. But the *red framed* side view reveals that in fact the bacteria are attached to the root surface, but at the lowest point of a root surface depression. The pictures are taken from Rothballer et al. (2003)

plane only (Schloter and Hartmann 1998). It is now widely accepted that the extent of internal colonization may depend on the bacterial strain, the plant species, and other yet unidentified factors (Bashan and Levanony 1990).

Von Bülow and Döbereiner (1975) reported high rates of nitrogen fixation by *Azospirillum* in association with maize roots using the acetylene reduction assay with excised roots. However, these findings were questioned because it turned out that this assay overestimates the actual amount of nitrogen fixed. Nevertheless, careful experiments with the ^{15}N dilution technique and N-balance estimations confirmed that in some cases biological nitrogen fixation could account for several percent of the total nitrogen in the plant. On the other hand, it was demonstrated in many field inoculation experiments worldwide that significant increases in yields of 5–30% could be achieved by inoculation with *Azospirillum* (Dobbelaere et al. 2001; Okon and Labandera-Gonzalez 1994). The yield increases were usually more pronounced at low fertility soils with low or medium rates of N-fertilizer application. The reason of these yield increases was a better development of the root architecture enabling the plant to explore more efficiently the soil and take up more nutrients and water. Therefore, the plant growth promoting effect of *Azospirillum* is mostly due to phytohormonal effects of auxins (e.g., indole-3-acetic acid) and other phytohormones like gibberellines produced by this bacterium as reviewed in Okon (1994) and Dobbelaere and Okon (2007). This phenotype of plant growth promotion is rather widespread also in other rhizosphere and endophytic bacteria. Probably, root-associated bacteria create a larger root surface area for colonization and interaction and thus enable the plant to reach more resources of eventually limiting nutrients and water in the soil.

3.2.3
Gluconacetobacter spp.

The response of a sugarcane crop to nitrogen fertilizer can be rather low, depending on the cultivar used. A sugarcane crop accumulates between 100–200 kg N ha^{-1} per season and most of the fixed nitrogen is removed from the field at harvest because the trash representing about 25% of the senescent leaves is almost always burned off before cutting and less than 10% of the fixed-N remains in the field (Oliveira et al. 1994). Thus, the continuous cropping of sugarcane should quickly deplete soil N and cane yields should decline. However, such effects are not usually observed even after decades or centuries of cane cropping. Therefore, it had to be postulated that sugarcane must have a significant N-input through biological nitrogen fixation (BNF) and indeed this could be demonstrated by the quantification of the nitrogen budget including ^{15}N-dilution analysis in sugarcane plants (Urquiaga et al. 1992). When grown with irrigation and ample phosphate fertilizer and molybdenum is applied as foliar spray (500 g/ha), some sugarcane varieties obtain more than 150 kg N ha^{-1} per year from BNF (Boddey et al. 2003). This is remarkable, because molybdenum is the key trace element required as essential cofactors in the nitrogenase enzyme necessary for N_2-fixation and for nitrate reductase in the N-assimilation pathway. Using the ^{15}N-natural abundance technique, 11 sites were studied in sugarcane plantations in Sao Paulo, Minas Gerais, and the Pernambuco States of Brazil. In nine of the sites, BNF inputs were significant and ranged from 25% to 60% of the plant nitrogen (Boddey et al. 2001). Therefore, diazotrophic bacteria need to effectively fix nitrogen in an endophytic association in sugarcane to create this amount of nitrogen nutrition.

The attempts in the late 1980s to isolate diazotrophic bacteria from the internal plant tissue of roots and the aerial parts of sugarcane varieties that are known to have high sugar concentrations (especially in the stem) soon resulted in the discovery of a new nitrogen-fixing species (Cavalcante and Döbereiner 1988). These endophytic bacteria were first named *Acetobacter diazotrophicus* (Gillis et al. 1989) and later renamed to *Gluconacetobacter diazotrophicus* (Yamada et al. 1997). This bacterium is not able to survive in soil (Baldani et al. 1997) and is transmitted from plant to plant mainly via plant cuttings (Reis et al. 1994). *G. diazotrophicus* was also found to be associated with spores of arbuscular mycorrhizal fungi (AM). AM fungi are regularly forming symbiotic interactions with the roots of vascular plants increasing the uptake of nutrients, mostly phosphate. Inoculation of plants with AM-fungi harboring *G. diazotrophicus* increased the translocation of diazotrophic endophytes into the plants (Döbereiner et al. 1993).

G. diazotrophicus is a true endophyte, because it was found inside the roots, stems, and leaves and trash of sugarcane. The bacteria have been localized in the xylem vessels (James et al. 1994, 2001; Sevilla et al. 2001) and in the apoplast space (Dong et al. 1994). Interestingly, it could also be isolated from the sugarcane mealy bug (*Saccharococcus sacchari*) (Ashbolt and Inkerman 1990). *G. diazotrophicus* was also found in *Pennisetum purpureum* (Reis et al. 1994), sweet potato (Paula et al. 1991), coffee (Jiménez-Salgado et al. 1997) pineapple (Tapia-Hernandez et al. 2000), a grass called *Eleusine coracana* (Loganathan et al. 1999) and several other plants such as tea (root), mango (fruit), banana (root) and ragi (root and stem) (Muthukumarasamy et al. 2002). Two other new diazotrophic *Gluconacetobacter* species, *G. johannae* and *G. azotocaptans*, could be isolated from coffee plants and pineapple (Fuentes-Ramirez et al. 2001a), but these could not be found in sugarcane. Specific PCR-primers are available to identify both *G. diazotrophicus* (Kirchhof et al. 1998) and *G. johannae* as well as *G. azotocaptans* (Fuentes-Ramirez et al. 2001b).

G. diazotrophicus is perfectly able to fix nitrogen with 10% added sucrose, and it still grows well in 30% sucrose (Cavalcante and Döbereiner 1988). *G. diazotrophicus* does not have an assimilatory nitrate reduction and thus continues to fix nitrogen even in the presence of high amounts of nitrate. *G. diazotrophicus* is not able to transport and respire sucrose, but it excretes the saccharolytic enzyme levan sucrase, which provides the bacterium with glucose for growth and fructose for the formation of the exopolysaccharide levan. It is hypothesized that the exopolysaccharide levan is the "glue" that holds microcolonies of *G. diazotrophicus* together at the colonization sites inside sugarcane (Reis et al. 2007). This could be of great importance for the protection of nitrogen-fixing micro-colonies against the oxygen damage of nitrogenase. During growth, the bacterium strongly acidifies the environment resulting in pH-values of 3 and below. Nevertheless, this bacterium continues to grow and fix nitrogen at this pH-level for several days (Stephan et al. 1991). In addition to BNF, phytohormone production was demonstrated in this bacterium. The auxin indole-3-acetic acid was shown to be produced in a defined culture medium (Fuentes-Ramirez et al. 1993). It was also shown that it produces a bacteriocin which inhibits *Xanthomonas albilineans*, the causal agent of leaf scald disease in sugarcane (Piñon et al. 2002). Furthermore, an antagonistic effect was demonstrated against the pathogenic fungus *Collelotrichum falcatum*, a causal agent of red-rot in sugarcane (Muthukumarasamy et al. 2002).

Detailed studies about the mechanism of plant growth stimulation and nitrogen fixation were carried out using a Nif$^-$ mutant of *G. diazotrophicus* PAL-5 (Sevilla et al. 2001). When fixed nitrogen was not limiting, growth promotion was present in plants inoculated with the wild type and the Nif$^-$ mutant MAd3a, arguing that a phytohormone effect was operative. However, under limiting conditions of fixed-N, plants inoculated with the wild type grew better and had a higher nitrogen content. This indicates a significant transfer of fixed N from *G. diazotrophicus* cells to sugarcane resulting in plant growth promotion. These results were confirmed using $^{15}N_2$ gas incorporation (Sevilla et al. 2001). Interestingly, it could be demonstrated that the colonization of sugar cane by *Gluconacetobacter diazotrophicus* is inhibited under high N-fertilization conditions (Fuentes-Ramirez et al. 1999). Following the inoculation of sugarcane by *G. diazotrophicus* specific gene expression was observed (Nogueira et al. 2001).

Inoculation experiments of *G. diazotrophicus* wild type and mutant strains were also performed with rice and resulted in stimulated growth of rice plantlets after inoculation with the wild type but only a little increase after inoculation with the Nif$^-$ mutant strain MAd3a (as reported in Reis et al. (2007)). When PAl5 wild type was inoculated to maize, enhanced plant growth independent of nitrogen fixation was shown, although it did not enter maize roots in significant numbers (Riggs et al. 2001).

3.3
Beta-Proteobacteria

Numerous plant-associated and endophytic beneficial as well as pathogenic bacteria belong to the beta-proteobacteria. The distribution of diazotrophy and plant endophytic as well as a symbiotic life style is more widely distributed in this bacterial subphylum than previously known.

3.3.1
Herbaspirillum spp.

The first species of the genus *Herbaspirillum* (Schmid et al. 2006) to be described was *H. seropedicae* (Baldani et al. 1986). These bacteria effectively colonize the roots, stems, and leaves of sugarcane endophytically (Baldani et al. 1996). Until now, *H. seropedica* has been reported to colonize 13 members of the *Gramineae*, usually the roots but also the stems of rice and maize (James and Olivares 1997; Olivares et al. 1996). The endophytic colonization of vascular tissues of leaves in *Sorghum bicolor* by some *H. seropedica* isolates was investigated in detail by James et al. (1997). The bacteria are curved rods with polar flagella and grow best on dicarboxylic acids. They fix N_2 over a rather wide pH range from 5.3 to 8.0 (Baldani et al. 1986). *H. seropedica* has a nitrate reductase. Its nitrogenase activity is only partially inhibited by up to 20 mM ammonium (Fu and Burris 1989). Direct evidence for a nitrogen-fixing endophytic association has been obtained from studies of *Herbaspirillum* sp. strain B501 in rice, *Oryza officinalis*, using *nifH:gfp* reporter fusions (Elbeltagy et al. 2001). In addition, it has been shown that *Herbaspirillum* inoculation increased growth and nitrogen content of aluminum tolerant rice varieties which are known for increased organic acid allocation to the roots and rhizospohere (Gyaneshwar et al. 2002).

In 1996, *Pseudomonas rubrisubalbicans*, a mild endophytic pathogen in some sugarcane varieties, was reclassified as *Herbaspirillum rubrisubalbicans* by Baldani et al. (1996). The mode of infection of mottled stripe diseased susceptible and resistant varieties of sugarcane by endophytic *Herbaspirillum* spp. was investigated in detail by Olivares et al. (Olivares et al. 1997). A third species, mostly harboring strains from clinical but also from plant origins was provisionally named *Herbaspirillum* species 3, but all of the strains were unable to fix nitrogen. More recently, *H. frisingense* was isolated from surface sterilized roots and stems of C4 fibre and energy plants *Pennisetum purpureum* in Brazil and *Miscanthus sinensis* in Germany (Kirchhof et al. 2001). Most recently, *H. hiltneri* was isolated from surface sterilized wheat roots, but these bacteria failed to harbor the *nif*-genes (Rothballer et al. 2006).

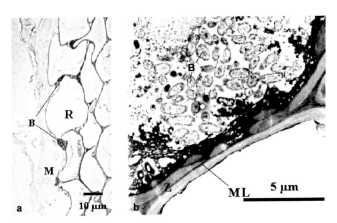

Fig. 2 Root cross section of a *Miscanthus sinensis* "Giganteus" root. Axenic roots were inoculated with *H. frisingense* Mb11. Roots were examined three days (**a**), and one week (**b**), after inoculation. Immunogold labeling with a polyclonal antiserum specific for strain Mb11 and subsequent enhancement with silver staining was performed. Numerous bacteria (B) have entered the intercellular spaces of the Rhizodermis (R) beneath the Mucigel layer (M) as can be seen in Fig. 2a. Intracellular colonization of strain Mb11 (B) in an intact xylem vessel in the central cylinder is shown in Fig. 2b; the middle lamella (ML) is not degraded

These bacteria are closely related to *H. lusitanum*, nitrogen fixing bacteria isolated from nodules of *Phaseolus vulgaris* (Valverde et al. 2003). The different *Herbaspirillum* spp. can be identified with a set of 16S-rRNA targeted oligonucleotide probes using the FISH-approach (Rothballer et al. 2006). Using these probes, the screening and characterization of new isolates from different sources is quickly possible avoiding the laborious genetic and phenotypic procedures. Using this screening, different *Herbaspirillum* spp. were detected in isolates from many other plants, including pineapple, banana, and rice (Magalhães-Cruz et al. 2001, Weber et al. 1999). Using immunological methods, as well as GFP-labeling and FISH-analysis, an efficient endophytic colonization of *Miscanthus* and wheat plants through roots was demonstrated using *H. frisingense* (Fig. 2) and *H. hiltneri* (Fig. 1), respectively.

3.3.2
Azoarcus spp.

Plant-associated *Azoarcus* spp. including the most studied strain BH72, *A. indigens, A. communis, Azovibrio restrictus, Azospira oryzae*, and *Azonexus fungiphilus* occur inside roots, on the root surface of grasses or in association with fungal resting stages. With the exception of *A. communis, Azoarcus* spp. have never been isolated from root-free soil. *Azoarcus* isolates were originally obtained from Kallar grass, *Leptochloa fusca* (L.) Kunth, a flood-tolerant halotolerant grass grown as a pioneer plant on salt-affected flooded low-fertility soils in the Punjab of Pakistan. Plant-associated *Azoarcus sensu lato* may be endemic and highly adapted to the host plant Kallar grass in saline-sodic soils. The plant-associated species *Azovibrio restrictus* and *Azospira oryzae*, however, showed a wider distribution and were isolated from surface sterilized roots of modern cultivars of *Oryza sativa* and wild rice species from the Philippines, Nepal, and Italy (Engelhard et al. 2000,) mostly from plants grown under flooded conditions as well as from fungal resting stages (sclerotia) (Hurek et al. 1997b).

The problem of isolating these bacteria more frequently may result from the fact that *Azoarcus* spp. shifts to an unculturable state in the association with the plant (Hurek et al. 2002). Another unique property of these bacteria is their apparent tight association with a fungus related to the *Basidiomycota* and the family *Ustilagomycetes*. It appeared that outside the plant, these bacteria survive in tight association with soil fungi (Hurek et al. 1997b). It was not possible to isolate Azoarcus spp. from sclerotium free sifted soil (Reinhold et al. 1986). Members of the genus *Ustilago* are plant-associated fungi and thus it is possible that the *Azoarcus*-harboring fungi also interact and infect plants. Thus, the colonization of resting stages of fungi may provide a shuttle for new infections of plants (Reinhold-Hurek and Hurek 2007). The interaction of *Azoarcus* sp. with AM-fungi, as has been observed in the case of *Gluconacetobacter diazotrophicus* in sugarcane, is not possible because of the environmental constraints for AM-fungi in both Kallar grass and rice fields caused by flooding.

Azoarcus sp. strain BH72 has been clearly shown by immunological and molecular marker techniques to colonize the roots of Kallar grass in the cortex region (aerenchyma) and to invade the stele of grasses through the vascular system (Hurek et al. 1991, 1994b; Reinhold-Hurek and Hurek 1998). Using in situ hybridization with fluorescent probes against the nitrogenase genes of Azoarcus sp. strain BH72 on sections of resin-embedded roots, nitrogen fixation was shown to be active in the aerenchyma of soil-grown Kallar grass (Hurek et al. 1997a). Apoplastic active nitrogen fixation in the cortex was also shown using nifH::gus and nifH::gfp fusions in rice seedlings in a gnotobiotic culture with Azoarcus sp. (Egener et al. 1999). Both infections and expression rates were higher when external carbon sources were added, but they were also detectable in their absence (Egener et al. 1998). Using $^{13}N_2$ gas, it could be shown that Azoarcus contributes fixed nitrogen to the plant in an unculturable state (Hurek et al. 2002). N-fertilization, the plant genotype, and other environmental conditions influence the nifH gene pools in rice roots (Tan et al. 2003).

In the case of Azoarcus sp. it has been clearly demonstrated, that endophytic colonization and the establishment of a plant growth promoting endophyte-grass-interaction are active processes. It could be shown that type IV pili, which are known as thin filamentous cell appendages of bacteria, are present also in Azoarcus sp. The respective genes pilAB were identified in Azoarcus sp. strain BH72 by Dörr et al. (1998). The emergence points of lateral roots are the primary sites of the colonization of grass roots by Azoarcus (Hurek et al. 1994b). The degree of colonization in pil^- mutants was strongly reduced, and thus type IV pili are essential for the efficient adhesion to, and entry into, roots. Most interestingly, a differential response by rice cultivars and an upregulation of jasmonate inducible defense proteins in some cultivars after inoculation with Azoarcus sp. strain BH72 was recently demonstrated (Miché et al. 2006).

An unique feature of the adaptation of Azoarcus sp. strain BH72 to extremely low oxygen tension (Hurek et al. 1994a) and the association with the fungus is the formation of intra-cytoplasmatic membrane stacks, called diazosomes (Hurek and Reinhold-Hurek 1995). Nitrogen fixing cells do not harbor these membranes under normal oxygen conditions (about $2\,\mu M$ of dissolved O_2). When these diazosomes are formed the cells become reprogrammed in terms of gene expression and protein composition. Interestingly, type IV pili are also involved in the interaction of Azoarcus with the Ascomycete fungus (related to Acremonium alternatum), which was isolated from surface sterilized roots of Kallar grass (Hurek and Reinhold-Hurek 1998). The pilAB-mutants adhere much less to the mycelium as compared to the Azoarcus wild type (Dörr et al. 1998). The decreased adhesion of the pilAB-mutant also resulted in decreased nitrogen-fixation activity and diazosome formation (Dörr et al. 1998). Once the *Azoarcus* cells have entered the plant roots, further progress in the apoplast is supported by the production of hydrolytic enzymes: an exoglycanase with cellobiohydrolase and β-glucosidase activities and an endoglucanase attacking oligosaccharides larger than cellobiose (Reinhold-Hurek et al. 1993, 2006). Thus, the colonization of the intercellular space consisting of mainly cellulose and hemicellulose and endophytic propagation is promoted. However, in contrast to most phytopathogens, *Azoarcus* sp. strain BH72 does not metabolize and grow on either the substrates or the breakdown products cellobiose and glucose (Reinhold-Hurek and Hurek 2000). In addition, the hydrolytic enzymes are not excreted into the environment but remain bound to the bacterial cell surface (Reinhold-Hurek et al. 1993). Therefore, the invasion by *Azoarcus* is much less aggressive as compared to the situation in plant pathogens. Mutants lacking these hydrolases are deficient in systemic spreading in the host and also showed a strong decrease in the intracellular infection of root epidermal cells of rice seedlings (Reinhold-Hurek and Hurek 2007). Further detailed insights into the physiological potentials of the N_2-fixing endophyte *Azoarcus* sp. strain BH72 was made possible by the presentation of the fully annotated genome (Krause et al. 2006).

3.3.3
Burkholderia spp.

The survey of endophytic diazotrophs in sugarcane and rice in the 1980s by the Döbereiner group at the EMPRAPA-center for Agrobiology in Seropedica, Rio de Janeiro, Brazil, also resulted in the isolation of a group of diazotrophs using LGIP semisolid nitrogen-free medium, usually applied for the enrichment and isolation of *G. diazotrophicus* (Reis et al. 1994). This group of isolates was provisionally named "isolates E" (Oliviera 1992) and were later characterized to belong to the genus *Burkholderia* (Hartmann et al. 1995). Diazotrophic isolates obtained from surface sterilized banana and pineapple material in Brazil (Magalhães-Cruz et al. 2001, Weber et al. 1999) as well as from maize and coffee plants in different climatic regions of Mexico (Estrada-De Los Santos et al. 2001) further documented an yet unknown richness of *Burkholderia* related bacteria associated with different plants. These observations led to the identification of *Burkholderia tropica* (Reis et al. 2004) including isolates from sugarcane plants collected from Pernambuco, Brazil, and South Africa and from maize and teosinte plants in Mexico. Another, closely related *Burkholderia* species, *B. unamae*, was identified covering the endophytic isolates of teosinte plants (Caballero-Mellado et al. 2004). A third endophytic group, *Cand. Burkholderia brasilensis*, combines the "bacteria E" isolates from rice plants with endophytic isolates of sweet potato (*Ipomea batatas*), cassava (*Manihot sculentum*) and sugarcane (Baldani V. L. D., pers. com.). *B. tropica* and *Cand. B. brasilensis* differ in their use of several carbon sources as well as their colony morphology on a medium containing mannitol at pH 5.5. They can also be differentiated by FISH-analysis based on 16S-rDNA-targeting oligonucleotide probes (Schmid and Hartmann 2007). The knowledge about the high diversity of endophytic *Burkholderia* spp. was even further extended by the characterization of the diversity of *Burkholderia* in field-grown maize and sugarcane (Perin et al. 2006). An endophytic colonization of *Vitis vinifera* L. by plant growth-promoting *Burkholderia* sp. strain PsJN was recently demonstrated by Compant et al. (2005).

Most interestingly, *Burkholderia* species have been isolated from the inside of nodules of tropical papilionoid legumes *Aspalathus carnosa* and *Machaerium lunatum* (Moulin et al. 2001) which harbor symbiotic *nifH* and *nodA* genes, closely resembling the *nifH* and *nodA* genes of Rhizobia. Therefore, these beta-subgroup nodule-associated bacteria are called "Beta-Rhizobia". These bacteria were phylogenetically characterized as *Burkholdera tuberum* STM678 and *B. phymatum* STM815 (Vandamme et al. 2002). Interestingly, *B. tuberum* is allocated to the same branch of the 16S-rDNA phylogenetic tree as cand. *B. brasilensis* and *B. kururiensis*. Bacteria closely related to *B. phymatum* had been already obtained from *Mimosa invisa* and *M. pudica* nodules in Papua, New Guinea by Trinick (1980). Another group of *Burkholderia* isolates from nodules of *Mimosa* spp. from Taiwan and South America were characterized as *B. mimosarum* (Chen et al. 2006), which is closely related to *B. tropica* and *B. unamae* in the 16S-rDNA phylogeny. It could be demonstrated that *B. phymatum* can also form highly effective nitrogen fixing nodules after infection of axenically grown *Mimosa* spp. (Elliott et al. 2007). In addition to *Burkholderia* spp., *Cupriovidus taiwanensis* (syn. *Ralstonia* or *Wautersia taiwanensis*) (Vandamme and Coenye 2004) have been isolated from nodules of *Mimosa pudica, M. diplotricha* and *M. pifra* in Taiwan (Chen et al. 2003) and from *M. pudica* nodules in north and south India (Verma et al. 2004a). Thus, it is quite possible that *Burkholderia* spp. which are endophytic in non-leguminous dicots or even monocots have some features in common with these nodule forming "Beta-Rhizobia".

Another fascinating facet of diazotrophic *Burkholderia* is the finding that an endosymbiotic bacterium has been discovered within spores of arbuscular mycorrhizal fungus *Gigaspora margarita*, which are closely related to *Burkholderia* (Jargeat et al. 2004) and to which *nif*-genes have been associated using molecular techniques (Minerdi et al. 2001).

3.4
Gamma-Proteobacteria

Nitrogen-fixing bacteria are well known among gamma-proteobacteria with a high diversity in enteric bacteria (Young 1992). However, only a few isolates such as *Klebsiella pneumoniae* 342 and *Pseudomonas stutzeri* A15 have been identified as endopyhtic bacteria up to this date.

3.4.1
Klebsiella pneumoniae

The diazotrophic endophyte *Klebsiella pneumoniae* 342 (Kp342) was originally isolated from the nitrogen-efficient maize line 342 from the CIM-MYT collection (Chelius and Triplett 2000). Various physiological tests and DNA:DNA hybridization assays, which showed 78.4% hybridization between DNA of Kp342 and DNA from the type strain of *K. pneumoniae* demonstrated that Kp342 belongs to the opportunistic human pathogen species *K. pneumoniae* (Dong et al. 2003a). However, using ribotyping, the strain Kp342 is allocated to the KpIII group, while most clinical isolates belong to the KpI group. Kp342 lacks type I pili, aerobactin production, and a ferric aerobactin receptor as virulence factors and expresses type 3 pili only weakly (Dong et al. 2003a). This strongly suggests that Kp342 is *not* an animal/human pathogen.

K. pneumoniae 342 very efficiently colonizes the following host plants endophytically: maize, *Triticum aestivum*, *Oryza sativa*, *Medicago trunctula*, *Medicago sativa*, and *Arabidopsis thaliana* (Dong et al. 2003b, 2003c). Most interestingly, only a single Kp342 cell was sufficient to give raise to a high internal colonization of the plant a few days after inoculation (Dong et al. 2003c). Optimal endophytic colonization was obtained with 10^3-10^4 inoculated cells per plant. The colonization efficiency was found to be different in dicots as compared to monocots as tested so far. While in dicots the cell numbers did not exceed 10^5 cells per gram, the bacterial colonization levels reached in monocots were considerably higher, ranging between 10^7 and 10^8 cells per g (Dong et al. 2003c).

Kp342 was found to produce the nitrogenase component NifH in planta when the plants were provided with sucrose. It partially relieved nitrogen-deficiency symptoms in greenhouse-grown wheat, whereas the Nif$^-$ mutant did not (Riggs et al. 2002). Nitrogen fixation could be demonstrated in wheat plants inoculated with Kp342 (Iniguez et al. 2004). Under standard agricultural conditions Kp342 inoculation enhances maize yields but could not relieve the nitrogen-deficiency symptoms in maize when fixed N was limiting (Riggs et al. 2002). This growth promotion independent of nitrogen fixation possibly involves the production of phytohormones by the endophytic bacterium (Triplett 2007), as in *Azospirillum* spp and other PGPR.

3.4.2
Pseudomonas stutzeri

In the early 1980s the team of C.B. You in China isolated diazotrophic bacteria from surface sterilized rice roots, which they identified as *Alcaligenes faecalis* (You and Zhou 1989). They also demonstrated that these bacteria were common in paddy soil and colonize rice roots endophytically. In particular, the *A. faecalis* strain A15 showed an efficient colonization of the intercellular spaces of the root epidermis and some cells even penetrated the plant's cell walls and colonized epidermal cells. Rice callus cultures inoculated with A15 could fix nitrogen, as shown with the ^{15}N tracer technique (You and Zhou 1989). The strain A15 was later assigned to the species *Pseudomonas stutzeri* A1501 (Vermeiren et al. 1999).

Nitrogen fixation was first shown in *Pseudomonas stutzeri* by Krotzky and Werner in 1987. The *nif*-operon and flanking genes have been characterized in detail and showed similarities in particular to *Azotobacter vinelandii* (Desnoues et al. 2003).

4
Conclusions and Outlook

When studying endophytic bacteria, the microscopic proof of the endophytic localization, together with an in situ identification of the endophytic microbial agent either by immunological, molecular biological (FISH), or gene marker (GFP) techniques, is absolutely fundamental. The fact that certain bacteria were isolated from surface sterilized specimen is not sufficient and can be regarded only as a single step towards the characterization of a true endophytic association. The direct proof of endophytic localization becomes even more important, as endophytes may turn to unculturable but physiologically active forms inside the plant host as a result of a supposedly efficient adaptation to the plant environment. For this reason, many more yet unknown endophytic diazotrophs may exist inside plant tissues which await discovery by culture independent molecular methods, and finally, isolation as pure freeliving cultures.

In some bacterial species, a plant endophytic life style is quite common or even the rule (e.g., in some species of the genera *Gluconacetobacter, Herbaspirillum* or *Azoarcus*), while in other cases only specific strains have endophytic potential (like in *Azospirillum*). Therefore, it is important to specify clearly the bacterial strain and the matching plant partner down to the cultivar level. It becomes apparent that the plant partner plays a major role in allowing or prohibiting the entry and distribution of a bacterium within its organs. The recognition processes in bacteria-plant interactions are quite well understood in the case of plant pathogens. Possibly, similar key determinants such as the lipopolysaccharide and exopolysaccharide structures, the flagellin type, and the type of signaling molecules produced by the bacteria, e.g., for quorum sensing, are potential elicitors of the plant's innate immune response.

Productive endophytic interactions of procaryotic and eucaryotic organisms are called symbiotic systems, although the detailed mechanisms of the interactions remain unclear in many cases so far. Furthermore, it must be closely observed whether or not, in some cases or under certain circumstances, a usually advantageous endophytic interaction may turn out to be harmful to the plant host. The presence of potentially pathogenic factors in endophytic bacteria is certainly one important issue which has to be researched. Genomic information in major endophytic diazotrophs such as *Azoarcus* sp. strain BH72, *Herbaspirillum seropedicae* Z87, *Klebsiella pneumoniae* 342, *Gluconacetobacter diazotrophicus* PAL5, *Azospirillum brasilense* Sp245, *A. lipoferum* 4B and *Pseudomonas stutzeri* A1501 is either complete or well on its way to becoming complete. The genomic information will also provide information about potential human health risks of endophytic bacteria and about their extent of relatedness to opportunistic human pathogens (e.g., *Burkholderia cepacia* and *Ochrobactrum antropi*). The rhizosphere is known as a habitat which harbors a diversity of bacteria which are closely related to or even undistinguishable from human pathogens (Berg et al. 2005). Although biotechnological applications are very tempting in many cases, the risks of endophytes to develop into or act as plant or human pathogens under certain circumstances need to be evaluated very carefully.

Acknowledgements This chapter is dedicated to Dr. Johanna Döbereiner in honor of her great achievements concerning diazotrophic bacteria in non-leguminous plants, and her pathfinding ideas about endophytic diazotrophs in *Gramineae*.

References

Amann RI, Krumholz L, Stahl DA (1990) Fluorescent-oligonucleotide probing of whole cells for determinative, phylogenetic, and environmental studies in microbiology. J Bacteriol 172:762–770

Amann RI, Ludwig W, Schleifer K-H (1995) Phylogenetic identification and in situ detection of individual microbial cells without cultivation. Microbiol Rev 59:143–169

Ashbolt NJ, Inkerman PA (1990) Acetic acid bacterial biota of the pink sugar cane mealybug, *Saccharococcus sacchari*, and its environs. Appl Environ Microbiol 56:707–712

Assmus B, Hutzler P, Kirchhof G, Amann R, Lawrence JR, Hartmann A (1995) In Situ Localization of *Azospirillum brasilense* in the rhizosphere of wheat with fluorescently labeled, rRNA-targeted oligonucleotide probes and scanning confocal laser microscopy. Appl Environ Microbiol 61:1013–1019

Baldani JI, Baldani VLD, Seldin L, Döbereiner J (1986) Characterization of *Herbaspirillum seropedicae* gen. nov., sp. nov., a new root-associated nitrogen-fixing bacterium. Int J Syst Bacteriol 36:86–93

Baldani JI, Pot B, Kirchhof G, Falsen E, Baldani VL, Olivares FL, Hoste B, Kersters K, Hartmann A, Gillis M, Döbereiner J (1996) Emended description of *Herbaspirillum*; inclusion of *Pseudomonas rubrisubalbicans*, a milk plant pathogen, as *Herbaspirillum rubrisubalbicans* comb. nov.; and classification of a group of clinical isolates (EF group 1) as *Herbaspirillum* species 3. Int J Syst Bacteriol 46:802–810

Baldani JI, Caruso L, Baldani VLD, Goi SR, Döbereiner J (1997) Recent advances in BNF with non-legume plants. Soil Biol Biochem 29:911–922

Bashan Y, Levanony H (1988) Interaction between *Azospirillum brasilense* Cd and wheat root cells during early stages of root colonization. In: Klingmüller W (ed) *Azospirillum* IV. Springer, Berlin, Germany, pp 166–173

Bashan Y, Levanony H (1990) Current status of *Azospirillum* inoculation technology. *Azospirillum* as a challenge for agriculture. Can J Microbiol 36:591–607

Becking JH (1963) Fixation of molecular nitrogen by an aerobic vibrio or spirillum. Antonie van Leeuwenhoeck 29:326

Becking JH (1982) *Azospirillum lipoferum*, a reappraisal. In: Klingmüller W (ed) *Azospirillum*, genetics, physiology, ecology, vol Experientia Suppl. 42. Birkhäuser Verlag, Basel, Switzerland, pp 130–149

Beijerinck MW (1925) Über ein Spirillum, welches freien Stickstoff binden kann? Zentralbl Bakteriol II Abt. 63:353–357

Berg G, Eberl L, Hartmann A (2005) The rhizosphere as a reservoir for opportunistic human pathogenic bacteria. Environ Microbiol 7:1673–1685

Bloemberg GV, Camacho Carvajal MM (2006) Microbial interactions with plants: A hidden world? In: Schulz B, Boyle C, Sieber TN (eds) Microbial Root Endophytes/Soil Biology, vol. 9. Springer, Berlin, Heidelberg, pp 321–336

Boddey RM, Chalk PM, Victoria RL, Matsui E, Döbereiner J (1983) The use of ^{15}N iusotope dilution technique to estimate the contribution of associated nitrogen fixation nutrition of *Paspalum notatum* cv. Batatais. Can J Microbiol 29:1036–1045

Boddey RM (1995) Biological nitrogen fixation in sugar cane: A key to energetically viable biofuel production. Crit Rev Plant Sci 14:263–279

Boddey RM, Polidoro JC, Resende AS, Alves BJR, Urquiaga S (2001) Use of the ^{15}N natural abundance technique for the quantification of the contribution of N_2 fixataion to sugar cane and other grasses. Austr J Plant Physiol 28:889–895

Boddey RM, Urquiaga S, Alves BJR, Reis VM (2003) Endophytic nitrogen fixation in sugarcane: Present knowledge and future applications. Plant Soil 252:139–149

Caballero-Mellado J, Martinez-Aguilar L, Paredes-Valdez G, Santos PE (2004) *Burkholderia unamae* sp. nov., an N_2-fixing rhizospheric and endophytic species. Int J Syst Evol Microbiol 54:1165–1172

Cavalcante VA, Döbereiner J (1988) A new acid tolerant nitrogen-fixing bacterium associated with sugarcane. Plant Soil 108:23–31

Chelius MK, Triplett EW (2000) Immunolocalization of dinitrogenase reductase produced by *Klebsiella pneumoniae* in association with *Zea mays* L. Appl Environ Microbiol 66:783–787

Chen W-M, James EK, Coenye T, Chou J-H, Barrios E, de Faria SM, Elliott GN, Sheu S-Y, Sprent JI, Vandamme P (2006) *Burkholderia mimosarum* sp. nov., isolated from root nodules of *Mimosa* spp. from Taiwan and South America. Int J Syst Evol Microbiol 56:1847–1851

Chen WM, James EK, Prescott AR, Kierans M, Sprent JI (2003) Nodulation of *Mimosa* spp. by the beta-proteobacterium *Ralstonia taiwanensis*. Mol Plant-Microbe Int 16:1051–1061

Chi F, Shen S-H, Cheng H-P, Jing Y-X, Yanni YG, Dazzo FB (2005) Ascending migration of endophytic *Rhizobia*, from roots to leaves, inside rice plants and assessment of benefits to rice growth physiology. Appl Environ Microbiol 71:7271–7278

Chowdhury SP, Schmid M, Hartmann A, Tripathi AK (2007) Identification of diazotrophs in the culturable bacterial community associated with roots of *Lasiurus scindicus* Henrard, a perennial grass of Thar desert, India. Microb Ecol (in press)

Compant S, Reiter B, Sessitsch A, Nowak J, Clement C, Ait Barka E (2005) Endophytic colonization of *Vitis vinifera* L. by plant growth-promoting bacterium *Burkholderia* sp. strain PsJN. Appl Environ Microbiol 71:1685–1693

Costacurta A, Vanderleyden J (1995) Synthesis of phytohormones by plant-associated bacteria. Crit Rev Microbiol 21:1–18

Daims H, Bruhl A, Amann R, Schleifer KH, Wagner M (1999) The domain-specific probe EUB338 is insufficient for the detection of all bacteria: development and evaluation of a more comprehensive probe set. Syst Appl Microbiol 22:434-444

Dalton DA, Kramer S, Azios N, Fusaro S, Cahill E, Kennedy C (2004) Endophytic nitrogen fixation in dune grasses (*Ammophila arenaria* and *Elymus mollis*) from Oregon. FEMS Microbiol Ecol 49:469-479

Dazzo FB, Yanni YG (2006) The natural *Rhizobium* - cereal crop association as an example of plant-bacteria interaction. In: Uphoff N, Ball AS, Fernandes E, Herren H, Husson O, Laing M, Palm C, Pretty J, Sanchez P (eds) Biological Approaches to Sustainable Soil Systems. Taylor & Francis Group, CRC Press, London, pp 109-126

Desnoues N, Lin M, Guo X, Ma L, Carreno-Lopez R, Elmerich C (2003) Nitrogen fixation genetics and regulation in a *Pseudomonas stutzeri* strain associated with rice. Microbiol Ecol 149:2251-2262

Dobbelaere S, Croonenborghs A, Amber T, Ptacek D, Vanderleyden J, Dutto P, Labandera-Gonzalez C, Caballero-Mellado J, Aguirre JF, Kapulnik Y, Shimon B, Burdman S, Kadouri D, Sarig S, Okon Y (2001) Responses of agronomically important crops to inoculation with *Azospirillum*. Austr J Plant Physiol 28:871-879

Dobbelaere S, Okon Y (2007) The plant growth-promoting effect and plant responses. In: Elmerich C, Newton WE (eds) Associative and endophytic nitrogen-fixing bacteria and cyanobacterial associations. Springer, Dordrecht, The Netherlands, pp 145-170

Döbereiner J, Day JM (1976) Associative symbioses in tropical grasses: Characterization of microorganisms and dinitrogen fixing sites. In: Newton WE, Nyman CJ (eds) Proceedings of the 1st international symposium on nitrogen fixation. Washington State University Press, Pullman, WA, pp 518-538

Döbereiner J (1992) History and new perspectives of diazotrophs in association with non-leguminous plants. Symbiosis 13:1-13

Döbereiner J, Reis VM, Paula MA, Olivares F (1993) Endophytic diazotrophs in sugar cane, cereal and tuber plants. In: Palacios R, Mora J, Newton WE (eds) New horizons in nitrogen fixation. Kluwer Academic Publishers, Dordrecht, The Netherlands, pp 671-676

Dong Y, Chelius MK, Brisse S, Kozyrovska N, Kovtunovych G, Podschun R, Triplett EW (2003a) Comparisons between two *Klebsiella*: The plant endophyte *K. pneumoniae* 342 and a clinical isolate *K. pneumoniae* MGH78578. Symbiosis 35:247-259

Dong Y, Iniguez AL, Ahmer BM, Triplett EW (2003b) Kinetics and strain specificity of rhizosphere and endophytic colonization by enteric bacteria on seedlings of *Medicago sativa* and *Medicago truncatula*. Appl Environ Microbiol 69:1783-1790

Dong Y, Iniguez AL, Triplett EW (2003c) Quantitative assessments of the host range and strain specificity of endophytic colonization by *Klebsiella pneumoniae* 342. Plant Soil 257:49-59

Dong Z, Canny MJ, McCully ME, Roboredo MR, Cabadilla CF, Ortega E, Rodes R (1994) A nitrogen-fixing endophyte of sugarcane stems (A new role for the apoplast). Plant Physiol 105:1139-1147

Dörr J, Hurek T, Reinhold-Hurek B (1998) Type IV pili are involved in plant-microbe and fungus-microbe interactions. Mol Microbiol 30:7-17

Eckert B, Weber OB, Kirchhof G, Halbritter A, Stoffels M, Hartmann A (2001) *Azospirillum doebereinerae* sp. nov., a nitrogen-fixing bacterium associated with the C4-grass *Miscanthus*. Int J Syst Evol Microbiol 51:17-26

Egener T, Hurek T, Reinhold-Hurek B (1998) Use of green fluorescent protein to detect expression of nif genes of *Azoarcus* sp. BH72, a grass-associated diazotroph, on rice roots. Mol Plant-Microbe Int 11:71-75

Egener T, Hurek T, Reinhold-Hurek B (1999) Endophytic expression of *nif* genes of *Azoarcus* sp. strain BH72 in rice roots. Mol Plant-Microbe Int 12:813–819

Elbeltagy A, Nishioka K, Sato T, Suzuki H, Ye B, Hamada T, Isawa T, Mitsui H, Minamisawa K (2001) Endophytic colonization and in planta nitrogen fixation by a *Herbaspirillum* sp. isolated from wild rice species. Appl Environ Microbiol 67:5285–5293

Elliott GN, Chen W-M, Chou J-H, Wang H-C, Sheu S-Y, Perin L, Reis VM, Moulin L, Simon MF, Bontemps C, Sutherland JM, Bessi R, de Faria SM, Trinick MJ, Prescott AR, Sprent JI, James EK (2007) *Burkholderia phymatum* is a highly effective nitrogen-fixing symbiont of *Mimosa* spp. and fixes nitrogen *ex planta*. New Phytol 173:168–180

Engelhard M, Hurek T, Reinhold-Hurek B (2000) Preferential occurrence of diazotrophic endophytes, *Azoarcus* spp., in wild rice species and land races of *Oryza sativa* in comparison with modern races. Environ Microbiol 2:131–141

Estrada-De Los Santos P, Bustillos-Cristales R, Caballero-Mellado J (2001) *Burkholderia*, a genus rich in plant-associated nitrogen fixers with wide environmental and geographic distribution. Appl Environ Microbiol 67:2790–2798

Fu H, Burris RH (1989) Ammonium inhibition of nitrogenase activity in *Herbaspirillum seropedicae*. J Bacteriol 171:3168–3175

Fuentes-Ramirez LE, Jimenez-Salgado T, Abarca-Ocampo IR, Caballero-Mellado J (1993) *Acetobacter diazotrophicus*, an indoleacetic acid producing bacterium isolated from sugarcane cultivars of Mexico. Plant Soil 154:145–150

Fuentes-Ramirez LE, Caballero-Mellado J, Sepulveda J, Martinez-Romero E (1999) Colonization of sugarcane by *Acetobacter diazotrophicus* is inhibited by high N-fertilization. FEMS Microbiol Ecol 29:117–128

Fuentes-Ramirez LE, Bustillos-Cristales R, Tapia-Hernandez A, Jimenez-Salgado T, Wang ET, Martinez-Romero E, Caballero-Mellado J (2001a) Novel nitrogen-fixing acetic acid bacteria, *Gluconacetobacter johannae* sp. nov. and *Gluconacetobacter azotocaptans* sp. nov., associated with coffee plants. Int J Syst Evol Microbiol 51:1305–1314

Fuentes-Ramirez LE, Bustillos-Cristales R, Tapia-Hernandez A, Jimenez-Salgado T, Wang ET, Martinez-Romero E, Caballero-Mellado J (2001b) Novel nitrogen-fixing acetic acid bacteria, *Gluconacetobacter johannae* sp. nov. and *Gluconacetobacter azotocaptans* sp. nov., associated with coffee plants. Int J Syst Evol Microbiol 51:1305–1314

Gillis M, Kersters K, Hoste B, Janssens D, Koppenstedt RM, Stephan MP, Teixeira KRS, Döbereiner J, De Ley J (1989) *Acetobacter diazotrophicus* sp. nov., a nitrogen fixing acetic acid bacterium associated with sugar cane. Int J Syst Bacteriol 39:361–364

Glick BR, Penrose DM, Li J (1998) A model for the lowering of plant ethylene concentrations by plant growth-promoting bacteria. J Theor Biol 190:63–68

Gutierrez-Zamora ML, Martinez-Romero E (2001) Natural endophytic association between *Rhizobium etli* and maize (*Zea mays* L.). J Biotechnol 91:117–126

Gyaneshwar P, James EK, Reddy PM, Ladha JK (2002) *Herbaspirillum* colonization increases growth and nitrogen accumulation in aluminium-tolerant rice varieties. New Phytol 154:131–145

Hallmann J, Berg G (2006) Spectrum and population dynamics of bacterial root endophytes. In: Schulz B, Boyle C, Sieber TN (eds) Microbial Root Endophytes/Soil Biology, vol. 9. Springer, Berlin, Heidelberg, pp 15–31

Hallmann J, Berg G, Schulz B (2006) Isolation procedures for endophytic microorganisms. In: Schulz B, Boyle C, Sieber TN (eds) Microbial Root Endophytes/Soil Biology, vol. 9. Springer, Berlin, Heidelberg, pp 299–319

Hartmann A, Baldani JI, Kirchhof G, Aßmus B, Hutzler P, Springer N, Ludwig W, Baldani VLD, Döbereiner J (1995) Taxonomic and ecoligic studies of diazotrophic rhizospere bacteria using phylogenetic probes. In: Fendrik I (ed) *Azospirillum* VI and related organisms. NATO ASI Series, vol. G 37. Springer, Berlin, Heidelberg, Germany, pp 415-427

Hartmann A, Baldani JI (2006) The genus *Azospirillum*. In: Dworkin M, Falkow S, Rosenberg E, Schleifer K-H, Stackebrandt E (eds) The Prokaryotes, vol. 5: Proteobacteria: Alpha and Beta Subclasses. Springer, New York, USA, pp 114-140

Hiltner L (1904) Über neuere Erfahrungen und Probleme auf dem Gebiete der Bodenbakteriologie unter besonderer Berücksichtigung der Gründüngung und Brache. Arbeiten der Deutschen Landwirtschaftlichen Gesellschaft 98:59-78

Hurek T, Reinhold-Hurek B, van Montagu M, Kellenberger E (1991) Infection of intact roots of Kallar grass and rice seedlings by *Azoarcus*. In: Polsinelli M, Materassi R, Vincenzini M (eds) Nitrogen fixation. Kluwer Academic Publishers, Dordrecht, The Netherlands, pp 235-242

Hurek T, Reinhold-Hurek B, Turner GL, Bergersen FJ (1994a) Augmented rates of respiration and efficient nitrogen fixation at nanomolar concentrations of dissolved O_2 in hyperinduced *Azoarcus* sp. strain BH72. J Bacteriol 176:4726-4733

Hurek T, Reinhold-Hurek B, Van Montagu M, Kellenberger E (1994b) Root colonization and systemic spreading of *Azoarcus* sp. strain BH72 in grasses. J Bacteriol 176:1913-1923

Hurek T, Reinhold-Hurek B (1995) Identification of grass-associated and toluene-degrading diazotrophs, *Azoarcus* spp., by analyses of partial 16S ribosomal DNA sequences. Appl Environ Microbiol 61:2257-2261

Hurek T, Egener T, Reinhold-Hurek B (1997a) Divergence in nitrogenases of *Azoarcus* spp., proteobacteria of the beta subclass. J Bacteriol 179:4172-4178

Hurek T, Wagner B, Reinhold-Hurek B (1997b) Identification of N_2-fixing plant- and fungus-associated *Azoarcus* species by PCR-based genomic fingerprints. Appl Environ Microbiol 63:4331-4339

Hurek T, Reinhold-Hurek B (1998) Interactions of *Azoarcus* sp. with rhizosphere fungi. In: Varma A, Hock B (eds) Mycorrhiza. 2nd. Springer, Berlin, Germany, pp 595-614

Hurek T, Handlley LL, Reinhold-Hurek B, Piche Y (2002) *Azoarcus* grass endophytes contribute fixed nitrogen to the plant in an unculturable state. Mol Plant Microbe Interact 15:233-242

Iniguez AL, Dong Y, Triplett EW (2004) Nitrogen fixation in wheat provided by *Klebsiella pneumoniae* 342. Mol Plant Microbe Interact 17:1078-1085

James EK, Reis VM, Olivares FL, Baldani JI, Döbereiner J (1994) Infection of sugar cane by the nitrogen-fixing bacterium *Acetobacter diazotrophicus*. J Exp Bot 45:757-766

James EK, Olivares FL (1997) Infection and colonization of sugarcane and other gramineous plants by endophytic diazotrophs. Crit Rev Plant Sci 17:77-119

James EK, Olivares FL, Baldani JI, Döbereiner J (1997) *Herbaspirillum*, an endophytic diazotroph colonising vascular tissue in leaves of *Sorghum bicolor* L. Moench. J Exp Bot 48:785-797

James EK, Olivares FL, de Oliveira AL, dos Reis FB Jr, da Silva LG, Reis VM (2001) Further observations on the interaction between sugar cane and *Gluconacetobacter diazotrophicus* under laboratory and greenhouse conditions. J Exp Bot 52:747-760

Jargeat P, Cosseau C, Ola'h B, Jauneau A, Bonfante P, Batut J, Becard G (2004) Isolation, free-living capacities, and genome structure of *Candidatus* Glomeribacter gigasporarum, the endocellular bacterium of the mycorrhizal fungus *Gigaspora margarita*. J Bacteriol 186:6876-6884

Jiménez-Salgado T, Fuentes-Ramirez LE, Tapia-Hernandez A, Mascarua-Esparza MA, Martinez-Romero E, Caballero-Mellado J (1997) *Coffea arabica* L., a new host plant for *Acetobacter diazotrophicus*, and isolation of other nitrogen-fixing acetobacteria. Appl Environ Microbiol 63:3676-3683

Khammas KM, Ageron E, Grimont PA, Kaiser P (1989) *Azospirillum irakense* sp. nov., a nitrogen-fixing bacterium associated with rice roots and rhizosphere soil. Res Microbiol 140:679–693

Kirchhof G, Baldani JI, Reis VM, Hartmann A (1998) Molecular assay to identify *Acetobacter diazotrophicus* and detect its occurrence in plant tissues. Can J Microbiol 44:12–19

Kirchhof G, Eckert B, Stoffels M, Baldani JI, Reis VM, Hartmann A (2001) *Herbaspirillum frisingense* sp. nov., a new nitrogen-fixing bacterial species that occurs in C4-fibre plants. Int J Syst Evol Microbiol 51:157–168

Kloepper JW, Ryu C-M (2006) Bacterial endophytes as elicitirs of induced systemic resistance. In: Schulz B, Boyle C, Sieber TN (eds) Microbial Root Endophytes/Soil Biology, vol. 9. Springer, Berlin, Heidelberg, pp 33–52

Krause A, Ramakumar A, Bartels D, Battistoni F, Bekel T, Boch J, Bohm M, Friedrich F, Hurek T, Krause L, Linke B, McHardy AC, Sarkar A, Schneiker S, Syed AA, Thauer R, Vorholter FJ, Weidner S, Puhler A, Reinhold-Hurek B, Kaiser O, Goesmann A (2006) Complete genome of the mutualistic, N_2-fixing grass endophyte *Azoarcus* sp. strain BH72. Nat Biotechnol 24:1385–1391

Loganathan P, Sunita R, Parida AK, Nair S (1999) Isolation and characterization of two genetically distant groups of *Acetobacter diazotrophicus* from a new host plant *Eleusine coracana* L. Appl Environ Microbiol 87:167–172

Magalhães-Cruz L, de Souza EM, Weber OB, Baldani JI, Döbereiner J, Pedrosa Fde O (2001) 16S ribosomal DNA characterization of nitrogen-fixing bacteria isolated from banana (*Musa* spp.) and pineapple (*Ananas comosus* (L.) Merril). Appl Environ Microbiol 67:2375–2379

Magalhaes FM, Baldani JI, Souto SM, Kuykendall JR, Döbereiner J (1983) A new acid-tolerant *Azospirillum* species. An Acad Bras Cien 55:417–430

McClung CR, van Berkum P, Davis RE, Sloger C (1983) Enumeration and localization of N_2-Fixing bacteria associated with roots of *Spartina alterniflora* Loisel. Appl Environ Microbiol 45:1914–1920

McCully ME (2001) Niches for bacterial endophytes in crop plants: A plant biologist's view. Austr J Plant Physiol 28:983–990

Miché L, Battistoni F, Gemmer S, Belghazi M, Reinhold-Hurek B (2006) Upregulation of jasmonate-inducible defense proteins and differential colonization of roots of *Oryza sativa* cultivars with the endophyte *Azoarcus* sp. Mol Plant Microbe Interact 19:502–511

Minerdi D, Fani R, Gallo R, Boarino A, Bonfante P (2001) Nitrogen fixation genes in an endosymbiotic *Burkholderia* strain. Appl Environ Microbiol 67:725–732

Miyamoto T, Kawahara M, Minamisawa K (2004) Novel endophytic nitrogen-fixing *Clostridia* from the grass *Miscanthus sinensis* as revealed by terminal restriction fragment length polymorphism analysis. Appl Environ Microbiol 70:6580–6586

Moulin L, Munive A, Dreyfus B, Boivin-Masson C (2001) Nodulation of legumes by members of the [beta]-subclass of proteobacteria. Nature 411:948–950

Muthukumarasamy R, Revathi G, Seshadri S, Lakshminarasimhan C (2002) *Gluconacetobacter diazotrophicus* (syn. *Acetobacter diazotrophicus*), a promising diazotrophic endophyte in the tropics. Curr Sci 83:137–145

Ngom A, Nakagawa Y, Sawada H, Tsukahara J, Wakabayashi S, Uchiumi T, Nuntagij A, Kotepong S, Suzuki A, Higashi S, Abe M (2004) A novel symbiotic nitrogen-fixing member of the *Ochrobactrum* clade isolated from root nodules of *Acacia mangium*. J Gen Appl Microbiol 50:17–27

Nogueira EdM, Vinagre F, Masuda HP, Vargas C, de Pádua VLM, de Silva FR, de Santos RV, Baldani JI, Ferreira PCG, Hemerly AS (2001) Expression of sugarcane genes induced by inoculation with *Gluconacetobacter diazotrophicus* and *Herbaspirillum rubrisubalbicans*. Genet Mol Biol 24:199–206

Okon Y (1994) *Azospirillum*-plant associations. *Azospirillum*-plant associations. CRC Press, Boca-Raton, Florida

Okon Y, Labandera-Gonzalez C (1994) Agronomic applications of *Azospirillum*: An evaluation of 20 years world-wide field inoculation. Soil Biol Biochem 26:1591–1601

Olivares FL, Baldani VLD, Reis VM, Baldani JI, Döbereiner J (1996) Occurrence of the endophytic diazotrophs *Herbaspirillum* spp. in roots, stems, and leaves, predominantly of Gramineae. Biol Fert Soils 21:197–200

Olivares FL, James EK, Baldani JI, Döbereiner J (1997) Infection of mottled stripe disease-susceptible and resistant sugar cane varieties by the endophytic diazotroph *Herbaspirillum*. New Phytol 135:723–737

Oliveira OC, Urquiaga S, Boddey RM (1994) Burning cane: The long term effects. Int Sug J 96:272–275

Oliviera E (1992) Estudo da associacao entre bacterias diazotroficas e arroz. Universidade Federal Rural do Rio de Janeiro. 135. Tese de Mestrado

Patriquin DG, Döbereiner J, Jain DK (1983) Sites and process of association between diazotrophs and grasses. Can J Microbiol 29:900–915

Paula MA, Reis VM, Döbereiner J (1991) Interactions of *Glomus clarum* with *Acetobacter diazotrophicus* in infection of sweet potato (*Ipomoea batatas*), sugarcane (*Saccharum* spp.) and sweet sorghum (*Sorghum vulgare*). Biol Fert Soils 11:111–115

Perin L, Martinez-Aguilar L, Paredes-Valdez G, Baldani JI, Estrada-de Los Santos P, Reis VM, Caballero-Mellado J (2006) *Burkholderia silvatlantica* sp. nov., a diazotrophic bacterium associated with sugar cane and maize. Int J Syst Evol Microbiol 56:1931–1937

Pinõn D, Casas M, Blanch M, Fontaniella B, Blanco Y, Vicente C, Solas MT, Legaz ME (2002) *Gluconacetobacter diazotrophicus*, a sugar cane endosymbiont, produces a bacteriocin against *Xanthomonas albilineans*, a sugar cane pathogen. Res Microbiol 153:345–351

Reinhold-Hurek B, Hurek T, Claeyssens M, van Montagu M (1993) Cloning, expression in *Escherichia coli*, and characterization of cellulolytic enzymes of *Azoarcus* sp., a root-invading diazotroph. J Bacteriol 175:7056–7065

Reinhold-Hurek B, Hurek T (1998) Interaction of gramineous plants with *Azoarcus* spp. and other diazotrophs: Identification, localization and perspectives to study their function. Crit Rev Plant Sci 17:29–54

Reinhold-Hurek B, Hurek T (2000) Reassessment of the taxonomic structure of the diazotrophic genus *Azoarcus* sensu lato and description of three new genera and new species, *Azovibrio restrictus* gen. nov., sp. nov., *Azospira oryzae* gen. nov., sp. nov. and *Azonexus fungiphilus* gen. nov., sp. nov. Int J Syst Evol Microbiol 50:649–659

Reinhold-Hurek B, Maes T, Gemmer S, Van Montagu M, Hurek T (2006) An endoglucanase is involved in infection of rice roots by the not-cellulose-metabolizing endophyte *Azoarcus* sp. strain BH72. Mol Plant Microbe Interact 19:181–188

Reinhold-Hurek B, Hurek T (2007) Endophytic associations of *Azoarcus* spp. In: Elmerich C, Newton WE (eds) Associative and endophytic nitrogen-fixing bacteria and cyanobacterial associations. Springer, Dordrecht, The Netherlands, pp 191–211

Reinhold B, Hurek T, Niemann E-G, Fendrik I (1986) Close association of *Azospirillum* and diazotrophic rods with different root zones of Kallar Grass. Appl Environ Microbiol 52:520–526

Reinhold B, Hurek T, Fendrik I, Pot B, Gillis M, Kersters K, Thielemans S, Deley J (1987) *Azospirillum halopraeferens* sp. nov, a nitrogen fixing organism associated with roots of kallar grass (*Leptochloa fusca* (L) Kunth). Int J Syst Bacteriol 37:43–51

Reinhold B, Hurek T (1988) Location of diazotrophs in the root interior with special attention to the Kallar grass association. Plant Soil 110:259–268

Reis VM, Olivares FL, Doebereiner J (1994) Improved methodology for isolation of *Acetobacter diazotrophicus* and confirmation of its endophytic habitat. World J Microbiol Biotech 10:101–104

Reis VM, de los Santos PE, Tenorio-Salgado S, Vogel J, Stoffels M, Guyon S, Mavingui P, Baldani VLD, Schmid M, Baldani JI, Balandreau J, Hartmann A, Caballero-Mellado J (2004) *Burkholderia tropica* sp. nov., a novel nitrogen-fixing, plant-associated bacterium. Int J Syst Evol Microbiol 54:2155–2162

Reis VM, Lee S, Kennedy C (2007) Biological nitrogen fixation in sugarcane. In: Elmerich C, Newton WE (eds) Associative and endophytic nitrogen-fixing bacteria and cyanobacterial associations. Springer, Dordrecht, The Netherlands, pp 213–232

Reiter B, Burgmann H, Burg K, Sessitsch A (2003) Endophytic *nifH* gene diversity in African sweet potato. Can J Microbiol 49:549–555

Riggs PJ, Chelius MK, Iniguez AL, Kaeppler SM, Triplett EW (2001) Enhanced maize productivity by inoculation with diazotrophic bacteria. Austr J Plant Physiol 28:829–836

Riggs PJ, Moritz RL, Chelius MK, Dong Y, Iniguez AL, Kaeppler SM, Casler MD, Triplett EW (2002) Isolation and characterization of diazotrophic endophytes from grasses and their effects on plant growth. In: Finan TR, O'Brian MR, Layzell DB, Vessey JK, Newton WE (eds) Nitrogen fixation: Global perspectives, Proceedings of the 13th International Congress on Nitrogen Fixation. CABI Publishing, Wallingford, UK, pp 263–267

Rosenblueth M, Martinez-Romero E (2006) Bacterial endophytes and their interactions with hosts. Mol Plant Microbe Interact 19:827–837

Rothballer M, Schmid M, Hartmann A (2003) *In situ* localization and PGPR-effect of *Azospirillum brasilense* strains colonizing roots of different wheat varieties. Symbiosis 34:261–279

Rothballer M, Schmid M, Klein I, Gattinger A, Grundmann S, Hartmann A (2006) *Herbaspirillum hiltneri* sp. nov., isolated from surface-sterilized wheat roots. Int J Syst Evol Microbiol 56:1341–1348

Schank SC, Smith RL, Weiser GC, Zuberer DA, Bouton JH, Quesenberry KH, Tyler ME, Milam JR, Littel RC (1979) Fluorescent antibody technique to identify *Azospirillum brasilense* associated with roots of grasses. Soil Biol Biochem 11:287–295

Schloter M, Wiehe W, Aßmus B, Steindl H, Becke H, Höflich G, Hartmann A (1997) Root colonization of different plants by plant-growth-promoting *Rhizobium leguminosarum* bv. *trifolii* R39 studied with monospecific polyclonal antisera. Appl Environ Microbiol 63:2038–2046

Schloter M, Hartmann A (1998) Endophytic and surface colonization of wheat roots (*Triticum aestivum*) by different *Azospirillum brasilense* strains studies with strain-specific monoclonal antibodies. Symbiosis 25:159–179

Schmid M, Rothballer M, Aßmus B, Hutzler P, Schloter M, Hartmann A (2004) Detection of microbes by confocal laser scanning microscopy. In: Kowalchuk GA, de Bruijn FH, Head IM, Akkermans ADL, van Elsas JD (eds) Molecular Microbial Ecology Manual II, vol 2. Kluwer Academic Publishers, Dordrecht, The Netherlands, pp 875–910

Schmid M, Baldani I, Hartmann A (2006) The genus *Herbaspirillum*. In: Dworkin M, Falkow S, Rosenberg E, Schleifer K-H, Stackebrandt E (eds) The Prokaryotes, vol 5: Proteobacteria: Alpha and Beta Subclasses. Springer, New York, USA, pp 141–150

Schmid M, Hartmann A (2007) Molecular phylogeny and ecology of root associated diazotrophic alpha- and beta-proteobacteria. In: Elmerich C, Newton WE (eds) Associative and endophytic nitrogen-fixing bacteria and cyanobacterial associations. Springer, Dordrecht, The Netherlands, pp 21-40

Schuhegger R, Ihring A, Gantner S, Bahnweg G, Knappe C, Vogg G, Hutzler P, Schmid M, van Breusegem F, Eberl L, Hartmann A, Langebartels C (2006) Induction of systemic resistance in tomato by N-acylhomoserine lactone-producing rhizosphere bacteria. Plant Cell Environ 29:909-918

Schulz B, Boyle C (2006) What are endophytes? In: Schulz B, Boyle C, Sieber TN (eds) Microbial Root Endophytes/Soil Biology, vol 9. Springer, Berlin, Heidelberg, pp 1-13

Sevilla M, Burris RH, Gunapala N, Kennedy C (2001) Comparison of benefit to sugarcane plant growth and $^{15}N_2$ incorporation following inoculation of sterile plants with *Acetobacter diazotrophicus* wild-type and Nif⁻ mutant strains. Mol Plant-Microbe Int 14:358-366

Sly LI, Stackebrandt E (1999) Description of *Skermanella parooensis* gen. nov., sp. nov. to accommodate *Conglomeromonas largomobilis* subsp. *parooensis* following the transfer of *Conglomeromonas largomobilis* subsp. *largomobilis* to the genus *Azospirillum*. Int J Syst Bacteriol 49:541-544

Starkey RL (1958) Interrelations between micro-organisms and plant roots in the rhizosphere. Bacteriol Review 22:154-172

Steenhoudt O, Vanderleyden J (2000) *Azospirillum*, a free-living nitrogen-fixing bacterium closely associated with grasses: genetic, biochemical and ecological aspects. FEMS Microbiol Rev 24:487-506

Stephan MP, Oliveira M, Teixeira KRS, Martinez-Drets G, Döbereiner J (1991) Physiology and nitrogen fixation of *Acetobacter diazotrophicus*. FEMS Microbiol Lett 77:67-72

Stoffels M, Castellanos T, Hartmann A (2001) Design and application of new 16S rRNA-targeted oligonucleotide probes for the *Azospirillum-Skermanella-Rhodocista*-cluster. Syst Appl Microbiol 24:83-97

Stone JK, Bacon CW, White JF (2000) An overview of endophytic microbes: endophytism defined. In: Bacon CW, White JF (eds) Microbial endophytes. Marcel Dekker, New York, pp 3-30

Sturz AV, Christie BR, Nowak J (2000) Bacterial endophytes: Potential role in developing sustainable systems of crop production. Crit Rev Plant Sci 19:1-30

Tan Z, Hurek T, Reinhold-Hurek B (2003) Effect of N-fertilization, plant genotype and environmental conditions on *nifH* gene pools in roots of rice. Environ Microbiol 5:1009-1015

Tapia-Hernandez A, Bustillos-Cristales MR, Jimenez-Salgado T, Caballero-Mellado J, Fuentes-Ramirez LE (2000) Natural endophytic occurrence of *Acetobacter diazotrophicus* in pineapple plants. Microb Ecol 39:49-55

Tarrand JJ, Krieg NR, Döbereiner J (1978) A taxonomic study of the *Spirillum lipoferum* group, with descriptions of a new genus, *Azospirillum* gen. nov. and two species, *Azospirillum lipoferum* (Beijerinck) comb. nov. and *Azospirillum brasilense* sp. nov. Can J Microbiol 24:967-980

Tombolini R, van der Gaag DJ, Gerhardson B, Jansson JK (1999) Colonization pattern of the biocontrol strain *Pseudomonas chlororaphis* MA 342 on barley seeds visualized by using green fluorescent protein. Appl Environ Microbiol 65:3674-3680

Trinick MJ (1980) Relationships amongst the fast -growing rhizobia of *Lablab purpureus*, *Leucaena leucocephala*, *Mimosa* spp., *Acacia farnesiana* and *Sesbania grandifora* and their affinities with other rhizobial groups. J Appl Bacteriol 49:39-53

Tripathi AK, Verma SC, Chowdhury SP, Lebuhn M, Gattinger A, Schloter M (2006) *Ochrobactrum oryzae* sp. nov., an endophytic bacterial species isolated from deepwater rice in India. Int J Syst Evolut Microbiol 56:1677-1680

Triplett EW (2007) Prospects for significant nitrogen fixation in grasses from bacterial endophytes. In: Elmerich C, Newton WE (eds) Associative and endophytic nitrogen-fixing bacteria and cyanobacterial associations. Springer, Dordrecht, The Netherlands, pp 303–314

Trujillo ME, Willems A, Abril A, Planchuelo AM, Rivas R, Ludena D, Mateos PF, Martinez-Molina E, Velázquez E (2005) Nodulation of *Lupinus* by strains of the new species *Ochrobactrum lupini* sp. nov. Appl. Environ Microbiol 71:1318–1327

Umali-Garcia M, Hubbell DH, Gaskins MH, Dazzo FB (1980) Association of *Azospirillum* with grass roots. Appl Environ Microbiol 39:219–226

Urquiaga S, Cruz KHS, Boddey RM (1992) Contribution of nitrogen fixation to sugarcane: nitrogen-15 and nitrogen balance estimates. Soil Sci Soc Am 56:105–114

Valverde A, Velazquez E, Gutierrez C, Cervantes E, Ventosa A, Igual J-M (2003) *Herbaspirillum lusitanum* sp. nov., a novel nitrogen-fixing bacterium associated with root nodules of *Phaseolus vulgaris*. Int J Syst Evol Microbiol 53:1979–1983

van Overbeek LS, van Vuurde J, van Elsas JD (2006) Application of molecular fingerprinting techniques to explore the diversity of bacterial endophytic communities. In: Schulz B, Boyle C, Sieber TN (eds) Microbial Root Endophytes/Soil Biology, vol 9. Springer, Berlin, Heidelberg, pp 1–13

Vandamme P, Goris J, Chen WM, de Vos P, Willems A (2002) *Burkholderia tuberum* sp. nov. and *Burkholderia phymatum* sp. nov., nodulate the roots of tropical legumes. Syst Appl Microbiol 25:507–512

Vandamme P, Coenye T (2004) Taxonomy of the genus *Cupriavidus*: a tale of lost and found. Int J Syst Evol Microbiol 54:2285–2289

Verma SC, Chowdhury SP, Tripathi AK (2004a) Phylogeny based on 16S rDNA and *nifH* sequences of *Ralstonia taiwanensis* strains isolated from nitrogen-fixing nodules of *Mimosa pudica*, in India. Can J Microbiol 50:313–322

Verma SC, Singh A, Chowdhury SP, Tripathi AK (2004b) Endophytic colonization ability of two deep-water rice endophytes, *Pantoea* sp. and *Ochrobactrum* sp. using green fluorescent protein reporter. Biotechnol Lett 26:425–429

Vermeiren H, Willems A, Schoofs G, de Mot R, Keijers V, Hai W, Vanderleyden J (1999) The rice inoculant strain *Alcaligenes faecalis* A15 is a nitrogen-fixing *Pseudomonas stutzeri*. Syst Appl Microbiol 22:215–224

von Bülow JFW, Döbereiner J (1975) Potential for nitrogen fixation in maize genotypes in Brazil Proc Natl Acad Sci USA 72:2389–2393

Wagner M, Horn M, Daims H (2003) Fluorescence *in situ* hybridisation for the identification and characterisation of prokaryotes. Curr Opin Microbiol 6:302–309

Weber OB, Baldani VLD, Teixeira KRS, Kirchhof G, Baldani JI, Döbereiner J (1999) Isolation and characterization of diazotrophic bacteria from banana and pineapple plants. Plant Soil 210:103–113

Webster G, Jain V, Davey MR, Gough C, Vasse J, Denarie J, Cocking EC (1998) The flavonoid naringenin stimulates the intercellular colonization of wheat roots by *Azorhizobium caulinodans*. Plant Cell Environ 21:373–383

Weller DM (1988) Biological control of soilborne plant pathogens in the rhizosphere with bacteria. Annu Rev Phytopathol 26:379–407

Yamada Y, Hoshino K, Ishikawa T (1997) The phylogeny of acetic acid bacteria based on the partial sequences of 16S ribosomal RNA: the elevation of the subgenus *Gluconoacetobacter* to the generic level. Biosci Biotech Biochem 61:1244–1251

Yanni YG, Rizk RY, Corich V, Squartini A, Ninke K, Philip-Hollingsworth S, Orgambide G, De Bruijn F, Stoltzfus J, Buckley D, Schmidt TM, Mateos PF, Ladha JK, Dazzo FB (1997) Natural endophytic association between *Rhizobium leguminosarum* bv.*trifolii* and rice roots and assessment of potential to promote rice growth. Plant Soil 194:99–114

You CB, Zhou FY (1989) Non-nodular endorhizospheric nitrogen fixation in wetland rice. Can J Microbiol 35:403–408

Young JPW (1992) Phylogenetic classification of nitrogen-fixing organisms. In: Stacey G, Burris RH, Evans HJ (eds) Biological Nitrogen Fixation. Chapman & Hall, New York, London, pp 43–86

Zhulin IB, Armitage JP (1993) Motility, chemokinesis, and methylation-independent chemotaxis in *Azospirillum brasilense*. J Bacteriol 175:952–958

Zhulin IB, Bespalov VA, Johnson MS, Taylor BL (1996) Oxygen taxis and proton motive force in *Azospirillum brasilense*. J Bacteriol 178:5199–5204

Zurdo-Pineiro JL, Rivas R, Trujillo ME, Vizcaino N, Carrasco JA, Chamber M, Palomares A, Mateos PF, Martinez-Molina E, Velázquez E (2007) *Ochrobactrum cytisi* sp. nov., isolated from nodules of *Cytisus scoparius* in Spain. Int J Syst Evol Microbiol 57:784–788

Subject Index

Abscisic acid 49
Accessory symbiotic genes 22
Acidothermus 114
Actinomycetes 104
Actinorhizal plants 48, 104, 155
N-Acylhomoserine lactone 276
AFLP 110
Agrobacterium tumefaciens 45
Akinetes, leaf cavities/sporocarps 249
Anabaena 185, 229
Anabaena/Nostoc 188
Ancestor, universal actinobacterial 116
Arachis hypogaea 51
Autoregulation 142–144
Auxin 142
Azoarcus spp. 274, 286
Azolla 211, 236
– hormogonia 247
– pockets 238
– pores 239
– primary branched hair 240
– simple hair cells 241
– symbiosis, third partner 255
– trichomes 240
Azolla-cyanobacterium interactions 235, 251
Azospirillum spp. 273, 280

Bacteriorhiza 274
Bactobionts 256
Banana 273
Bartonella 72
Betulaceae 128
Biological control, endophytic bacteria 275
Blasia pustilla 189
Blastococcus 111
Bowenia 226
Bradyrhizobiaceae 46

Bradyrhizobium 6, 31
Brucella 45, 72
Bryophytes, hormogonium-inducing factor (HIF) 187
Burkholderia spp. 9, 288

Calothrix spp. 185, 230
Capsule 159
Carbon metabolism 232
– nodules 163
– non-symbiotic *Frankia* 162
Cassava 288
Casuarinaceae 128
Ceanothus 128–132
Ceratozamia 226
Chamaecrista 28
Chemotaxis 188
Chigua 226
Chlorogloeopsis spp. 185
Chromosomes 11
Citrulline 231
Coevolution 113
Coffee 273, 288
Colletia 128
Comptonia 107
Coralloid roots 227
Coriariaceae 128, 133, 134
Cortex 158
Cospecialization 116
Crack entry 51
– signaling 61
Cross-inoculation 5, 106
Cucurbitales 128–132, 134
Cyanobacteria 182
– *Azolla* symbiosis, structural characteristics 237
– symbiotic, specificity/diversity 229
Cyanobacterial partners 185
Cyanobiont, structural aspects 246

Cyanolichens 265
Cycads 226
Cytokinin 49, 142

Daidzein 53
Datisca cannabina 134
Datiscaceae 128, 132–134
Dermatophilus 111
Diazotrophs 4
Dinitrogenase 116, 155
Dioon 226
DNA homology 107, 108

Ecotypes 12
Elaeagnaceae 128, 132
Encephalartos 226
Endophytes 106
– diazotrophic 276
Endophytic bacteria 274
Envelopes 159, 238
Ethylene 144
Eubacteria 258
Evolution, rate 114

Fabaceae 5, 46
Fagales 128–134
Flavonoid co-inducers 13
Flavonoid inducers 80
Flavonoid perception 52
Flavonoids 53
Formononetin 53
Frankia 4, 104, 127–143, 145, 146
– diversity 117
– evolution 113
– fossil record 114
– galls 104
– host infectivity 107, 108
– host range 106
– hyphae 131
– infectivity 115
– neighbors 111
– pure cultures 104
– vesicles 104, 130, 133–139, 155
Frankia alni 106
Frankiaceae 111
Frankinae, phylogeny 113
Fungi 107
– parasitic 107

Genbank 111

Gene expression profiling 77
Gene transfer, horizontal 116
Genome, rhizobial, structure 11
Genospecies 108
Geodermatophilus 111
Glands, entry points 216
glnII 113
Gluconacetobacter diazotrophicus 273, 278, 282
Glutamine synthetase/glutamate synthase 193, 231
Glycine soja 139
Gramineae 273
Gunneraceae 183
Gunnera–Nostoc symbiosis 181, 207

Herbaspirillum spp. 9, 273, 284
Heterocysts 182
– differentiation 195
– filaments 247
High throughput 90
Horizontal gene transfer (HGT) 11
Hormogonia 186
– *Azolla* 247
– *Nostoc* 181
– differentiation 187, 195
– movement 188
Hormogonium repressing factor (HRF) 189
Hormogonium-inducing factor (HIF) 187
Hornworts 184, 191
Host, plasma membrane 159
– promiscuous 107, 110
Host defence responses 215
Host preference 275
Housekeeping genes 21
Hydrogen metabolism 170
Hypernodulation 49
Hyphae 104

Induced systemic resistance (ISR) 276
Infection, inter-/intracellular 127, 130–134
Infection threads 51, 132–135, 139, 142
Infection thread-like structure 131–135, 143
Inositol triphosphates 58
Intercellular penetration 130
Interfacial matrix 159
Isoflavonoids 53

Subject Index

Kallar grass (*Leptochloa fusca*)   274, 277, 288
Kentrothamnus   128
Klebsiella pneumoniae   290, 342 274
Koch's postulates   105

Lateral root organ defective (*latd*) mutant   49
Leaf cavity, *Azolla*   237
Legumes   5, 46, 128–134
– taxonomy/phylogeny   27
Lepidozamia   226
Lipid laminae   159
Lipo-chitin oligosaccharide (LCOs)   13
Lotus japonicus   48, 140, 141
Luteolin   53

Macrozamia spp.   185, 191, 226
Maize   273, 288
Medicago sativa   139, 142, 144
Medicago truncatula   48, 144
Medicarpin   53
Metabolites   119
Metabolomes   89
Methanothermococcus thermolithothrophicus   22
Microarrays   77
Microcycas   226
Micromonospora   116
Miscanthus   273
MLST   110, 119
Molecular clock   29, 114
Molecular interactions/dialogue   13, 52
Myc factors   139, 140
Mycorrhiza   5, 274
Myricaceae   128
Myrothamnus   208

Naringenin   54
nifD-nifK, intergenic   118
Nitrogen, transfer   231
Nitrogen fixation   166, 230
– genes   22, 113
Nitrogen metabolism   230
– nodules   165
– non-symbiotic *Frankia*   163
Nod factors   13, 16, 46, 54, 139–142, 146
– bacterial infection   59
– perception by host plant   55

Nod factor entry receptor   57, 59
Nod factor receptors   57
Nod factor signaling (NFS) genes   56
NodD   52
– / flavonoid complex   16
Nodulation genes   13
Nodulators, spontaneous   49
Nodules, anatomy   158
– apical meristem   159
– hyphae   158
– lenticels   158
– lobe meristem   158
– periderm   158
– roots   158
– sporangia   158
– vesicles   158
Nostoc   267
– growth/metabolism   190
– nitrogen fixation   193
– strains, lichen-forming/plant-associated   267
Nutrient sources   83

Ochrobactrum spp.   278
Omics   72
2-Oxoglutarate   195
Oxygen dilemma   170
Oxygen protection   129, 137, 139

Parasponia   5, 47, 128
Paspalum notatum   277
Peltigera spp.   266
Penicillium nodositatum   141
Pennisetum   273
Peptide mass fingerprinting (PMF)   86
Perisymbiont matrix/membrane   159
pH   117
Phospholipases   58
Photosymbiodeme   266
Physiological adaptation   186, 199
Phytohormones, auxin   142
– cytokinin   142
– ethylene   144
Pineapple   273
Plant growth promotion agents   275
Plant partners   184
Plasmids   11
Plasmodium   107
Pre-coralloids   227
Prenodules   51, 131–134, 143, 145

Proteobacteria 6, 29
– alpha- 278
– beta- 284
– gamma- 289
Proteomes 85
Pseudomonas stutzeri 274, 290

recA 111
Rhamnaceae 128
Rhizobia 5, 11, 45, 278
– diversity/taxonomy 18
– host specificity 15
– species concept 11
Rhizobia–legume symbiosis 13
– evolution/coevolution 28
Rhizobiaceae 46
Rhizobial genomes 72
– structure 11
Rhizobium leguminosarum bv. *trifolii* 279
Rhizobium spp. 278
Rhizosphere 118, 274
Rice 273
RNA, 16S rRNA 104
Root hair infection 130
Root nodules 46
– actinorhizal 104
Roots, adventitious (shoot-associated) 217
– coralloid 227
Rosaceae 128, 132, 133, 136
Rosales 128, 129, 134
Rosid clade 115
Rubisco activity 192, 232

Semantides 104
Sinorhizobium 3, 8, 15, 18, 52

Specificity 184, 198, 274
Sporangia, multilocus 111
Spores 104
Stangeriaceae 226
Stress conditions 83
Sugarcane 273, 277, 288
Surface sterilization 275
Sweet potato 273, 288
Symbiont, obligate 106
Symbiosis 107, 116
Symbiosomes 50
Symbiotic association, establishment 187
Symbiotic conditions 79
Symbiotic filament 230
Symbiotic regulatorM (SyrM) 52
SymRKs 140, 141
Systemic acquired resistance (SAR) 276

Talguenea 128
Teat cells 239
Thermal springs 111
Transcriptomes 77
Trichomes 240
– *Azolla* 240

Ulmaceae 47

Vesicles 104, 130, 133–139, 155

Wheat 273

Zamia 226
Zamiaceae 226